Holt McDougal Mathematics

Course 1
Kentucky Mathematics Worktext

HOLT McDOUGAL
a division of Houghton Mifflin Harcourt

Printed in the United States of America

ISBN: 978-0-55-400406-8
ISBN: 0-55-400406-2
7 8 9 10 2266 14
4500507614

About the Worktext and Table of Contents

The Grade 6 **Kentucky Mathematics Worktext** is designed to help all students master the mathematics standards found in the Grade 6 **Combined Curriculum Document,** including the **Program of Studies** and the **Core Content for Assessment Version 4.1.** Each standard is taught in one or more lessons, providing plenty of practice to help you build confidence in your math skills. A correlation showing which worktext lessons address each standard appears on p. xv.

Each chapter opens with a hands-on Project and closes with Chapter Summary discussion questions and KCCT Prep items. All Worktext lessons contain Exploration activities, Review for Mastery worksheets, and Homework and Practice worksheets. The column at right states which pages in ***Holt McDougal Mathematics** Course 1* support Worktext lessons.

Chapter 1 · Whole Numbers and Patterns

Lessons denoted with (*) include Hands-On Labs. Lessons with (†) include Technology Labs.

Chapter 2 • Introduction to Algebra

Lessons denoted with (*) include Hands-On Labs. Lessons with (†) include Technology Labs.

Chapter 3 • Decimals

Lessons denoted with (*) include Hands-On Labs. Lessons with (†) include Technology Labs.

Chapter 4 • Number Theory and Fractions

	Worktext	Page	Program of Studies	Core Content	HM Math Course 1
Project	Down the Drain	106			
4A	Divisibility	108	MA-6-NPO-S-PNO1	MA-06-1.5.1	4-1
4B	Factors and Prime Factorization	112	MA-6-NPO-S-PNO1	MA-06-1.5.1	4-2, Lab 4-2
4C	Greatest Common Factor	115	MA-6-NPO-S-PNO1	MA-06-1.5.1	4-3, Lab 4-3
4D	Decimals and Fractions	118	MA-6-NPO-S-NS1, MA-6-NPO-S-NS5	MA-06-1.1.1, MA-06-1.1.2, MA-06-1.1.3	4-4, Lab 4-4
4E	Equivalent Fractions	122	MA-6-NPO-S-PNO2, MA-6-NPO-S-NS1	MA-06-1.1.2	4-5, Lab 4-5A
4F	Mixed Numbers and Improper Fractions*	125	MA-6-NPO-S-NS1	MA-06-1.1.1, MA-06-1.1.2	4-6
4G	Comparing and Ordering Fractions*†	130	MA-6-NPO-S-NS1, MA-6-NPO-S-NS5	MA-06-1.1.2, MA-06-1.1.3	4-7
4H	Adding and Subtracting with Like Denominators	137	MA-6-NPO-S-NS2, MA-6-NPO-S-NO1, MA-6-NPO-S-NO2	MA-06-1.1.2, MA-06-1.3.1	4-8
4I	Estimating Fraction Sums and Differences	140	MA-6MNPO-S-E1, MA-6MNPO-S-E2	MA-06-1.1.2, MA-06-1.2.1	4-9
Review	Chapter Summary	143			
Practice	KCCT Prep	144			

Lessons denoted with (*) include Hands-On Labs. Lessons with (†) include Technology Labs.

Chapter 5 • Fraction Operations

	Worktext	Page	Program of Studies	Core Content	HM Math Course 1
Project	Paint the Town	146			
5A	Least Common Multiple†	148	MA-6-NPO-S-PNO1, MA-6-NPO-S-PNO3	MA-06-1.5.1	5-1
5B	Adding and Subtracting with Unlike Denominators†	153	MA-6-NPO-S-NS2, MA-6-NPO-S-NO1, MA-6-NPO-S-NO2	MA-06-1.1.2, MA-06-1.3.1	5-2, Lab 5-2
5C	Adding and Subtracting Mixed Numbers†	158	MA-6-NPO-S-NS2, MA-6-NPO-S-NO1, MA-6-NPO-S-NO2	MA-06-1.3.1	5-3
5D	Regrouping to Subtract Mixed Numbers	164	MA-6-NPO-S-NS2, MA-6-NPO-S-NO1, MA-6-NPO-S-NO2, MA-6-AT-S-VEO2	MA-06-1.1.2, MA-06-1.3.1, MA-06-5.2.1	5-4, Lab 5-4
5E	Solving Fraction Equations: Addition and Subtraction	167	MA-6-NPO-S-NO3, MA-6-AT-S-VEO1, MA-6-AT-S-VEO3, MA-6MAT-S-EI1, MA-6MAT-S-EI3	MA-06-1.3.2, MA-06-5.2.2, MA-06-5.3.1	5-5
5F	Multiplying Fractions by Whole Numbers†	170	MA-6-NPO-S-NS2, MA-6-NPO-S-NO1, MA-6-NPO-S-NO2, MA-6-AT-S-VEO2	MA-06-1.1.2, MA-06-1.3.1, MA-06-5.2.1	5-6
5G	Multiplying Fractions	175	MA-6-NPO-S-NS2, MA-6-NPO-S-NO1, MA-6-NPO-S-NO2, MA-6-AT-S-VEO2	MA-06-1.1.2, MA-06-1.3.1, MA-06-5.2.1	5-7, Lab 5-7
5H	Multiplying Mixed Numbers	178	MA-6-NPO-S-NS2, MA-6-NPO-S-NO1, MA-6-NPO-S-NO2	MA-06-1.3.1	5-8
5I	Dividing Fractions and Mixed Numbers	181	MA-6-NPO-S-NS2, MA-6-NPO-S-NO1, MA-6-NPO-S-NO2	MA-06-1.1.2, MA-06-1.3.1	5-9, Lab 5-9
5J	Solving Fraction Equations: Multiplication and Division	184	MA-6-NPO-S-NO3, MA-6-AT-S-VEO1, MA-6-AT-S-VEO3, MA-6MAT-S-EI1, MA-6MAT-S-EI3	MA-06-1.3.2, MA-06-5.2.2, MA-06-5.3.1	5-10
Review	Chapter Summary	187			
Practice	KCCT Prep	188			

Lessons denoted with (*) include Hands-On Labs. Lessons with (†) include Technology Labs.

Chapter 6 • Collecting and Displaying Data

	Worktext	Page	Program of Studies	Core Content	HM Math Course 1
Project	Enjoying Nature	190			
6A	Make a Table	192	MA-6-DAP-S-DR4, MA-6-DAP-S-ES1	MA-06-4.1.1, MA-06-5.1.5	6-1
6B	Mean, Median, Mode, and Range*	195	MA-6-DAP-S-CD2	MA-06-4.2.1	6-2, Lab 6-2
6C	Bar Graphs*	202	MA-6-DAP-S-DR2, MA-6-DAP-S-DR4	MA-06-4.1.1, MA-06-4.1.4	6-4, Lab 6-4
6D	Line Plots, Frequency Tables, and Histograms*	210	MA-6-DAP-S-DR2, MA-6-DAP-S-DR4	MA-06-4.1.1, MA-06-4.1.4	6-5
6E	Line Graphs†	216	MA-6-DAP-S-DR2, MA-6-DAP-S-DR4	MA-06-4.1.1, MA-06-4.1.4	6-7
6F	Stem-and-Leaf Plots	222	MA-6-DAP-S-DR2	MA-06-4.1.1	6-9
6G	Venn Diagrams	225	---	MA-06-4.1.1, MA-06-4.1.4	4-Ext.
6H	Choosing an Appropriate Display*	228	MA-6-DAP-S-DR1, MA-6-DAP-S-DR3, MA-6-DAP-S-ES1	MA-06-4.1.2, MA-06-4.1.4	6-10
Review	Chapter Summary	234			
Practice	KCCT Prep	235			

Lessons denoted with (*) include Hands-On Labs. Lessons with (†) include Technology Labs.

Chapter 7 • Proportional Relationships

Lessons denoted with (*) include Hands-On Labs. Lessons with (†) include Technology Labs.

Chapter 8 • Geometric Relationships

	Worktext	Page	Program of Studies	Core Content	HM Math Course 1
Project	Tangram Tangle	289			
8A	Building Blocks of Geometry	291	MA-6-G-S-SR3	MA-06-3.1.1	8-1
8B	Measuring and Classifying Angles	294	MA-6-M-S-MPA2, MA-6-M-S-MPA4, MA-6-G-S-SR3	MA-06-3.1.1	8-2
8C	Angle Relationships*†	298	MA-6-M-S-MPA2, MA-6-M-S-MPA4, MA-6-G-S-SR3	MA-06-3.1.1	8-3
8D	Classifying Lines	306	MA-6-G-S-SR2	MA-06-3.1.1	8-4, Lab 8-4
8E	Triangles	309	MA-6-G-S-SR1, MA-6-G-S-SR4	MA-06-3.1.2	8-5, Lab 8-5, Lab 8-5
8F	Quadrilaterals	313	MA-6-G-S-SR4	MA-06-3.1.2	8-6
8G	Polygons*	316	MA-6-G-S-SR1	MA-06-3.1.2	8-7
8H	Geometric Patterns*	323	MA-G-AT-S-PRF1	MA-06-5.1.1	8-8
8I	Congruence	328	MA-6-G-S-SR6	MA-06-3.1.4	8-9
Review	Chapter Summary	331			
Practice	KCCT Prep	332			

Lessons denoted with (*) include Hands-On Labs. Lessons with (†) include Technology Labs.

Chapter 9 • Measurement and Geometry

Lessons denoted with (*) include Hands-On Labs. Lessons with (†) include Technology Labs.

Chapter 10 • Measurement: Area and Volume

	Worktext	Page	Program of Studies	Core Content	HM Math Course 1
Project	The Architecture of Flowers	380			
10A	Estimating and Finding Area*	382	MA-6-M-S-MPA3	MA-06-2.1.1	10-1
10B	Area of Triangles and Trapezoids	388	MA-6-M-S-MPA3, MA-6-AT-S-EI2	MA-06-2.1.1, MA-06-2.1.3	10-2, Lab 10-2
10C	Area of Composite Figures	392	MA-6-M-S-MPA3	MA-06-2.1.1	10-3
10D	Comparing Perimeter and Area*	395	MA-6-M-S-MPA1, MA-6-M-S-MPA2, MA-6-M-S-MPA6	MA-06-2.1.3	10-4
10E	Area of Circles	402	MA-6-AT-S-EI2	--	10-5, Lab 10-5
10F	Three-Dimensional Figures*	405	MA-6-G-S-SR5	MA-06-3.1.3	10-6, Lab 10-6
Review	Chapter Summary	410			
Practice	KCCT Prep	411			

Lessons denoted with (*) include Hands-On Labs. Lessons with (†) include Technology Labs.

Chapter 11 • Integers, Graphs, and Functions

	Worktext	Page	Program of Studies	Core Content	HM Math Course 1
Project	Graphing Earth	413			
11A	Tables and Functions†	415	MA-6-G-S-CG2, MA-6-AT-S-PRF1, MA-6-AT-S-PRF2, MA-6-AT-S-PRF4	MA-06-4.1.2, MA-06-5.1.1, MA-06-5.1.2, MA-06-5.1.3, MA-06-5.1.4, MA-06-5.1.5	11-9
11B	Ordered Pairs	422	MA-6-AT-S-PRF3, MA-6-G-S-CG1	MA-06-3.3.1	6-6
11C	Graphing Functions†	425	MA-6-G-S-CG2, MA-6-AT-S-PRF2, MA-6-AT-S-PRF3, MA-6-AT-S-PRF4	MA-06-4.1.2, MA-06-5.1.2, MA-06-5.1.3, MA-06-5.1.4, MA-06-5.1.5	11-10
11D	Reflections and Translations	432	MA-6-G-S-TS2	MA-06-3.2.2	8-10
11E	Rotations	435	MA-6-G-S-TS3	MA-06-3.2.3	8-10
11F	Line Symmetry	438	MA-6-G-S-TS1	MA-06-3.2.1	8-11
Review	Chapter Summary	441			
Practice	KCCT Prep	442			

Lessons denoted with (*) include Hands-On Labs. Lessons with (†) include Technology Labs.

Chapter 12 • Probability

	Worktext	Page	Program of Studies	Core Content	HM Math Course 1
Project	Financial Planning	444			
12A	Introduction to Probability	446	MA-6-DAP-S-P5	--	12-1
12B	Experimental Probability†	449	MA-6-DAP-S-P4	MA-06-4.4.2	12-2, Lab 12-2
12C	Counting Methods and Sample Spaces*	454	MA-6-DAP-S-P1, MA-6-DAP-S-P2, MA-6-DAP-S-P5	MA-06-4.4.1	12-3
12D	Theoretical Probability*	461	MA-6-DAP-S-P5	MA-06-4.4.3	12-4
12E	Compound Events†	468	MA-6-DAP-S-P1, MA-6-DAP-S-P2, MA-6-DAP-S-P5	MA-06-4.4.1	12-5, Lab 12-5
12F	Making Predictions†	473	MA-6-DAP-S-CD1, MA-6-DAP-S-P3, MA-6-DAP-S-P5	MA-06-4.4.2	12-6
12G	Sample Size and Reliability	478	MA-6-DAP-S-ES2	MA-06-2.1.1	--
Review	Chapter Summary	481			
Practice	KCCT Prep	482			

Lessons denoted with (*) include Hands-On Labs. Lessons with (†) include Technology Labs.

Kentucky Program of Studies and Core Content

The table below lists the standards in the Grade 6 Program of Studies and the Core Content for Assessment and shows which Worktext lessons address each standard.

Grade 6 Program of Studies		Worktext Lessons
MA-6-NPO-S-E1	Students will estimate and mentally compute to solve real-world and/or mathematical problems with whole numbers, fractions, decimals and percents, checking for reasonable and appropriate computational results.	1B, 1E, 3B, 3C, 3D, 4I, 7J
MA-6-NPO-S-E2	Students will estimate large and small quantities of objects.	1B, 3B, 4I
MA-6-NPO-S-NO1	Students will develop addition, subtraction, multiplication and division of common fractions and decimals with manipulatives and symbols (e.g., mental computation, paper and pencil, calculators).	1E, 3C, 3D, 3E, 3F, 3H, 4H, 5B, 5C, 5D, 5F, 5G, 5H, 5I
MA-6-NPO-S-NO2	Students will add, subtract, multiply, divide and apply order of operations with whole numbers, fractions and decimals to solve real-world problems.	1D, 3C, 3D, 3E, 3F, 3G, 4H, 5B, 5C, 5D, 5F, 5G, 5H, 5I
MA-6-NPO-S-NO3	Students will explain and/or demonstrate inversely-related operations (addition and subtraction; multiplication and division).	2E, 2F, 2G, 2H, 3H, 5E, 5J
MA-6-NPO-S-NS1	Students will continue to develop number sense using fractions, decimals and percents, including percents greater than 100% and improper fractions.	3A, 3G, 4D, 4E, 4F, 4G, 7G, 7H
MA-6-NPO-S-NS2	Students will extend applications of operations (+, −, ×, ÷) to include fractions and decimals.	3C, 3D, 3E, 3F, 4H, 5B, 5C, 5D, 5F, 5G, 5H, 5I
MA-6-NPO-S-NS3	Students will develop place value of large and small numbers, including decimals.	1A, 3A

Grade 6 Program of Studies		Worktext Lessons
MA-6-NPO-S-NS4	Students will explore positive integral exponents (e.g. squares, cubes).	1C
MA-6-NPO-S-NS5	Students will compare, order and convert between whole numbers, fractions, decimals and percents using concrete materials, drawings or pictures and mathematical symbols (e.g., $<$, $\leq$, $>$, $\geq$, $=$, $\neq$, order on a number line).	1A, 3A, 4D, 4G, 7G, 7H
MA-6-NPO-S-PNO1	Students will determine prime numbers, composite numbers, prime factorization, factors, multiples, greatest common factor and least common multiple.	4A, 4B, 4C, 5A
MA-6-NPO-S-PNO2	Students will simplify fractions and determine equivalent fractions.	4E, 5B, 5C
MA-6-NPO-S-PNO3	Students will use prime numbers, composite numbers, factors, multiples and divisibility to solve problems.	5A
MA-6-NPO-S-PNO4	Students will explore and/or demonstrate how applications of properties (e.g., commutative, associative, inverse and identity for addition and multiplication) show relationships among numbers and operations.	1E
MA-6-NPO-S-RP1	Students will develop meaning of percent and how to determine a percentage.	7G, 7H, 7I, 7J
MA-6-NPO-S-RP2	Students will develop meaning of ratio (e.g., describe and compare two sets of data using ratios and appropriate notations: 3:5, $\frac{3}{5}$, 3 to 5).	7A, 7B
MA-6-NPO-S-RP3	Students will define and apply ratios to solve real-world problems.	7A, 7B, 7C, 7E, 7F

Grade 6 Program of Studies		Worktext Lessons
MA-6-M-S-MPA1	Students will find perimeter of regular and irregular polygons in metric and U.S. customary units.	10D, 9E
MA-6-M-S-MPA2	Students will read and use measurement tools (e.g., rulers, scales, protractors, angle rulers).	10D, 7D, 8B, 8C, 9A, 9B
MA-6-M-S-MPA3	Students will find area of plane figures composed of triangles, squares and rectangles by subdividing and measuring; use square units appropriately.	10A, 10B, 10C
MA-6-M-S-MPA4	Students will estimate and find angle measures and segment measures.	8B, 8C, 9A, 9B
MA-6-M-S-MPA5	Students will estimate measurements in standard units, including fractions and decimals.	9A, 9B, 9E
MA-6-M-S-MPA6	Students will explain how measurements and measurement formulas are related or different (e.g., compare the perimeter with the area of a rectangle).	10D
MA-6-M-S-SM1	Students will describe and provide examples of U.S. Customary and metric units of measurement; use these units to solve real-world and/or mathematical problems.	9A, 9B, 9C, 9D

	Grade 6 Program of Studies	Worktext Lessons
MA-6-M-S-SM2	Students will estimate, compare and convert (meaning to make ballpark comparisons/not memorize conversion factors between U.S. and metric) units of measurement for length, weight/mass and volume/capacity within the U.S. customary system and within the metric system: • length (e.g., parts of an inch, inches, feet, yards, miles, millimeters, centimeters, meters, kilometers); • weight/mass (e.g., pounds, tons, grams, kilograms); • volume/capacity (e.g., cups, pints, quarts, gallons, milliliters, liters).	9A, 9B, 9C, 9D
MA-6-G-S-CG1	Students will identify and graph ordered pairs on a positive coordinate system, identifying the origin, axes and ordered pairs.	11B
MA-6-G-S-CG2	Students will apply graphing in the positive coordinate system to solve real-world and mathematical problems.	11A, 11C
MA-6-G-S-SR1	Students will formulate and use the rules for the sum of angle measures in a triangle (180°) and in a quadrilateral (360°).	8E, 8G
MA-6-G-S-SR2	Students will identify and use relationships among lines (e.g., parallel, perpendicular).	8D
MA-6-G-S-SR3	Students will identify, describe and provide examples of the basic geometric elements (points, rays, lines, segments, angles [acute, right, obtuse], planes, radius, diameter, circumference).	8A, 8B, 8C, 9F

Grade 6 Program of Studies		Worktext Lessons
MA-6-G-S-SR4	Students will identify, describe and provide examples and properties of two-dimensional figures (circles, triangles [acute, right, obtuse, scalene, isosceles, equilateral], quadrilaterals, regular polygons); apply these properties and figures to solve real-world problems.	8E, 8F, 9F
MA-6-G-S-SR5	Students will describe, provide examples of and identify properties (e.g., vertices, angles, faces, edges, congruent parts) of common three-dimensional figures (spheres, cones, cylinders, prisms and pyramids).	10F
MA-6-G-S-SR6	Students will describe and provide examples of congruent and similar plane figures; apply congruent and similar plane figures to solve real-world problems.	7D, 7E, 8I
MA-6-G-S-TS1	Students will determine lines of symmetry for a plane figure, sketch plane figures with multiple lines of symmetry and apply line symmetry to real-world and/or mathematical situations.	11F
MA-6-G-S-TS2	Students will transform (translate and reflect across a horizontal or vertical line) figures in the first quadrant of the coordinate plane and determine new coordinates of the shape after transformation.	11D
MA-6-G-S-TS3	Students will explore the rotation of a figure in a plane in the first quadrant, with and without manipulatives.	11E
MA-6-DAP-S-CD1	Students will make predictions, draw conclusions and verify results from statistical data and probability experiments.	12F

	Grade 6 Program of Studies	Worktext Lessons
MA-6-DAP-S-CD2	Students will determine and apply measures of distribution (mean, median, mode, range).	6B
MA-6-DAP-S-DR1	Students will select an appropriate graph to represent given data and justify the selection.	6H
MA-6-DAP-S-DR2	Students will collect, organize, construct, analyze and interpret data in a variety of graphical methods, including line plots, line graphs, circle graphs, bar graphs and stem-and-leaf plots.	6C, 6D, 6E, 6F
MA-6-DAP-S-DR3	Students will compare data from various types of graphs.	6H
MA-6-DAP-S-DR4	Students will relate different representations of data (e.g., tables, graphs, diagrams, plots).	6A, 6C, 6D, 6E
MA-6-DAP-S-ES1	Students will pose questions; collect, organize and display data.	6A, 6H
MA-6-DAP-S-ES2	Students will explore how sample size affects the reliability of the outcome.	12G
MA-6-DAP-S-P1	Students will describe or determine (e.g., tables, tree diagrams) the sample space of an event.	12C, 12E
MA-6-DAP-S-P2	Students will investigate solutions to probability problems using counting techniques, tree diagrams, charts and tables.	12C, 12E
MA-6-DAP-S-P3	Students will make predictions, draw conclusions and verify results from statistical data and probability experiments.	12F
MA-6-DAP-S-P4	Students will determine simple probabilities based on the results of an experiment and make inferences based on the data.	12B

Grade 6 Program of Studies		Worktext Lessons
MA-6-DAP-S-P5	Students will explore the role of probability in decision making.	12A, 12C, 12D, 12E, 12F
MA-6-AT-S-EI1	Students will use concrete and/or informal methods to solve equations with one variable that model real-world situations.	2D, 2E, 2F, 2G, 2H, 3H, 5E, 5J
MA-6-AT-S-EI2	Students will solve problems involving simple formulas (e.g., $A = lw$, $D = rt$).	10B, 10E, 9F
MA-6-AT-S-EI3	Students will model and solve real-world problems with one variable equations and inequalities (e.g., $8x = 4$, $x + 2 > 5$).	2E, 2F, 2G, 2H, 2I, 3H, 5E, 5J
MA-6-AT-S-PRF1	Students will recognize, create and extend patterns (give an informal description of the continuation of a pattern and/or generalize a pattern through a verbal rule).	11A, 8H
MA-6-AT-S-PRF2	Students will represent, interpret and describe function relationships through tables, graphs and verbal rules.	11A, 11C
MA-6-AT-S-PRF3	Students will organize input-output coordinate pairs into tables and plot points in the first quadrant of a coordinate (Cartesian) system/grid.	11B, 11C
MA-6-AT-S-PRF4	Students will explain how the change in one quantity affects change in another quantity (e.g., in tables or graphs, input/output tables).	11A, 11C
MA-6-AT-S-VEO1	Students will explore the use of variables in expressions and equations.	2D, 2E, 2F, 2G, 2H, 3H, 5E, 5J
MA-6-AT-S-VEO2	Students will substitute numerical values for variables and evaluate algebraic expressions.	2A, 5D, 5F, 5G
MA-6-AT-S-VEO3	Students will describe, define and provide examples of algebraic expressions based on real-world and/or mathematical situations.	2B, 2C, 5E, 5J

Grade 6 Core Content for Assessment Version 4.1		Worktext Lessons
MA-06-1.1.1	Students will provide examples of and identify fractions, decimals and percents.	3A, 4D, 4F, 7G, 7H
MA-06-1.1.2	Students will describe and provide examples of representations of numbers (whole numbers, fractions in simplest form, mixed numbers, decimals, percents) and operations in a variety of equivalent forms using models, diagrams, and symbols (e.g., number lines, 10 by 10 grids, rectangular arrays, number sentences), based on real-world and mathematical problems.	3A, 3C, 3D, 4D, 4E, 4F, 4G, 4H, 4I, 5B, 5D, 5F, 5G, 5I, 7G, 7I
MA-06-1.1.3	Students will convert between any two of the following numbers: fractions, decimals, and percents (less than or equal to 100%); and will compare and order these numbers.	3A, 4D, 4G, 7G, 7H
MA-06-1.2.1	Students will estimate to solve real-world and mathematical problems with whole numbers, fractions, decimals and percents, checking for reasonable and appropriate computational results.	1B, 3B, 3C, 3D, 3G, 4I, 7J
MA-06-1.3.1	Students will add, subtract, multiply and divide whole numbers, fractions and decimals to solve real-world problems and apply order of operations to simplify numerical expressions.	1C, 1D, 3C, 3D, 3E, 3F, 4H, 5B, 5C, 5D, 5F, 5G, 5H, 5I
MA-06-1.3.2	Students will explain how operations (addition and subtraction; multiplication and division) are inversely related.	2E, 2F, 2G, 2H, 3H, 5E, 5J
MA-06-1.4.1	Students will describe and apply ratios to solve real-world problems.	7A, 7B, 7C, 7E, 7F, 7G

Grade 6 Core Content for Assessment Version 4.1		Worktext Lessons
MA-06-1.5.1	Students will identify and apply prime numbers, composite numbers, prime factorization, factors, multiples and divisibility to solve real-world and mathematical problems (e.g., prime factorization to determine a least common multiple [LCM] or greatest common factor [GCF]).	4A, 4B, 4C, 5A
MA-06-1.5.2	Students will identify the use of properties (commutative properties of addition and multiplication, the associative properties of addition and multiplication and the identity properties for addition and multiplication) to simplify numerical expressions.	1E
MA-06-2.1.1	Students will measure lengths (to the nearest eighth of an inch or the nearest centimeter) and will determine and use in real-world and mathematical problems: • area and perimeter of triangles; • area and perimeter of quadrilaterals (rectangles, squares); (using the Pythagorean theorem will not be required as a strategy) and • area and perimeter of compound figures composed of triangles and quadrilaterals.	10A, 10B, 10C, 12G, 9E
MA-06-2.1.2	Students will estimate measurements in standard units including fractions and decimals.	9A, 9B, 10B
MA-06-2.1.3	Students will explain how measurements and measurement formulas are related or different (perimeter and area of rectangles).	10D
MA-06-2.2.1	Students will convert units within the same measurement system and use these units to solve real-world problems.	9C, 9D

Grade 6 Core Content for Assessment Version 4.1		Worktext Lessons
MA-06-3.1.1	Students will describe and provide examples of the basic geometric elements (points, rays, lines, segments, angles [acute, right, obtuse], planes, radius, diameter, circumference).	8A, 8B, 8C, 8D, 9F
MA-06-3.1.2	Students will describe, and provide examples of the elements (e.g., sides, vertices, angles, congruent parts) of two-dimensional figures (circles, triangles, quadrilaterals, regular polygons), and will apply these elements and figures to solve real-world and mathematical problems.	8E, 8F, 8G, 9F
MA-06-3.1.3	Students will describe, provide examples of, and identify elements (e.g., vertices, angles, faces, edges, congruent parts) of common three-dimensional figures (spheres, cones, cylinders, prisms, and pyramids).	10F
MA-06-3.1.4	Students will identify and describe congruent figures, and will apply congruent figures to solve real-world and mathematical problems.	8I
MA-06-3.1.5	Students will identify similar figures and apply similar figures to solve real-world and mathematical problems.	7D, 7E
MA-06-3.2.1	Students will describe, provide examples of, and apply line symmetry to real-world and mathematical problems.	11F
MA-06-3.2.2	Students will: • reflect figures across a horizontal or vertical line in the first quadrant; • translate figures in a plane in the first quadrant and • determine the coordinates of the image after transformation in the first quadrant.	11D
MA-06-3.2.3	Students will identify rotations of figures in the plane (90° and 180°).	11E

Grade 6 Core Content for Assessment Version 4.1		Worktext Lessons
MA-06-3.3.1	Students will identify and graph ordered pairs on a positive coordinate system (Quadrant I), correctly identifying the origin, axes and ordered pairs; and will apply graphing in the coordinate system to solve real-world and mathematical problems.	11B
MA-06-4.1.1	Students will analyze and make inferences from data displays (drawings, tables/charts, pictographs, bar graphs, circle graphs, line plots, Venn diagrams, line graphs, stem-and-leaf plots).	6A, 6C, 6D, 6E, 6F, 6G, 7B, 9G
MA-06-4.1.2	Students will explain how different representations of data (e.g., tables, graphs, diagrams, plots) are related.	11A, 11C, 6H
MA-06-4.1.4	Students will determine and construct appropriate data displays (bar graphs, line plots, Venn diagrams, tables, line graphs), and will explain why the type of display is appropriate for the data.	6C, 6D, 6E, 6G, 6H
MA-06-4.2.1	Students will determine and apply the mean, median, mode and range of a set of data.	6B
MA-06-4.4.1	Students will describe or determine (e.g., tables, tree diagrams) the sample space of an event for a real-world or mathematical situation.	12C, 12E
MA-06-4.4.2	Students will determine single event probabilities based on the results of an experiment and will make inferences based on the data.	12B, 12F
MA-06-4.4.3	Students will explore the theoretical probability of simple events.	12D
MA-06-5.1.1	Students will extend, describe rules for patterns and find a missing term in a pattern from real-world and mathematical problems.	11A, 1F, 8H

Grade 6 Core Content for Assessment Version 4.1		Worktext Lessons
MA-06-5.1.2	Students will create tables for functions and will apply the tables to solve real-world problems.	11A, 11C
MA-06-5.1.3	Students will describe, define, provide examples of, and apply to real-world and mathematical problems functions using tables, graphs and verbal rules.	11A, 11C
MA-06-5.1.4	Students will explain how tables, graphs and patterns relate to each other.	11A, 11C
MA-06-5.1.5	Students will explain how the change in one quantity affects change in another quantity (e.g., in tables or graphs, input/output tables).	6A, 11A, 11C
MA-06-5.2.1	Students will substitute values for variables (up to two different variables) and evaluate algebraic expressions.	2A, 5D, 5F, 5G
MA-06-5.2.2	Students will describe, define and provide examples of variables and expressions with a missing value based on real-world and mathematical problems.	2B, 2C, 3H, 5E, 5J
MA-06-5.3.1	Students will model and solve real-world and mathematical problems with simple equations and inequalities (e.g., $8x = 4$, $x + 2 = 5$)	2D, 2E, 2F, 2G, 2H, 2I, 3H, 5E, 5J

CHAPTER 1

Project
Whole Numbers: Feed the Animals

Career: Veterinary Technician

Do you like caring for animals? Veterinary technicians perform many of the same tasks for veterinarians as nurses do for doctors. Veterinary technicians also do research that can help animals. To care for animals, technicians must know what the animals need to eat and how they behave with other types of animals. Large plant-eating animals, many of which live in Africa, need to eat specific kinds of grasses and trees.

Before you start the project, work through the Animal Food Intake worksheet on the next page.

In this chapter project . . .

You design a zoo habitat.

What to do:

- Research three animals of interest to you. Learn about the climates and terrain they live in, as well as the varieties and amounts of food they eat.
- Choose one of the three animals to feature in your habitat.
- Determine the size of the habitat your animal requires.
- Estimate the volume and weight of food your animal typically eats in one month.
- Determine the cost of feeding and caring for your animal for one year.

What to present:

- An overview of your animal's habitat. Include information on the size and characteristics of the habitat.
- A budget for feeding and caring for your animal for one year.

Where's the math?

- Calculate the approximate number of square feet, acres, or hectares required for your habitat.
- Determine the storage space required for the amount of food you plan to keep for your animal.
- Estimate the cost of feeding and caring for your animal for one year.

go.hrw.com
Chapter Project Online
KEYWORD: MR7 Ch1

Name ______________________ Date __________ Class __________

CHAPTER 1

Project Recording Sheet

Whole Numbers and Patterns

Animal Food Intake

How does the amount of food that these animals eat compare to one another? Complete the chart to find the amount of food each animal would eat in a year. Make an estimate first; then make the calculations.

Food Intake

Animal	**Weight** (lb)	**Daily Food Intake** (lb)	**Estimated Food Intake per Year** (lb)	**Calculated Food Intake per Year** (lb)
Elephant	11,000	660		
Hippopotamus	5500	90		
Giraffe	2500	75		
Buffalo	1500	45		
Zebra	950	30		

Does the weight of the animal determine how much it eats per pound?

____

You can also compare the animals by using the relationship below:

Daily Food Intake/Weight

For example, the comparison of an African buffalo's food needs compared to its weight would be $\frac{45 \text{ lb}}{1500 \text{ lb}}$ or 0.03 or $\frac{3}{100}$. This is read as "a buffalo needs three hundredths of a pound of food for every pound of body weight."

Use the chart above to complete the following chart. Round each answer to the nearest hundredth.

Which animal needs the most food per pound of body weight? ______________

Extension: What things determine the amount of food that an animal needs?

____

____

____

____

Animal	**Daily Food Intake/Weight**
Elephant	
Hippopotamus	
Giraffe	
Buffalo	
Zebra	

EXPLORATION

Program of Studies and Core Content
MA-6-NPO-S-NS3, MA-6-NPO-S-NS5

1A Comparing and Ordering Whole Numbers

Values **increase** as you move **right** on a number line.

Values **decrease** as you move **left** on a number line.

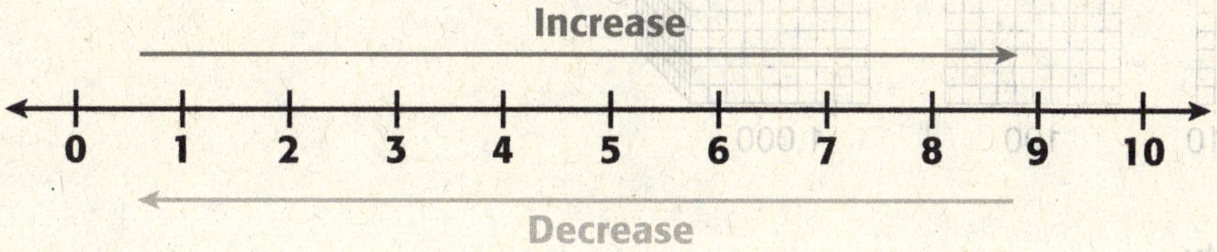

The number line below shows some large whole numbers between 1,000 and 1,100.

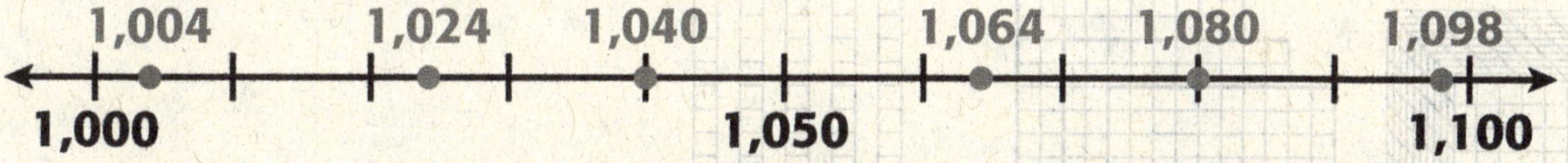

Compare the numbers from the number line above.
Write < or > .

1. 1,024 ☐ 1,080

2. 1,024 ☐ 1,004

3. 1,064 ☐ 1,040

4. 1,004 ☐ 1,040

5. 1,040 ☐ 1,024

6. 1,064 ☐ 1,098

Think and Discuss

7. Explain how to use a number line to compare whole numbers.

8. Describe how to use place value to compare the numbers 1,004 and 1,040.

Name ______________________ Date __________ Class __________

LESSON **1A**

Hands-On Lab

Comparing Whole Numbers

You can create and compare whole numbers using base 10 blocks.

KEY

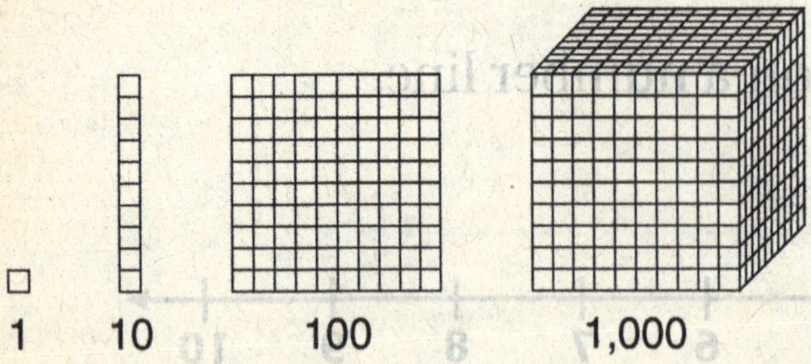

Activity

Which is greater 1,350 or 1,172? Use base 10 blocks to show your answer.

Create 1,350 with base 10 blocks.

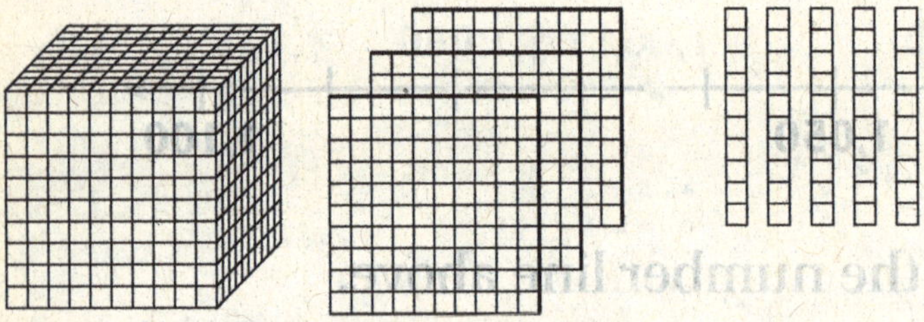

Create 1,172 with base 10 blocks.

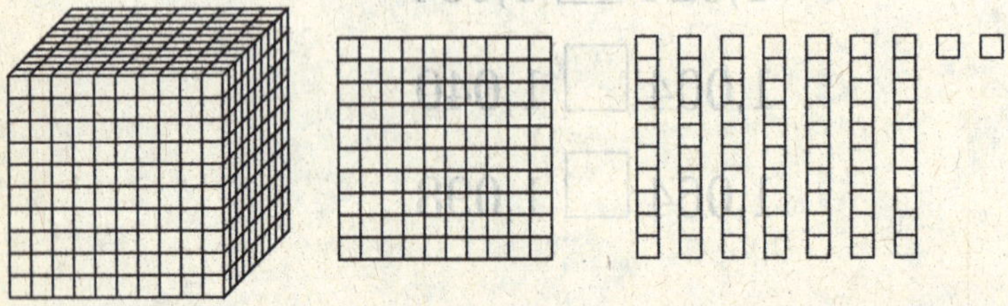

Compare the numbers starting with the thousands.

Each number has 1 thousand block, so look at the hundreds.
1,350 has 3 hundred squares, but 1,172 has 1 hundred square.

1,350 is greater than 1,172.

Name ______________________________ Date ____________ Class ____________

LESSON 1A

Hands-On Lab

Comparing Whole Numbers, continued

Think and Discuss

1. Explain how to compare numbers with different numbers of digits. For example, how do you compare 670 and 2,824?

____

____

____

Try This

1. Use base 10 blocks to represent 321 and 148. Draw the numbers below.

321

148

Which number is greater? ____________________

Name ______________________________ Date ____________ Class ____________

LESSON 1A

Hands-On Lab

Comparing Whole Numbers, continued

2. Use base 10 blocks to represent 2,010 and 1,535. Sketch a model of each number below.

2,010

1,535

Which number is greater? ______________________

3. Use base 10 blocks to represent 729 and 1,004. Sketch a model of each number below.

729

1,004

Which number is greater? ______________________

Name ______________________________ Date ____________ Class ____________

LESSON **1A**

Review for Mastery

Comparing and Ordering Whole Numbers

You can use place value to compare or order whole numbers.
Use < or > to compare the numbers.

289,865 ☐ 289,765

Thousands			Ones			Compare the digits from left to right.
H	T	O	H	T	O	
2	8	9	**8**	6	5	First Number
2	8	9	**7**	6	5	Second Number

$8 > 7$
So, $289{,}865 > 289{,}765$

Write < or > to compare the numbers.

1.

Thousands			Ones		
H	T	O	H	T	O
		3	5	4	7
		3	5	3	2

3,547 ☐ 3,532

2.

Thousands			Ones		
H	T	O	H	T	O
		9	5	3	6
		9	6	3	5

9,536 ☐ 9,635

Write the numbers in order from least to greatest.
976; 859; 924

Ones		
H	T	O
9	7	6
8	5	9
9	2	4

Compare the numbers in pairs.
$976 > 859$, $976 > 924$, and $859 < 924$.
So the numbers from least to greatest are
859; 924; 976.

Write the numbers in order from least to greatest.

3.

Ones		
9	5	4
9	4	5
9	6	9

____

4.

Ones		
3	4	3
3	3	4
4	3	4

____

5.

Ones		
8	9	4
8	9	2
9	6	5

____

Name ______________________ Date ____________ Class ____________

LESSON 1A

Homework and Practice

Comparing and Ordering Whole Numbers

Write < or > to compare the numbers.

1. 75 ☐ 57 **2.** 212 ☐ 221 **3.** 780 ☐ 1,003

4. 6,789 ☐ 6,802 **5.** 31,090 ☐ 31,900 **6.** 243,603 ☐ 243,036

Write the numbers from least to greatest.

7. 61; 106; 96 **8.** 389; 396; 369 **9.** 85; 158; 58

______________ ______________ ______________

10. 2,101; 210; 2,010 **11.** 2,649; 916; 2,496 **12.** 7,119; 71,119; 719

______________ ______________ ______________

Write the numbers from greatest to least.

13. 46; 64; 16 **14.** 916; 619; 169 **15.** 248; 419; 189

______________ ______________ ______________

16. 606; 777; 515 **17.** 6,100; 6,216; 6,126 **18.** 1,551; 5,115; 1,515

______________ ______________ ______________

19. The smallest two great lakes are lake Erie and lake Ontario. Lake Erie covers 32,630 square miles and Lake Ontario covers 34,850 square miles. What is the smallest great lake?

____

20. The smallest two major natural lakes in the world are Lake Urmia in Asia and lake Manitoba in North America. Lake Urmia covers 1,815 square miles and Lake Manitoba covers 1,799 square miles. What major natural lake covers the smallest area?

____

EXPLORATION

Program of Studies and Core Content

MA-6-NPO-S-E1, MA-6-NPO-S-E2, MA-06.1.2.1

1B Estimating with Whole Numbers

Harvard Middle School is collecting aluminum cans to recycle for a fund-raiser. The number of cans that each grade collected is shown in the bar graph.

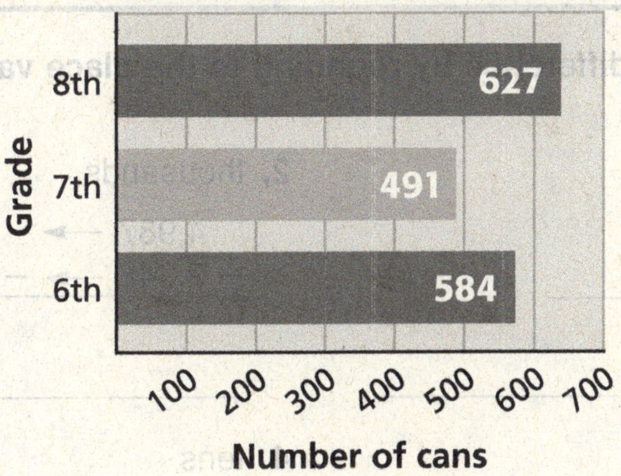

1. Estimate the total number of cans that Harvard Middle School collected.

The principal announced that nearly 2,000 cans were collected. The newspaper reported that over 1,700 cans were collected.

2. Is either of these reports correct?

3. Why are the reports different?

Think and Discuss

4. **Describe** how you reached your estimate.

5. **Identify** some words that indicate whether an amount is an estimate or approximation.

Name ______________________________ Date __________ Class __________

LESSON **1B**

Review for Mastery

Estimating with Whole Numbers

In mathematics, you can find an estimate when an exact answer is not needed. An estimate is close to the exact answer.

You can use rounding to estimate sums and differences.

A. Estimate the sum by rounding to the hundreds.

3,478	→	3,500
+ 7,136	→	+ 7,100
		10,600

B. Estimate the difference by rounding to the thousands.

23,848	→	24,000
− 16,132	→	− 16,000
		8,000

Estimate each sum or difference by rounding to the place value indicated.

1. hundreds

789 →
+ 453 → +

____

2. thousands

4,987 →
− 2,348 → −

____

3. tens

456 →
+ 875 → +

____

4. tens

876 →
− 432 → −

____

5. hundreds

6,898 →
+ 2,671 → +

____

6. thousands

1,857 →
+ 3,598 → +

____

7. hundreds

8,813 →
− 2,384 → −

____

8. thousands

9,128 →
− 4,716 → −

____

Name ______________________ Date ____________ Class ____________

LESSON 1B

Review for Mastery

Estimating with Whole Numbers, continued

You can use rounding and basic facts to estimate products. Count the number of zeros in your rounded numbers. They will appear to the right of the basic fact in your estimate.

Estimate 8 × 532.

8 × 532
↓ ↓
8 × 500 Round each factor.
↓ ↓ two zeros
4,000

Use rounding to estimate each product.

9. 28 × 5
↓ ↓
____

10. 78 × 11
↓ ↓
____

11. 67 × 19
↓ ↓
____

12. 93 × 7
↓ ↓
____

Compatible numbers are numbers that are easy to compute mentally. One compatible number divides evenly into the other.

Estimate the quotient of 553 ÷ 8.

Step 1: What are the multiples of 8?
8 16 24 32 40 48 56 64
Which multiple is closest to 55?
56 is close to 55.
8 and 560 are compatible numbers.

Step 2: Divide. 560 ÷ 8 = 70

Use compatible numbers to estimate each quotient.

13. 748 ÷ 25 __________

14. 557 ÷ 8 __________

15. 417 ÷ 7 __________

16. 241 ÷ 3 __________

Name ______________________ Date __________ Class __________

LESSON 1B

Homework and Practice

Estimating with Whole Numbers

Estimate each sum or difference.

1. 51 + 13 ________

2. 619 − 190 ________

3. 88 − 19 ________

4. 1,497 + 3,809 ________

5. 160 + 201 ________

6. 5,002 − 2,005 ________

7. 27,169 + 16,167 ________

8. 81,169 − 49,136 ________

9. 389,000 + 213,695 ________

Estimate each product or quotient.

10. 75 ÷ 7 ________

11. 63 × 7 ________

12. 69 ÷ 5 ________

13. 8 × 19 ________

14. 163 ÷ 8 ________

15. 49 × 6 ________

16. 316 ÷ 8 ________

17. 266 ÷ 9 ________

18. 69 × 6 ________

19. Sebastian can read 105 pages in an hour. About how many pages can he read in 4 hours?

____

20. A professional basketball court is 94 feet long. If a player runs the length of the court 6 times, about how far did he run?

____

21. Cassandra, a gardener, is preparing to fertilize a lawn. The lawn is 40 yards by 35 yards. One bag of fertilizer will cover an area of 100 square yards. How many bags of fertilizer does Cassandra need to buy?

____

Program of Studies and Core Content
MA-6-NPO-S-NS4,
MA-06-1.3.1

1C Exponents

1. Maria won the grand prize on a game show. She will be given $2 the first month, $4 the second month, $8 the third month, and so on as her payment is doubled each month for one year.

a. Complete the table.

b. How much will Maria receive in the fifth month?

c. How much will Maria receive in the eighth month?

d. Use a calculator to determine how much Maria will receive in the last month of the year.

Month	Amount ($)
1	2
2	4
3	8
4	
5	
6	
7	
8	

Think and Discuss

2. Describe the pattern in the table.

3. Explain how the values in the table compare with the values 2, 2^2, 2^3, 2^4, and so on.

Name ______________________ Date ____________ Class ____________

LESSON 1C

Technology Lab

Evaluating Exponents Using Your Calculator

When a number is written with a base and an exponent it is in exponential form. The number 5^3 is in exponential form. 5 is the base and 3 is the exponent.

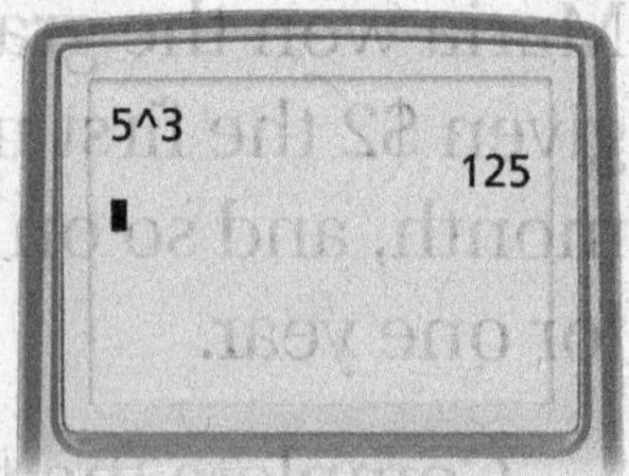

Evaluate 5^3 on your graphing calculator by pressing 5, then pressing ^ and 3.

Press ENTER.

What happens?

The calculator displays **125.**

Activity

1. Evaluate 4^8 on a calculator. Write the product in standard form.

   The product is **65,536.**
   $4 \times 4 \times 4 \times 4 \times 4 \times 4 \times 4 \times 4 = 65{,}536$

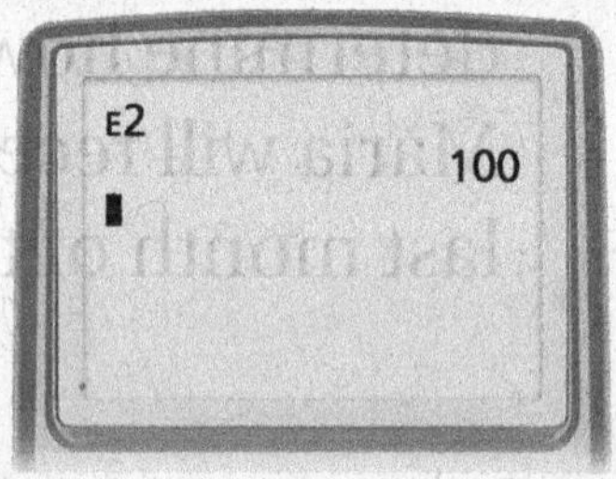

2. Use the EE function (the second function of the comma key) to enter 10^2 into a calculator using scientific notation.

   To enter 10^2, press 2nd, ,, 2, ENTER.

   The product is 100.
   $10 \times 10 = 100$

Think and Discuss

1. Tell what happens if you enter 10^0 into your calculator.

____

Try This

Use your calculator to evaluate each expression.

1. 12^1 ______________ **2.** 6^5 ______________ **3.** 50^2 ______________

Name ______________________ Date __________ Class __________

LESSON 1C

Review for Mastery

Exponents

You can write a number in exponential form to show repeated multiplication. A number written in exponential form has a base and an exponent. An exponent tells you how many times a number, called the base, is used as a factor.

8^4 ← exponent

↑

base

Write the expression in exponential form.

$6 \times 6 \times 6$

6 is used as a factor 3 times.

$6 \times 6 \times 6 = 6^3$

Write each expression in exponential form.

1. $8 \times 8 \times 8 \times 8 \times 8$ **2.** 3×3 **3.** $5 \times 5 \times 5 \times 5$ **4.** $7 \times 7 \times 7$

________ ________ ________ ________

You can find the value of expressions in exponential form.

Find the value.

2^5

Step 1: Write the expression as repeated multiplication.

$2^5 = 2 \times 2 \times 2 \times 2 \times 2$

Step 2: Multiply.

$2 \times 2 \times 2 \times 2 \times 2 = 32$

$2^5 = 32$

Find each value.

5. 12^3 **6.** 9^3 **7.** 10^4 **8.** 4^6

________ ________ ________ ________

Name ________________ Date __________ Class __________

LESSON 1C

Homework and Practice

Exponents

Write each expression in exponential form.

1. $2 \times 2 \times 2 \times 2$ ______

2. $3 \times 3 \times 3 \times 3 \times 3$ ______

3. $9 \times 9 \times 9$ ______

4. 6×6 ______

5. $5 \times 5 \times 5 \times 5 \times 5 \times 5 \times 5$ ______

6. $7 \times 7 \times 7 \times 7 \times 7 \times 7$ ______

Find each value.

7. 5^2 ______

8. 5^4 ______

9. 3^4 ______

10. 2^6 ______

11. 121^1 ______

12. 4^5 ______

13. 12^2 ______

14. 29^0 ______

Compare using <, >, or =.

15. 3^4 ☐ 9^2

16. 2^4 ☐ 3^3

17. 5^3 ☐ 10^2

18. 2^2 ☐ 16^1

19. 6^2 ☐ 2^5

20. 8^2 ☐ 2^6

21. What whole number equals 49 when it is squared and 343 when it is cubed?

______

22. Use exponents to write the number 16 three different ways?

______

EXPLORATION

Program of Studies and Core Content
MA-6-NPO-S-NO2, MA-06-1.3.1

1D Order of Operations

Calculators are programmed to perform operations in a certain order. Each keystroke sequence below results in 17.

3 [×] 5 [ENTER] [+] 2 [ENTER]   2 [+] 3 [×] 5 [ENTER]

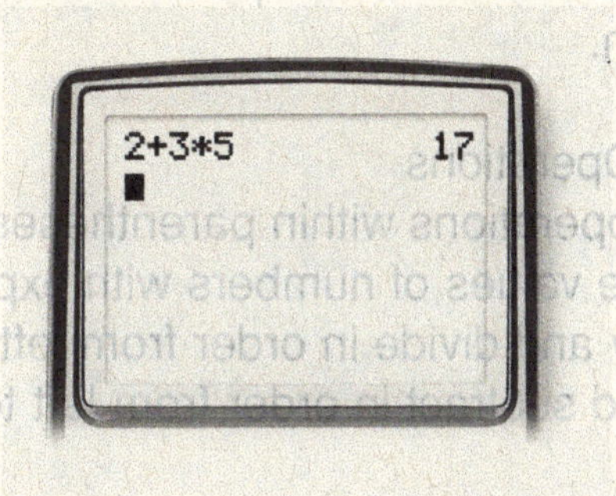

For each keystroke sequence, determine the order of operations the calculator follows.

1. [(] 2 [+] 3 [)] [×] 5 [ENTER]
2. 2 [∧] 3 [−] 1 [×] 4 [ENTER]
3. 2 [∧] [(] 3 [−] 1 [)] [×] 4 [ENTER]

Write the keystroke sequence for each expression.

4. $5 - 2^2$
5. $(2 - 3)^3 + 2$

Think and Discuss

6. **Explain** why there needs to be a rule for the order of operations.

Name ______________________ Date __________ Class __________

LESSON 1D

Review for Mastery

Order of Operations

A mathematical phrase that includes only numbers and operations is called a numerical expression.

$9 + 8 \times 3 \div 6$ is a numerical expression.

To evaluate a numerical expression, you find its value.

You can use the order of operations to evaluate a numerical expression.

Order of Operations
1. Do all operations within parentheses.
2. Find the values of numbers with exponents.
3. Multiply and divide in order from left to right.
4. Add and subtract in order from left to right.

Evaluate the expression.

$60 \div (7 + 3) + 3^2$	
$60 \div 10 + 3^2$	Do all operations within parentheses.
$60 \div 10 + 9$	Find the values of numbers with exponents.
$6 + 9$	Multiply and divide in order from left to right.
15	Add and subtract in order from left to right.

Evaluate each expression.

1. $7 \times (12 + 8) - 6$

$7 \times$ ______ $- 6$

______ $- 6$

____

2. $10 \times (12 + 34) + 3$

$10 \times$ ______ $+ 3$

______ $+ 3$

____

3. $10 + (6 \times 5) - 7$

$10 +$ ______ $- 7$

______ $- 7$

____

4. $2^3 + (10 - 4)$

____

5. $7 + 3 \times (8 + 5)$

____

6. $36 \div 4 + 11 \times 8$

____

7. $5^2 - (2 \times 8) + 9$

____

8. $3 \times (12 \div 4) - 2^2$

____

9. $(3^3 + 10) - 2$

____

Name ______________________ Date __________ Class __________

LESSON 1D

Homework and Practice

Order of Operations

Evaluate each expression.

1. $7 + 3 \times 4$

2. $(45 - 10) \times 2$

3. $(20 + 16) \div 2 + 10$

4. $2^2 \times 3 \div 4$

5. $9 \times (7 + 1) \div 2$

6. $(9 - 2) + 8 - (1 + 3)$

7. $12 + 36 \div 2^2$

8. $7 + 90 \div 9 + 10$

9. $24 + (3^4 - 1) \div 10$

Insert parentheses so that each equation is correct.

10. $16 - 5 \times 2 - 2 = 20$

11. $2^5 - 25 \times 7 = 49$

12. $30 \div 3 \times 5 - 1 = 1$

13. $21 - 5 + 6 = 10$

14. $8 + 4 \cdot 2 + 7 \div 3 = 20$

15. $6 + 4^2 - 9 = 13$

16. $3 \times 21 - 10 + 3 = 36$

17. $3 \times 100 - 5^2 = 225$

18. $14 \div 2 + 7 \times 2 = 21$

19. Tiffany jogged 3 miles a day for the first week of her exercise plan. Then she jogged 5 miles a day for the next 10 days. How many miles did Tiffany jog in all?

____

20. Juan's father bought 2 tickets to the baseball game and 2 large bags of peanuts. Each ticket cost \$25 and each bag of peanuts cost \$2. he paid with a \$100 bill. How much change did he get back?

____

Program of Studies and Core Content

MA-6-NPO-S-NO1, MA-6-NPO-S-E1, MA-6-NPO-S-PNO4, MA-06-1.5.2

1E Mental Math

1. Choose one expression from each pair to evaluate using mental math.

a.	$25 \cdot 24$	$(25 \cdot 4) \cdot 6$
b.	$5 \cdot 22$	$(5 \cdot 20) + (5 \cdot 2)$
c.	$13 + 44 + 27$	$44 + (13 + 27)$
d.	$41 + 32 + 9 + 18$	$(41 + 9) + (32 + 18)$

2. What makes the expressions you chose easier to evaluate?

3. What makes the expressions you did not choose more difficult to evaluate?

Think and Discuss

4. **Discuss** the mental math strategies you used.

5. **Compare** the first expression with the second expression in each pair. How are they alike? How are they different?

Name ______________________ Date ____________ Class ____________

LESSON **1E**

Review for Mastery

Mental Math

Commutative Property

Changing the order of addends does not change the sum.

$21 + 13 = 13 + 21$

Changing the order of factors does not change the product.

$5 \times 7 = 7 \times 5$

Associative Property

Changing the grouping of addends does not change the sum.

$(3 + 8) + 4 = 3 + (8 + 4)$

Changing the grouping of factors does not change the product.

$2 \times (7 \times 4) = (2 \times 7) \times 4$

Distributive Property

When you multiply a number by a sum, you can

- Find the sum and then multiply. $3 \times (8 + 4) = 3 \times 12 = 36$

or

- Multiply the number by each addend and then find the sum.
  $3 \times (8 + 4) = (3 \times 8) + (3 \times 4) = 24 + 12 = 36$

Identify the property shown.

1. $3 \times (2 \times 6) = (3 \times 2) \times 6$ ______________

2. $7 + 18 = 18 + 7$ ______________

3. $4 \times (8 + 5) = 4 \times 13$ ______________

4. $11 \times 8 = 8 \times 11$ ______________

5. $3 \times (8 + 4) = (3 \times 8) + (3 \times 4)$ ______________

6. $(3 + 8) + 4 = 3 + (8 + 4)$ ______________

Identify the property shown and the missing number in each equation.

7. $9 + 16 = y + 9$ ______________

8. $4 \times (3 \times 2) = (4 \times n) \times 2$ ______________

9. $3 \times (11 + 4) = 3 \times a$ ______________

10. $6 \times (9 + 14) = b \times 23$ ______________

Name ______________________ Date __________ Class __________

LESSON 1E

Review for Mastery

Mental Math, continued

Find each sum or product.

A. 8 + 9 + 22 + 31

8 + 22 + 9 + 31 Use the Commutative Property.

(8 + 22) + (9 + 31) Use the Associative Property.

30 + 40 Use mental math to add.

70

B. 5 × 7 × 4

7 × 5 × 4 Use the Commutative Property.

7 × (5 × 4) Use the Associative Property.

7 × 20 Use mental math to multiply.

140

Find each sum or product. Identify any properties you used.

11. 3 + 58 + 27 + 22 __________

12. 8 × 3 × 5 __________

13. 5 × 3 × 4 __________

14. 54 + 32 + 78 + 106 __________

15. 84 + 11 + 26 + 39 __________

16. 10 × 3 × 7 __________

Find the product.

6 × 34

Step 1: Write one factor as a sum of two numbers.
6 × 34 = 6 × (30 + 4)

Step 2: Use the Distributive Property.
6 × (30 + 4) = (6 × 30) + (6 × 4)

Step 3: Use mental math to multiply and add.
(6 × 30) + (6 × 4) = 180 + 24 = 204

Use the Distributive Property to find each product.

17. 6 × 43 __________

18. 12 × 34 __________

19. 53 × 4 __________

20. 74 × 8 __________

Name ______________________ Date __________ Class __________

LESSON 1E

Homework and Practice

Mental Math

Find each sum or product. Identify any properties you used.

1. 26 + 14 + 11 + 9

2. 5 × 5 × 4

3. 38 + 27 + 13 + 12

4. 5 × 6 × 3

5. 11 + 32 + 58 + 19

6. 8 × 10 × 4

7. 23 + 7 + 8 + 42

8. 4 × 15 × 4

9. 12 × 2 × 6

Use the Distributive Property to find each product.

10. 9 × 19

11. 8 × 41

12. 7 × 52

13. 5 × 46

14. 4 × 35

15. 6 × 26

16. 3 × 63

17. 2 × 79

18. 11 × 15

19. 10 × 24

20. 12 × 13

21. 14 × 21

22. Each DVD costs $21. How much will it cost to buy 5 DVDs? Which property did you use to solve this problem with mental math?

23. Mrs. White bought three cases of erasers. Each case has 10 boxes. In each box there are 25 erasers. Use mental math to find how many erasers Mrs. White bought.

24. It took Marcus one hour to drive 55 miles. Use mental math to find how many miles Marcus could drive in 4 hours.

Program of Studies and Core Content
MA-06-5.1.1

1F Patterns and Sequences

1. Examine the sequence of figures below and look for a pattern.

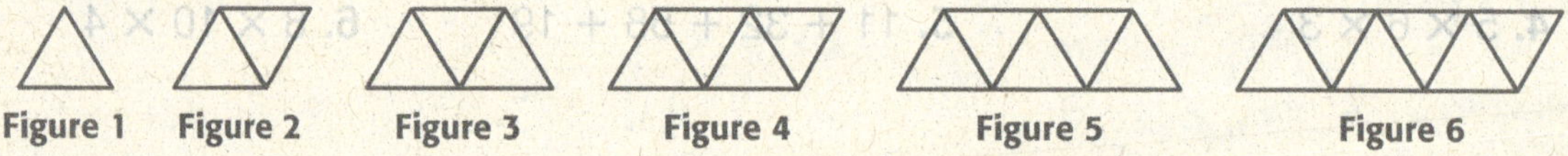

a. Sketch the next two figures in your pattern. Count the number of line segments it takes to draw each.

b. Copy and complete the table for your pattern.

Figure Number	1	2	3	4	5	6	7	8	9	10
Number of Line Segments	3	5	7							

Find the next three numbers in each sequence.

2. 1, 3, 5, 7, ____, ____, ____, ...

3. 96, 84, 72, ____, ____, ____, ...

4. 1, 3, 6, 10, 15, ____, ____, ____, ...

Think and Discuss

5. Describe the pattern you noticed in the sequence of triangles.

6. Explain how you found the next three numbers in exercises 2–4 above.

Name ______________________ Date ____________ Class ____________

LESSON **1F**

Hands-On Lab

Explore Patterns with Perfect Squares

An arrangement of smaller squares can be used to make bigger squares. The number of smaller squares that make up a bigger square follows a pattern.

Activity

Use square algebra tiles to make the shapes below. Each shape is a square. The number of tiles in each shape is listed below it.

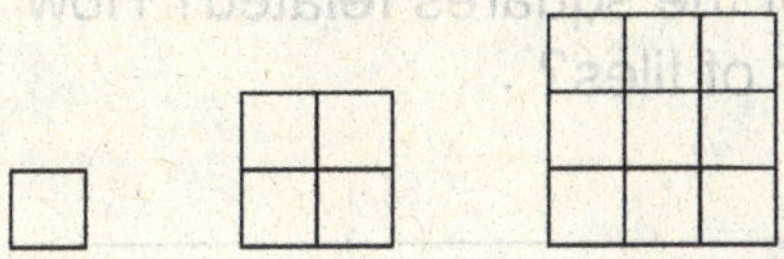

1 tile   4 tiles   9 tiles

Count the number of square tiles in each of the larger squares below.

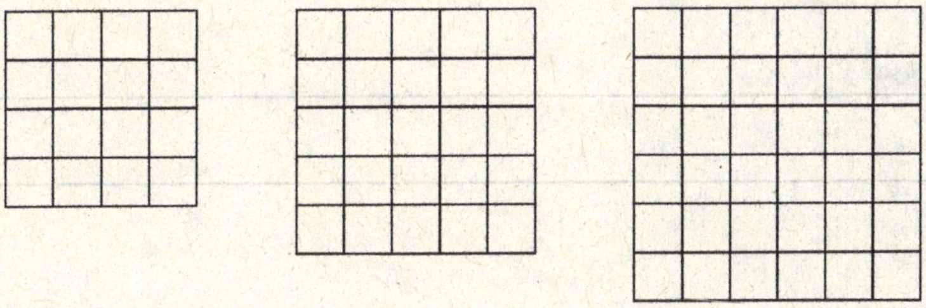

1. ________ **2.** ________ **3.** ________

Continue to make larger squares. Make one that is 7 tiles wide and 7 tiles high; then make one that is 8 wide and 8 high. Count the number of squares in each shape.

4. 7 by 7 ________ **5.** 8 by 8 ________

Name ______________________ Date ____________ Class ____________

LESSON 1F

Hands-On Lab

Explore Patterns with Perfect Squares, continued

Think and Discuss

1. The numbers 1, 4, 9, 16, 25, etc. are known as perfect squares. Why do you think they are called perfect squares?

____

____

2. How are the width and the height of the squares related? How are they related to the total number of tiles?

____

____

3. How could you find the next numbers that are perfect squares without tiles?

____

____

Try This

1. Write down the first 10 perfect squares, starting with 0.

0 1 4 9 16 25 36 ___

2. Now subtract each number from the one after it.

1 3 5 ___ ___ ___ ___ ___ ___

3. What pattern do you see in the difference between two perfect squares?

____

____

4. Use your answer from Exercise 3 above to find all perfect squares less than 200.

____

____

Name ______________________________ Date ____________ Class ____________

LESSON 1F

Review for Mastery

Patterns and Sequences

Find the next three numbers in the sequence.

8, 12, 16, 20, 24, □, □, □, ...

Step 1: Look at pairs of numbers to find the pattern.

8, 12, 16, 20, 24, □, □, □, ...

8 + 4 = 12 12 + 4 = 16 16 + 4 = 20

The pattern is to add 4.

Step 2: Use the pattern to name the next three numbers.

24 + 4 = 28 28 + 4 = 32 32 + 4 = 36

The next three numbers are 28, 32, and 36.

Find the next three numbers in each sequence.

1. 5, 8, 6, 9, 7 □, □, □, ...

____

2. 90, 80, 70, 60, □, □, □, ...

____

3. 2, 8, 4, 16, □, □, □, ...

____

4. 10, 14, 18, 22, □, □, □, ...

____

5. 13, 21, 29, 37, □, □, □, ...

____

6. 24, 12, 16, 8, □, □, □, ...

____

7. 14, 12,10, 8, □, □, □, ...

____

8. 1, 7, 13, 19, □, □, □, ...

____

9. 1, 3, 6, 10, □, □, □, ...

____

10. 40, 38, 36, □, □, □, ...

____

11. 54, 45, 36, □, □, □, ...

____

12. 10, 25, 40, □, □, □, ...

____

13. 36, 29, 22, □, □, □, ...

____

14. 18, 36, 72, □, □, □, ...

____

Name ______________________ Date ____________ Class ____________

LESSON 1F

Homework and Practice

Find a Pattern

Identify a pattern in each sequence, and name the next three terms.

1. 3, 9, 27, 81, __, __, __,...

2. 1, 2, 4, 8, __, __, __,...

3. 97, 90, 83, 76, __, __, __,...

4. 160, 80, 40, 20, __, __, __,...

5. 6, 11, 16, 21, __, __, __,...

6. 110, 101, 92, 83, 74, __, __, __,...

7. 2, 8, 32, 128, __, __, __,...

8. 109, 106, 103, 100 __, __, __,...

9. 15, 22, 29, 36, __, __, __,...

10. 72, 63, 54, 45 __, __, __,...

Identify a pattern in each sequence, and name the missing terms.

11. 3, 12, __, 192, __, 3,072,...

12. 3, 22, __, 60, __, __, 117, 136,...

13. 160, 145, __, __, 100, 85, __, 55,...

14. 704, 352, __, __, 44, __, 11,...

15. 6, 26, 46, __, __, 106, 126, __,...

16. 86, 80, __, 68, __, 56, __, 44,...

17. Tonya recorded the time it took her to run a mile each day for the past 3 days. It took Tonya 11 minutes 5 seconds the first day, 11 minutes the second day, and 10 minutes 55 seconds the third day. If this pattern continues, how long will it take Tonya to run a mile on the 5th day?

CHAPTER 1

Chapter Summary

Answer these question to summarize the important concepts from Chapter 1 in your own words.

1. Explain how to compare two whole numbers.

2. Explain how to evaluate 8^4.

3. List the steps when evaluating an expression with more than one operation.

4. How can you use mental math to find the product of 7 and 53?

5. Explain how to find the rule in a numerical pattern.

For more review of Chapter 1:

- Complete the Chapter 1 Study Guide and Review on pages 42–44 of your textbook.
- Complete the Ready to Go On quizzes on pages 18 and 38 of your textbook.

Name ______________________ Date __________ Class __________

CHAPTER 1

KCCT Prep

Multiple Choice

1. There are 19 boxes of fruit bars on a supermarket shelf. Each box contains 12 fruit bars. Which is the best estimate of the total number of fruit bars on the shelf?

Ⓐ 30 Ⓒ 200
Ⓑ 10 Ⓓ 300

2. Javier correctly simplified the expression 18 − 3 × 2 + □ and got a result of 30. What is the missing number in the expression?

Ⓐ 0 Ⓒ 12
Ⓑ 1 Ⓓ 18

3. Which pattern has a rule that involves two different operations?

Ⓐ 8, 16, 32, 64, 128, ...
Ⓑ 3, 11, 19, 27, 35, ...
Ⓒ 2, 10, 9, 17, 16, 24, ...
Ⓓ 5, 6, 8, 11, 15, ...

4. Use estimation to determine which of these is closest to 1,000.

Ⓐ 6 × 21 Ⓒ 51 × 18
Ⓑ 19 × 5 Ⓓ 22 × 499

5. Which of these is an example of the Associative Property?

Ⓐ 6 + 0 = 6
Ⓑ 6 + 8 = 8 + 6
Ⓒ 6 + (8 + 2) = (6 + 8) + 2
Ⓓ 8 × 1 = 8

6. The number of daily flights from an airport to Chicago is 37, to Denver is 11, to Los Angeles is 42, and to Memphis is 9. Which is the best estimate of the total number of flights to these destinations in 5 days?

Ⓐ 100 Ⓒ 500
Ⓑ 400 Ⓓ 1,000

7. Which of these terms will NOT appear if you continue the following pattern? 3, 2, 6, 5, 9, 8, 12, 11, 15, ...

Ⓐ 60 Ⓒ 66
Ⓑ 61 Ⓓ 69

8. The expression 6 × 11 has the same value as 11 × 6 because of which property?

Ⓐ Associative Property
Ⓑ Commutative Property
Ⓒ Identity Property of Addition
Ⓓ Identity Property of Multiplication

9. When the expression $25 - (8 \square 2)^2$ is simplified, the result is 9. What is the missing operation in the expression?

Ⓐ + Ⓒ ×
Ⓑ − Ⓓ ÷

10. One pair of parentheses is added to the expression 8 + 2 × 3 + 7. When the expression is simplified, which of these results is NOT possible?

Ⓐ 21 Ⓒ 37
Ⓑ 28 Ⓓ 100

11. The equation $9 + (15 + 2) = (9 + 15) + 2$ is an example of which property?

Ⓐ Associative Property
Ⓑ Commutative Property
Ⓒ Identity Property of Addition
Ⓓ None of these

12. Which sequence has 180 as one of its terms?

Ⓐ 1, 2, 4, 8, 16, 32, ...
Ⓑ 7, 14, 21, 28, 35, ...
Ⓒ 2, 3, 2, 5, 2, 7, 2, 9, ...
Ⓓ 5, 4, 10, 9, 15, 14, ...

13. Marc writes additional terms to continue each sequence. He stops when he gets to a term that is greater than 100. Which sequence requires the fewest additional terms?

Ⓐ 5, 10, 15, 20, ...
Ⓑ 2, 15, 4, 15, 6, 15, 8, ...
Ⓒ 1, 2, 4, 8, 16, ...
Ⓓ 65, 67, 69, 71, ...

14. Meg drives at an average speed of 52 miles per hour. Which is the best estimate of the time it takes her to drive 297 miles?

Ⓐ 4 hours
Ⓑ 5 hours
Ⓒ 6 hours
Ⓓ 8 hours

15. In which expression is it possible to add one set of parentheses so that the new expression has a value of 12?

Ⓐ $6 \div 2 + 8$
Ⓑ $1 + 6 \times 2$
Ⓒ $4 + 2 \times 2$
Ⓓ $2 \times 3 + 4$

Open Response

16. A wall contains a large pattern consisting of rows of tiles. The first row has 15 tiles. The second row has 18 tiles, the third row has 21 tiles, the fourth row has 24 tiles, and so on.

a. Suppose the pattern continues. How many tiles are in the 6th row?

b. Describe the pattern.

c. Does any row of the pattern have exactly 400 tiles? Why or why not?

17. Ann and Ben start saving money at same time. The table shows their total savings for the first five weeks.

Week	1	2	3	4	5
Ann's Savings ($)	8	17	26	35	44
Ben's Savings ($)	19	27	35	43	51

a. Ann and Ben continue to save at the same rate. How much will they each have saved after 7 weeks?

b. Ben claims that he will always have more savings than Ann. Do you agree or disagree? Why?

18. Gary is having clothing printed with a school logo for a graduation ceremony. A T-shirt costs $9, a baseball jacket costs $37, and a baseball cap costs $4.

a. Gary orders 22 t-shirts. Show how to estimate the total cost of the shirts.

b. Gary orders 11 baseball jackets. Show how to estimate total cost of the jackets.

c. Gary has a total of $650 to spend on the clothes. He would like to order some baseball caps. About how many baseball caps can he order and stay within his $650 budget? Explain.

CHAPTER 2

Project
Algebra: Traffic Jam

Career: Traffic Engineer

Have you ever wondered why traffic moves quickly through one intersection but slowly through another? Traffic engineers program stoplights so that vehicles can move smoothly through intersections. There are many variables at a traffic intersection—the number of vehicles that pass, the time of day, and the direction in which each vehicle travels are examples. Traffic engineers use this information to control the timing of stoplights.

Before you start the project, work through the Vehicle Travel worksheet on the next page.

In this chapter project . . .

You predict how a public transit system would change traffic.

What to do:

- Choose a busy intersection in your town and record the number of cars traveling in each direction during the morning rush hour.
- Use your data to find out which direction most of the traffic travels. Using a map of your town, make a drawing of what a train line for commuters might look like.
- Write equations to show how many cars would pass through your intersection if one third, one half, and three quarters of the drivers took the train. Would the train affect traffic in all directions the same way? Explain why or why not.

What to present:

- A table that shows how many cars pass through the intersection and how these numbers might change if there were a train.
- The equations you wrote to model the change in traffic patterns.

Where's the math?

- Gather data on traffic patterns at an intersection.
- Use equations to predict how traffic patterns might change if there were a train.

go.hrw.com
Chapter Project Online
KEYWORD: MR7 Ch2

Name ______________________ Date __________ Class __________

CHAPTER 2

Project Recording Sheet

Introduction to Algebra

Vehicle Travel

Traffic engineers use mathematics and computers to program the lights at intersections so traffic moves as smoothly as possible. One way to compare vehicles moving in different directions is to see how many pass in a given period of time such as a minute. We can write this mathematically as:

vehicles per minute • total minutes = total vehicles

Use the information above to write equations and complete the table below:

Street	Total Vehicles	Time (Minutes)	Equation	Vehicles/ Minute
West Spain	120	60		
Second Street East	30	3		
Fifth Street West	60	20		
Broadway	180	45		
MacArthur	100			5

Answer the following questions using the table above.

- Which is the busiest street? __________
- Which is the quietest street? __________
- Which street(s) have less than four vehicles per minute? __________
- How long (on the average) would you have to wait on Fifth Street West for six vehicles to come by? __________
- How does the wait on Fifth Street West for six vehicles compare to the wait on Second Street East? __________
- If you counted from eight in the morning to five at night, about how many vehicles would you expect to see on Broadway? __________
- There are five more cars per minute on State Street than on Second Street East. Write an equation to find the number of vehicles on State Street. __________

Extension: What things determine the average number of vehicles on a street? __________

EXPLORATION

2A Variables and Expressions

Program of Studies and Core Content

MA-6-AT-S-VE02, MA-06-5.2.1

1. Look at the sequence of connected squares.

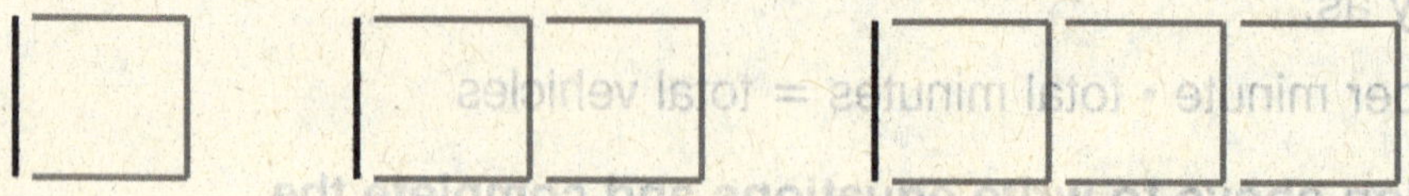

1 + 3   1 + 3 + 3   1 + 3 + 3 + 3

a. Sketch the next two squares.

b. To complete the table for the connected squares, count the number of segments it takes to draw each square.

Number of Connected Squares	1	2	3	4	5	10	20	100
Number of Segments	4	7						

c. How can you find the number of segments if you know the number of squares?

Think and Discuss

2. Explain the reasoning you used to find the number of segments in one hundred connected squares.

3. Explain the reasoning you could use to find the number of segments in one thousand connected squares.

Name ______________________________ Date ____________ Class __________

LESSON **2A**

Technology Lab 1

Variables and Expressions

You can use the memory function on your calculator to help you compute total cost.

Activity

It's fun going places, such as the movies, with a group of friends. Use your calculator to figure out how much money you will need for you and groups of friends to go to the movies.

MOVIE CHOICES

Option A	COST of movie: $6.00
Option B	COST of movie with popcorn and drink: $8.00
Option C	COST of movie, popcorn, drink, and candy: $9.00

Your calculator uses the [STO>] button to store values. You enter a value, press the [STO>] key, then choose a letter to represent the value. After choosing a letter, it can take the place of the value.

1. Compute Option A: You and 6 friends see just the movie.

   Press 6 (for $6.00) and then [STO>] [ALPHA] A [ENTER]. (This will keep the cost in the memory of the calculator so that you do not have to enter it every time you change the number of friends going to the movies.)

2. Next, press [ALPHA] A [×] and then enter the number of people who will be going to the movies. In this case it will be 7, you and 6 friends.

3. To get the total cost for 7 people to go to the movies, press the [ENTER] key. Record the total in the table.

4. Clear the display. Determine the cost for you and 8 friends, then the cost for you and 2 friends. Enter the total costs in the table.

5. Compute Option B: $8.00 for a movie ticket with popcorn and a drink. Use the [STO>] key to store that value as B, and then compute the cost for each number of Movie Goers.

6. Continue calculating the total cost for you and friends to see a movie, have popcorn and a drink, and get a box of candy too. Write the costs in the table.

Name ______________________ Date ____________ Class ____________

LESSON **2A**

Technology Lab 1

Variables and Expressions, continued

Movie Choice	Number of Movie Goers	TOTAL Cost
Option A1	you and 6 friends	
Option A2	you and 8 friends	
Option A3	you and 2 friends	
Option B1	you and 6 friends	
Option B2	you and 8 friends	
Option B3	you and 2 friends	
Option C1	you and 5 friends	
Option C2	you and 2 friends	
Option C3	you choose how many friends	

Think and Discuss

1. How do you decide which value to store in memory as a constant and which one to use as the variable that is entered each time?

____

____

____

2. When in your everyday life would the calculator's memory function be helpful to you?

____

____

____

____

Try This

Store each constant in your calculator. Calculate the total for each variable.

1. 15 in memory × 5 = ____________

   × 7 = ____________

   × 9 = ____________

2. 18.3 in memory + 12 = ____________

   + 20 = ____________

   + 4 = ____________

Name ______________________ Date __________ Class __________

LESSON **2A**

Technology Lab 2

Evaluate Expressions

You can use a graphing calculator to evaluate algebraic expressions. A graphing calculator is especially helpful when you are evaluating an expression for many values of the variable.

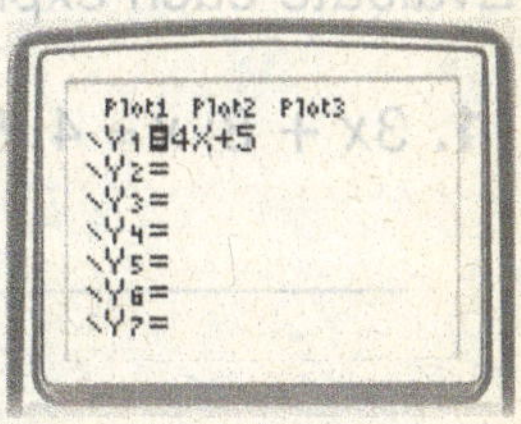

Activity

Evaluate $4x + 5$ for $x = 0, 1, 2, 3, 4, 5, 6, 7, 8, 9,$ and 10.

Enter $4x + 5$ into the **Y** = menu. [Y=] 4 [X,T,θ,n] [+] 5.

Press [2nd] [WINDOW] (TBLSET) to access the

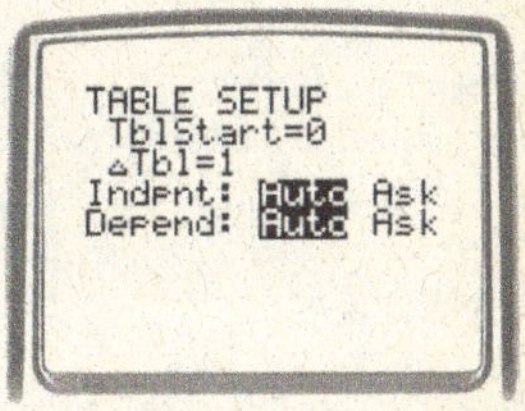

TABLE SETUP menu. TblStart tells which value of x the table should begin with. Make sure this is set to 0, because 0 is the smallest value of x for which you must evaluate the expression. **ΔTbl** gives the difference between successive x-values. **ΔTbl = 1** means that the x-values in the table will increase by 1.

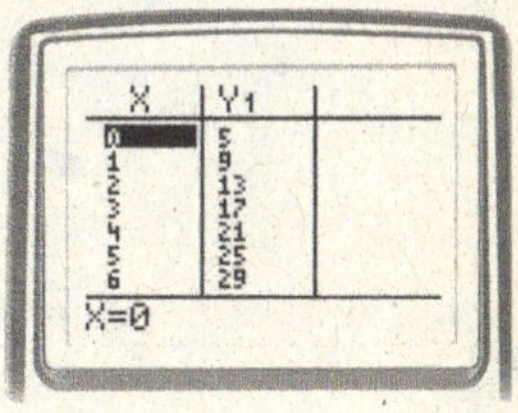

[2nd] [GRAPH] (TABLE) The **Y1** column shows the value of $4x + 5$ for several x-values. For example, when $x = 0$, $4x + 5 = 5$. Use the arrow keys to find the answers for other x-values.

Think and Discuss

1. In the table, one row has 5 in the **X** column and 25 in the **Y1** column. What does this mean?

____

2. When does $y = 41$?

____

3. Would you use a calculator to evaluate $x + 2$ for $x = 5$? Explain why or why not.

____

____

Name ______________________________ Date ____________ Class ____________

LESSON 2A

Technology Lab 2

Evaluate Expressions, continued

Try This

Evaluate each expression for the given *x*-values. Show your work.

1. $3x + 8$; $x = 4, 5,$ and 9

____

2. $45x + 67$; $x = 8, 10,$ and 12

____

3. $4x + 7$; $x = 0, 1, 2, 3, 4,$ and 5

____

4. $30 + 25x$; $x = 5, 10, 15,$ and 20

____

Name ______________________ Date __________ Class __________

LESSON **2A**

Review for Mastery

Variables and Expressions

A variable is a letter or a symbol that stands for a number that can change. A constant is an amount that does not change.

A mathematical phrase that contains at least one variable is an algebraic expression. In the algebraic expression $x + 5$, x is a variable and 5 is a constant.

When you evaluate an algebraic expression, substitute a number for the variable and then find the value.

To evaluate the algebraic expression $m - 8$ for $m = 12$, first replace the variable m in the expression with 12.

$m - 8$

$12 - 8$

Then find the value of the expression.

$12 - 8 = 4$

The value of $m - 8$ is 4 when $m = 12$.

Evaluate each expression for the given value of the variable.

1. $x + 5$, for $x = 6$ ____________

2. $3p$, for $p = 5$ ____________

3. $z \div 4$, for $z = 24$ ____________

4. $w - 7$, for $w = 15$ ____________

To find the missing values in a table, use the given values of the variable.

x	**4*x***
3	12
4	■
5	■

Think: $x = 3$, so $4x = 4 \cdot 3 = 12$

Think: $x = 4$, so $4x = 4 \cdot 4 = 16$

Think: $x = 5$, so $4x = 4 \cdot 5 = 20$

Evaluate each expression to find the missing values in the tables.

5.

x	*x* + 7
3	10
5	
7	

6.

y	*y* – 2
9	
10	
14	

Name ______________________ Date ____________ Class ____________

LESSON 2A

Homework and Practice

Variables and Expressions

Evaluate each expression to complete the table.

1.

n	*n* − 7
60	53
51	
42	
33	

2.

n	*n* ÷ 2
24	12
30	
36	
42	

3.

n	*n* + 11
19	30
13	
7	
3	

Find an expression for each table.

4.

n	
2	6
6	18
10	30
14	42

5.

n	
81	27
60	20
48	16
21	7

6.

n	
12	7
34	29
56	51
78	73

Evaluate each expression for the indicated values of the variables.

7. $3x - 4y$; $x = 4$ and $y = -3$

8. $6x \cdot y$; $x = -1$ and $y = 2$

9. $5x + y + 7$; $x = 1$ and $y = 6$

10. $2y - x + 1$; $x = 3$ and $y = 3$

11. Ann pays $300 for membership to a local gym. She is allowed to bring one guest on any visit. John pays Ann $5 to go to the gym with her occasionally. Describe what the expression 300 − 5*t* could represent. Then evaluate the expression for *t* = 5, 10, 15, and 20.

____

____

____

EXPLORATION

Program of Studies and Core Content
MA-6-AT-S-VE03, MA-06-5.2.2

2B Translate Between Words and Math

Drawing pictures and using formulas can help you translate between words and math.

A basketball court is 50 ft wide by 94 ft long. What is its area?

Formula for area

$A = \text{length} \times \text{width}$

$A = 94 \times 50$

$A = 4{,}700$

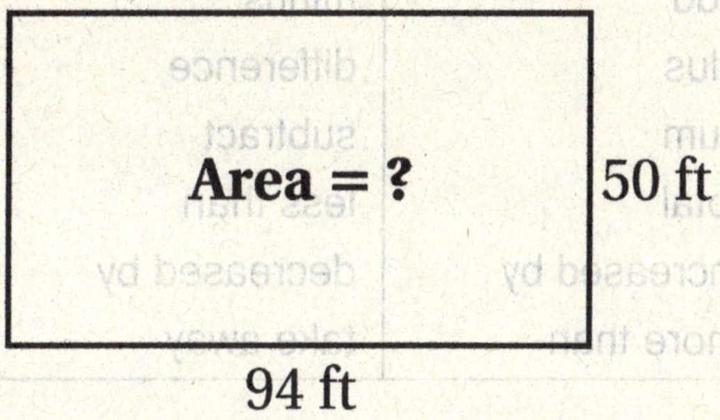

The area is 4,700 ft^2.

In some word problems, word order may be confusing. For example, the following problems can be translated in at least two different ways.

Rewrite each problem to make it clearer.

	Word Problem	Possible Translations	Better Word Problem
1.	Write the expression "4 times *x* plus 6."	$4x + 6$ or $4(x + 6)$	
2.	Translate "the square root of *n* minus 3."	$\sqrt{n} - 3$ or $\sqrt{n - 3}$	

Think and Discuss

3. **Explain** what you did to rewrite numbers 1 and 2 to make them easier to translate into math.

Name ______________________ Date __________ Class __________

LESSON **2B**

Review for Mastery

Translate Between Words and Math

There are key words that tell you which operations to use for mathematical expressions.

Addition (combine)	**Subtraction** (less)	**Multiplication** (put together groups of equal parts)	**Division** (separate into equal groups)
add	minus	product	quotient
plus	difference	times	divide
sum	subtract	multiply	
total	less than		
increased by	decreased by		
more than	take away		

You can use key words to help you translate between word phrases and mathematical phrases.

A. 3 plus 5 — $3 + 5$
B. 3 times *x* — $3x$
C. 5 less than *p* — $p - 5$
D. *h* divided by 6 — $h \div 6$

Write each phrase as a numerical or algebraic expression.

1. 4 less than 8 ________
2. *q* divided by 3 ________
3. *f* minus 6 ________
4. *d* multiplied by 9 ________

You can use key words to write word phrases for mathematical phrases.

A. $7k$
- the product of 7 and k
- 7 times *k*

B. $5 - 2$
- 5 minus 2
- 2 less than 5

Write a phrase for each expression.

5. $z \div 4$ ________
6. $5 \cdot 6$ ________
7. $m - 6$ ________
8. $s + 3$ ________

Name ______________________________ Date __________ Class __________

LESSON 2B

Homework and Practice

Translate Between Words and Math

Write an expression.

1. Steven has twice as many CDs as he has DVDs. If *d* represents the number of DVDs Steven has, write an expression for the number of CDs Steven has. ______________

2. The longest snake is a reticulated python. The reticulated python is 35 feet long. The next longest snake is the anaconda. Let *a* represent the length of an anaconda. Write an expression for the difference between the two snake lengths. ______________

Write each phrase as a numerical or algebraic expression.

3. 26 added to 10 ______________

4. 9 less than *b* ______________

5. 5 times 6 ______________

6. the product of 8 and *z* ______________

7. the difference of 100 and *c* ______________

8. the quotient of 200 and *a* ______________

9. 9 plus *g* ______________

10. *K* divided by 19 ______________

11. 28 multiplied by *p* ______________

Write two phrases for each expression.

12. $w - 2$ ______________ ______________ ______________

13. $15 + 5$ ______________ ______________ ______________

14. 96(69) ______________ ______________ ______________

15. Michelle has 10 dollars more than Amy does. If *a* represents the amount of money Amy has, what expression represents the amount of money Michelle has? ______________

16. Adrian has half as much money as Eric does. If *e* represents the amount of money Eric has, what expression shows the amount of money Adrian has? ______________

EXPLORATION

Program of Studies and Core Content
MA-6-AT-S-VE03,
MA-06-5.2.2

2C Translating Between Tables and Expressions

You can explore geometric patterns to help you write algebraic expressions. Consider this sequence of figures.

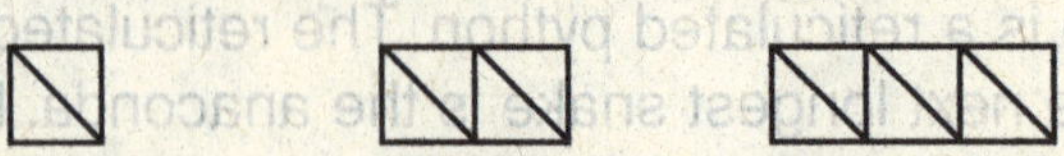

Figure 1 Figure 2 Figure 3

The first figure has 5 segments, the second figure has 9 segments, and the third figure has 13 segments.

Figure Number	1	2	3
Number of Segments	5	9	13

1. Draw the next two figures in the pattern.
2. Complete the table.

Figure Number	1	2	3	4	5
Number of Segments	5	9	13		

3. Describe any patterns you notice in the table.

Think and Discuss

4. **Explain** how you can find the number of segments in the 6th figure without drawing it.
5. **Explain** Explain how you can find the number of segments in any figure in the pattern if you know the number of the figure.

Name ______________________________ Date ____________ Class ____________

LESSON 2C

Review for Mastery

Translating Between Tables and Expressions

You can write an expression for data in a table.
The expression must work for all of the data.

Cats	Legs
1	4
2	8
3	12
c	?

Think: When there is 1 cat, there are 4 legs. $4 \times 1 = 4$
When there are 2 cats, there are 8 legs. $4 \times 2 = 8$
When there are 3 cats, there are 12 legs. $4 \times 3 = 12$

So, when there are c cats, there are $4c$ legs.

You can write an expression for the sequence in a table.
Find a rule for the data in the table that works for the whole sequence.

Position	1	2	3	4	5	n
Value of Term	4	5	6	7	8	?

Step 1 Look at the value of the term in position 1.
4 is **3 more** than 1.

Step 2 Try the rule for position 2.
5 is **3 more** than 2.

Step 3 Try the rule for the rest of the positions.
6 is **3 more** than 3, 7 is **3 more** than 4, and 8 is **3 more** than 5.

So, the expression for the sequence is $n + 3$.

Write an expression for the missing value in each table.

1.

People	Legs
1	2
2	4
3	6
p	

2.

Yoko's Age	Mel's Age
9	19
10	20
11	21
y	

Write an expression for the sequence in the table.

3.

Position	1	2	3	4	5	n
Value of Term	3	6	9	12	15	

Name ______________________ Date ____________ Class __________

LESSON 2C

Homework and Practice

Translating Between Tables and Expressions

Fill in the missing values or expressions in each table.

1.

Lessons	Weeks
6	2
12	
18	6
	w

2.

Cups	Calories
	110
2	220
3	330
c	

The table at the right shows data from Mr. Yang's last history quiz. Use the table for Exercises 3-6.

Correct	Incorrect	Score
17	3	85
15	5	75
18	2	90
19	1	95
18	2	90
14	6	60
c		

3. What is the relationship between the data in the first two columns?

____

4. What is the relationship between the data in the first and last columns?

____

5. Fill in the missing expressions for the last row of the table.

6. Suppose Amanda answered *x* questions incorrectly on the quiz. Write an expression for her score.

____

Read each of the following situations. Fill in the table based on what you read.

7. Wendy has no money in her savings account. She decides to start saving $40 per week to purchase a canoe next summer.

Wendy's Account

Week	Savings
1	
3	
5	
w	

8. Dirk is training to run a marathon. He would like to increase the miles per week ran in April by 5 miles per week in May, for each week.

Miles per Week

April	May
22	
23	
28	
m	

EXPLORATION

Program of Studies and Core Content

MA-6-AT-S-VE01, MA-6-AT-S-EI1, MA-06-5.3.1

2D Equations and Their Solutions

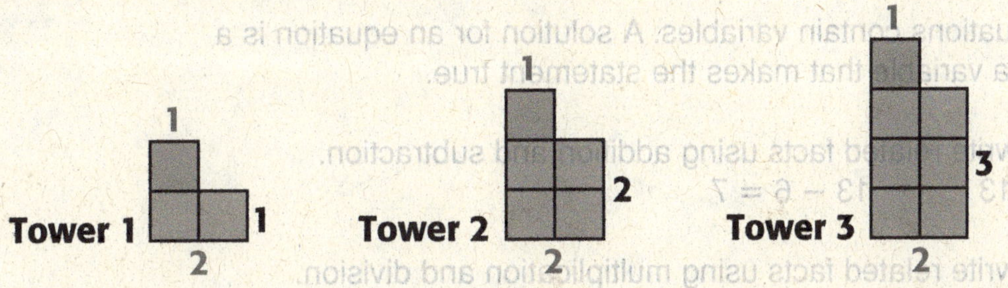

In the sequence of towers, the base of each tower is always 2 squares wide. The heights of the towers vary. If we call the height (not including the square on top) of each tower h, we can represent this pattern with the following expression:

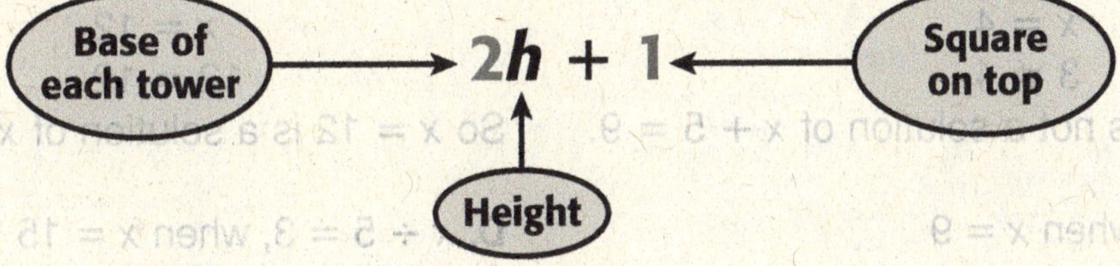

1. Use the pattern in the sequence of towers to draw a tower with 11 squares. Which tower number is it in the sequence?

2. Use the pattern to solve the equation $2h + 1 = 21$.

3. Look at the sequence of grids and draw a picture of the grid that has 10 shaded squares.

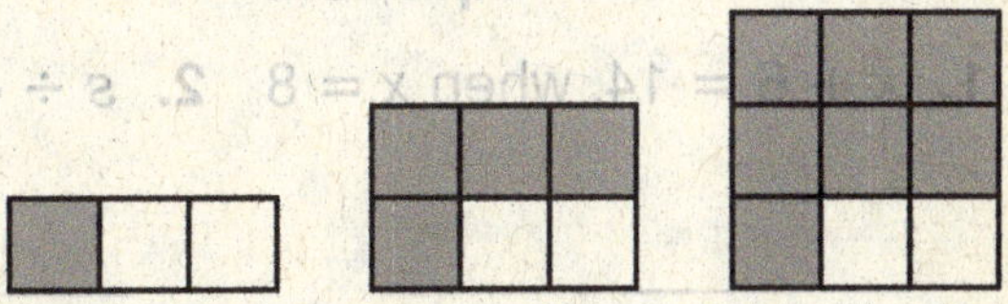

   a. Where in the sequence does this grid occur?

   b. Write an equation for the problem in **3a.**

Think and Discuss

4. **Discuss** what is meant by "a solution of an equation."

Name ______________________ Date __________ Class __________

LESSON 2D

Review for Mastery

Equations and Their Solutions

An equation is a mathematical sentence that says that two quantities are equal.

Some equations contain variables. A solution for an equation is a value for a variable that makes the statement true.

You can write related facts using addition and subtraction.
$7 + 6 = 13$ $\quad$ $13 - 6 = 7$

You can write related facts using multiplication and division.
$3 \cdot 4 = 12$ $\quad$ $12 \div 4 = 3$

You can use related facts to find solutions for equations. If the related fact matches the value for the variable, then that value is a solution.

A. $x + 5 = 9$, when $x = 3$
Think: $9 - 5 = x$
$x = 4$
$3 \neq 4$
So $x = 3$ is not a solution of $x + 5 = 9$.

B. $x - 7 = 5$, when $x = 12$
Think: $5 + 7 = x$
$x = 12$
$12 = 12$
So $x = 12$ is a solution of $x - 7 = 5$.

C. $2x = 14$, when $x = 9$
Think: $14 \div 2 = x$
$x = 7$
$9 \neq 7$
So $x = 9$ is not a solution for $2x = 14$.

D. $x \div 5 = 3$, when $x = 15$
Think: $3 \cdot 5 = x$
$x = 15$
$15 = 15$
So $x = 15$ is a solution for $x \div 5 = 3$.

Use related facts to determine whether the given value is a solution for each equation.

1. $x + 6 = 14$, when $x = 8$ ____________
2. $s \div 4 = 5$, when $s = 24$ ____________
3. $g - 3 = 7$, when $g = 11$ ____________
4. $3a = 18$, when $a = 6$ ____________
5. $26 = y - 9$, when $y = 35$ ____________
6. $b \cdot 5 = 20$, when $b = 3$ ____________
7. $15 = v \div 3$, when $v = 45$ ____________
8. $11 = p + 6$, when $p = 5$ ____________
9. $6k = 78$, when $k = 12$ ____________

Name ______________________ Date __________ Class __________

LESSON **2D**

Homework and Practice

Equations and Their Solutions

Determine whether the given value of the variable is a solution.

1. $26 - w = 18$ for $w = 8$ ______

2. $a \div 3 = 19$ for $a = 58$ ______

3. $42 - g = 15$ for $g = 26$ ______

4. $k \times 6 = 108$ for $k = 18$ ______

5. $32 \div z = 8$ for $z = 4$ ______

6. $u + 23 = 41$ for $u = 17$ ______

7. $\frac{d}{3} = 4$ for $d = 15$ ______

8. $13y = 65$ for $y = 5$ ______

9. $36 + m = 81$ for $m = 45$ ______

10. $7 \times j = 84$ for $j = 12$ ______

11. $v - 31 = 13$ for $v = 45$ ______

12. $69 + c = 121$ for $c = 52$ ______

13. $4r = 107$ for $r = 27$ ______

14. $\frac{91}{q} = 13$ for $q = 7$ ______

15. $n - 0 = 70$ for $n = 70$ ______

16. $h + 16 = 73$ for $h = 58$ ______

Replace each [?] with a number that makes the equation correct.

17. $9 - 5 = 6 -$ [?] ___

18. $2 \times$ [?] $= 4 \times 4$ ___

19. [?] $+ 7 = 5 + 8$ ___

20. $35 \div 5 = 49 \div$ [?] ___

21. [?] $- 2 = 10 - 3$ ___

22. $21 \times 1 =$ [?] $\times 7$ ___

23. Erin has 5 CD towers of CDs, for a total of 475 CDs. Each tower holds the same number of CDs. Write an equation using the variable c to model this situation. What does your variable represent?

____

____

24. Mr. Wright and Mr. Miller took their classes on a field trip. They took a total of 49 students. Mr. Wright had 25 students in his class. Write an equation to model this situation using the variable s.

____

____

EXPLORATION

Program of Studies and Core Content

MA-6-NPO-S-NO3, MA-6-AT-S-VE01, MA-6-AT-S-EI1, MA-6-AT-S-EI3, MA-06-1.3.2, MA-06-5.3.1

2E Addition Equations

How much change from a dollar do you get when you buy something that costs 51 cents?

This problem can also be expressed as **what number** plus **51** is **100**?

$$n + 51 = 100$$

$$49 + 51 = 100$$

$$n = 49$$ The change is 49¢.

Find the value of *n* in each equation.

1. $4 + n = 100$ $n =$ ________
2. $n + 45 = 100$ $n =$ ________
3. $19 + n = 100$ $n =$ ________
4. $n + 65 = 100$ $n =$ ________
5. $100 = 41 + n$ $n =$ ________

Think and Discuss

6. **Discuss** your strategies for solving the equations.
7. **Explain** how you can mentally find the solution to $n + 125 = 500$.

Name ______________________ Date ___________ Class ___________

LESSON 2E

Hands-On Lab

Solving Addition Equations

You can use algebra tiles to help you solve equations.

KEY

Activity

To solve the equation $x + 1 = 4$, you need x by itself on one side of the equal sign. Show this with algebra tiles. You can add or remove tiles as long as you add the same amount or remove the same amount on both sides.

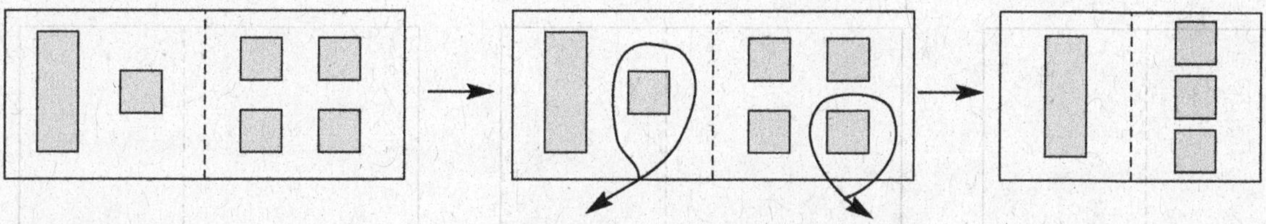

$x + 1 \quad = \quad 4$ Remove 1 from each side. $x = 3$

1. Circle the tiles to remove to solve $x + 3 = 5$. Then draw the tiles in the last box.

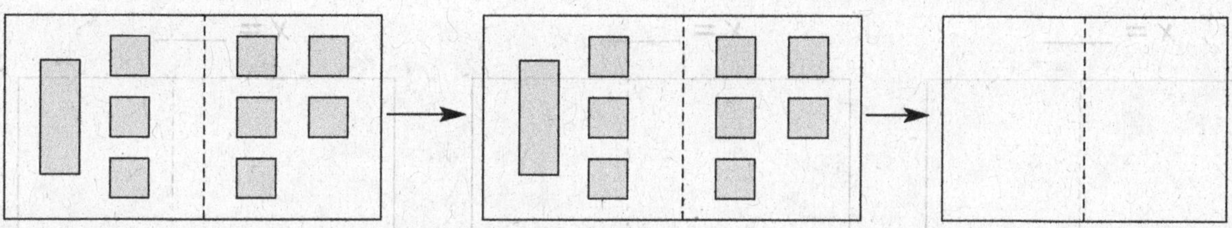

$x + 3 \quad = \quad 5$ Remove ___ tiles from each side. $x =$ ___

Name ______________________________ Date ____________ Class ____________

LESSON 2E

Hands-On Lab

Solving Addition Equations, continued

Think and Discuss

1. When you remove tiles, what operation are you representing?

____

2. Give an example of an equation that would require you to remove four tiles to solve.

____

Try This

Draw algebra tiles to model each equation. Then solve the equation.

1. $x + 2 = 7$

   $x =$ ____

2. $x + 1 = 5$

   $x =$ ____

3. $x + 5 = 8$

   $x =$ ____

4. $3 + x = 4$

   $x =$ ____

5. $x + 4 = 9$

   $x =$ ____

6. $2 + x = 2$

   $x =$ ____

Name ______________________ Date __________ Class __________

LESSON 2E

Review for Mastery

Addition Equations

To solve an equation, you need to get the variable alone on one side of the equal sign.

You can use tiles to help you solve addition equations.

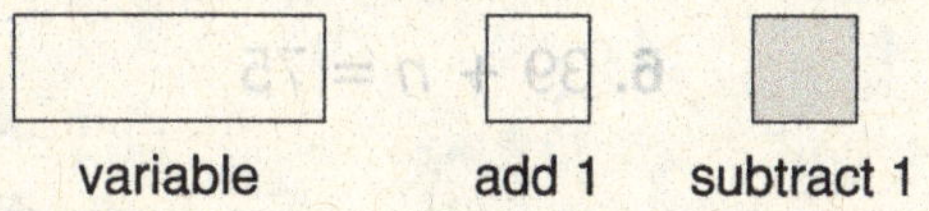

Subtraction undoes addition, so you can use subtraction to solve addition equations.

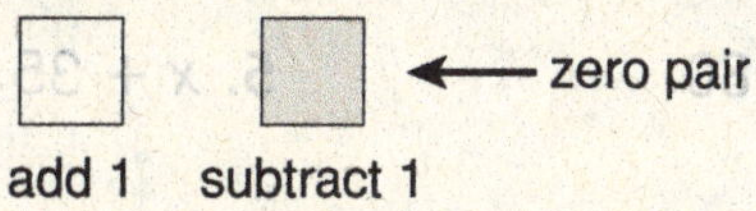

One positive tile and one negative tile is called a zero pair because together they have a value of zero.

To solve $x + 3 = 5$, first use tiles to model the equation.

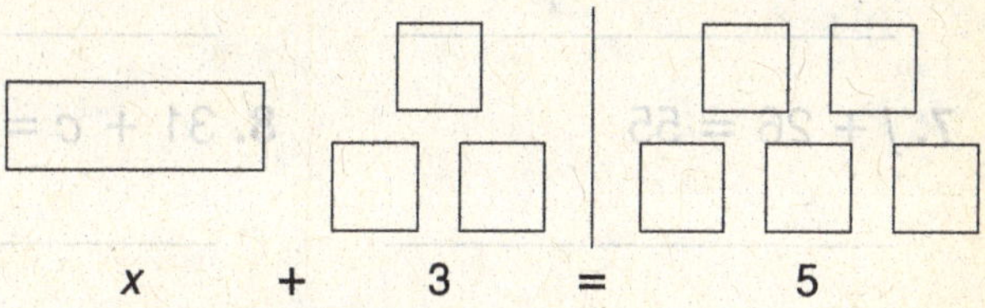

Next, add enough subtraction tiles to get the variable alone. Then add the same number of subtraction tiles to the other side of the equal sign.

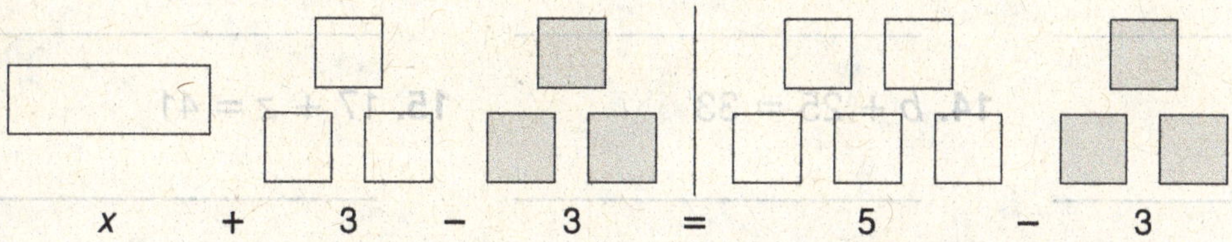

Then remove the greatest possible number of zero pairs from each side of the equal sign.

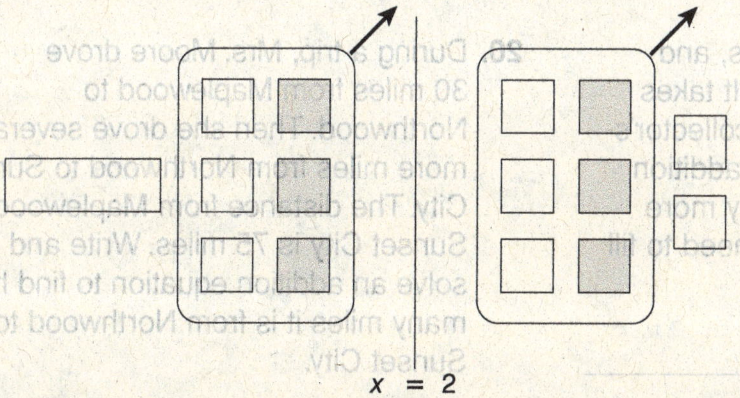

Check: $x + 3 = 5$

$2 + 3 \stackrel{?}{=} 5$

$5 \stackrel{?}{=} 5$ ✓

The remaining tiles represent the solution. $x = 2$

Use tiles to solve each equation. Then check each answer.

1. $x + 8 = 11$ ____________

2. $x + 4 = 9$ ____________

3. $x + 7 = 13$ ____________

Name ______________________ Date ____________ Class ____________

LESSON 2E

Homework and Practice

Solving Addition Equations

Solve each equation. Check your answers.

1. $13 + g = 30$

2. $m + 9 = 22$

3. $s + 11 = 42$

4. $61 + y = 80$

5. $x + 35 = 79$

6. $39 + n = 75$

7. $j + 26 = 55$

8. $31 + c = 46$

9. $w + 49 = 67$

Solve each equation.

10. $f + 5 = 9$

11. $22 + d = 31$

12. $v + 16 = 22$

13. $36 + k = 43$

14. $b + 25 = 33$

15. $17 + z = 41$

16. $p + 7 = 21$

17. $16 + q = 31$

18. $26 + a = 62$

19. Mike has 35 baseball cards, and Ira has 42 baseball cards. It takes 60 baseball cards to fill a collector's book. Write and solve two addition equations to find how many more baseball cards they each need to fill one collector's book.

20. During a trip, Mrs. Moore drove 30 miles from Maplewood to Northwood. Then she drove several more miles from Northwood to Sunset City. The distance from Maplewood to Sunset City is 75 miles. Write and solve an addition equation to find how many miles it is from Northwood to Sunset City.

EXPLORATION

Program of Studies and Core Content

MA-6-NPO-S-NO3, MA-6-AT-S-VE01, MA-6-AT-S-EI1, MA-6-AT-S-EI3, MA-06-1.3.2, MA-06-5.3.1

2F Subtraction Equations

After spending \$11, Jane has \$3 left in her purse. How much did she have to begin with?

Beginning amount **\$11 spent** **Amount left**

$n - 11 = 3$

$14 - 11 = 3$

$n = 14$ She had \$14 to begin with.

Find the value of n in each equation.

1. $n - 25 = 75$ $n =$ ________
2. $n - 4 = 19$ $n =$ ________
3. $n - 7 = 35$ $n =$ ________
4. $n - 14 = 21$ $n =$ ________
5. $n - 20 = 83$ $n =$ ________

Think and Discuss

6. **Describe** your strategies for solving the subtraction equations.
7. **Explain** how you can find the solution to $n - 125 = 375$.

Name ______________________ Date ____________ Class ____________

LESSON 2F

Review for Mastery

Subtraction Equations

To solve an equation, you need to get the variable alone on one side of the equal sign.

You can use tiles to help you solve subtraction equations.

Addition undoes subtraction, so you can use addition to solve subtraction equations.

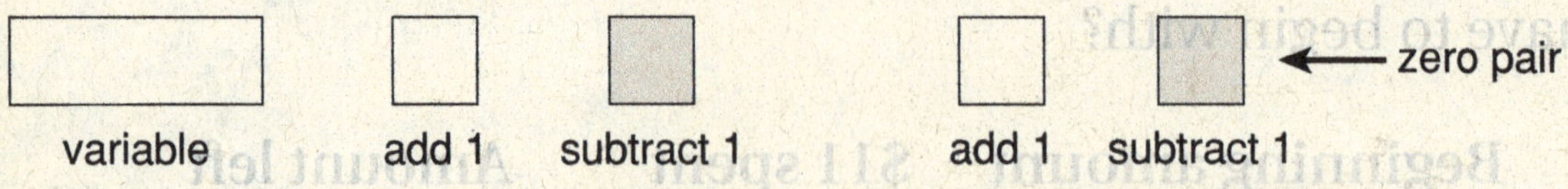

One positive tile and one negative tile is called a zero pair because together they have a value of zero.

To solve $x - 4 = 2$, first use tiles to model the equation.

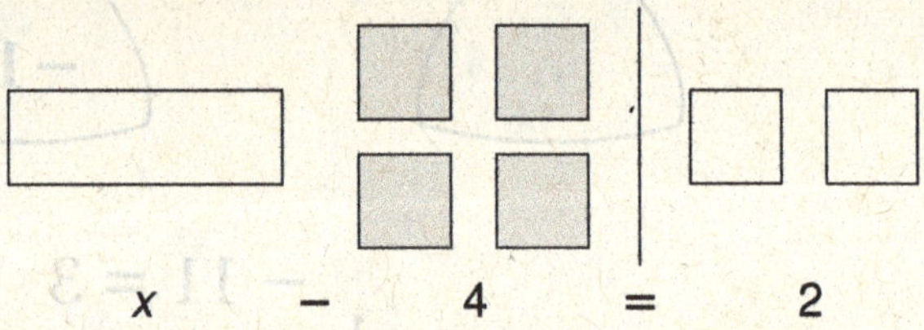

Next, add enough addition tiles to get the variable alone. Then add the same number of addition tiles to the other side of the equal sign.

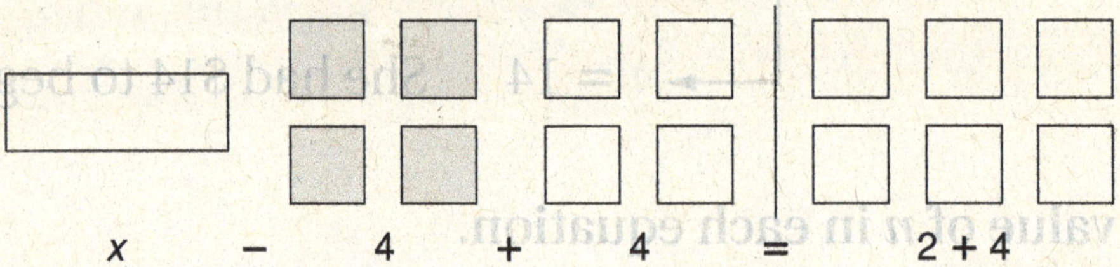

Then remove the greatest possible number of zero pairs from each side of the equal sign.

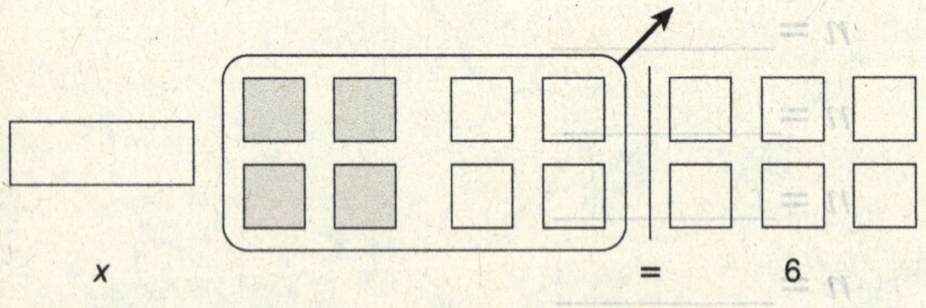

Check: $x - 4 = 2$

$6 - 4 \stackrel{?}{=} 2$

$2 \stackrel{?}{=} 2$ ✓

The remaining tiles represent the solution.

$x = 6$

Use tiles to solve each equation. Then check each answer.

1. $x - 5 = 3$ ____________

2. $x - 2 = 5$ ____________

3. $x - 6 = 4$ ____________

4. $x - 8 = 1$ ____________

5. $x - 3 = 9$ ____________

6. $x - 7 = 3$ ____________

Name ______________________ Date __________ Class __________

LESSON 2F

Homework and Practice

Solving Subtraction Equations

Solve each equation. Check your answers.

1. $k - 7 = 23$
2. $z - 13 = 32$
3. $11 = m - 2$
4. $c - 19 = 10$
5. $34 = x - 12$
6. $r - 31 = 25$
7. $w - 19 = 9$
8. $y - 7 = 29$
9. $12 = a - 54$

Solve each equation.

10. $9 = n - 6$
11. $p - 13 = 25$
12. $26 = z - 9$
13. $21 = b - 21$
14. $44 = f - 8$
15. $s - 19 = 13$
16. $q - 91 = 3$
17. $h - 17 = 17$
18. $56 = g - 29$

19. Shelly took 22 pictures at the zoo. She now has 10 pictures left on the roll. Write and solve a subtraction equation to find out how many photos Shelly had when she went to the zoo.

20. Shelly bought a poster for $15 while she was at the zoo. She now has $7. Write and solve a subtraction equation to find out how much money Shelly took to the zoo.

EXPLORATION

Program of Studies and Core Content

MA-6-NPO-S-NO3, MA-6-AT-S-VEO1, MA-6-AT-S-EI1, MA-6-AT-S-EI3, MA-06-1.3.2, MA-06-5.3.1

2G Multiplication Equations

Bill bought 7 tickets to a basketball game for \$21. How much did each ticket cost?

t	t	t	t	t	t	t	= \$21

$$7 \cdot t = 21$$

$$7 \cdot 3 = 21$$

$$t = 3$$ Each ticket cost \$3.

Find the value of n in each equation.

1. $4 \cdot n = 36$ $n =$ ________
2. $100 \cdot n = 500$ $n =$ ________
3. $50 = 5 \cdot n$ $n =$ ________
4. $24 \cdot n = 48$ $n =$ ________
5. $10 \cdot n = 240$ $n =$ ________

Think and Discuss

6. **Discuss** your strategies for solving the equations.
7. **Explain** how you can find the solution to $4 \cdot n = 200$.

Name ____________________ Date __________ Class __________

LESSON 2G

Hands-On Lab

Solving Multiplication Equations

You can use algebra tiles to help you solve equations.

KEY

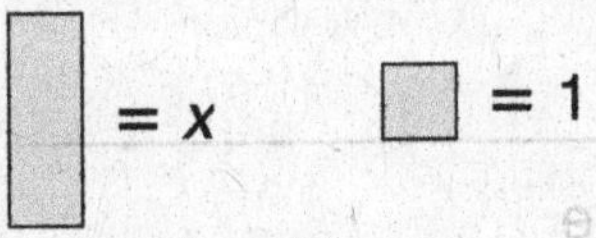

Activity

To solve the equation $3x = 6$, you need one x by itself on one side of the equal sign. Show this with algebra tiles. Divide each side into equal parts.

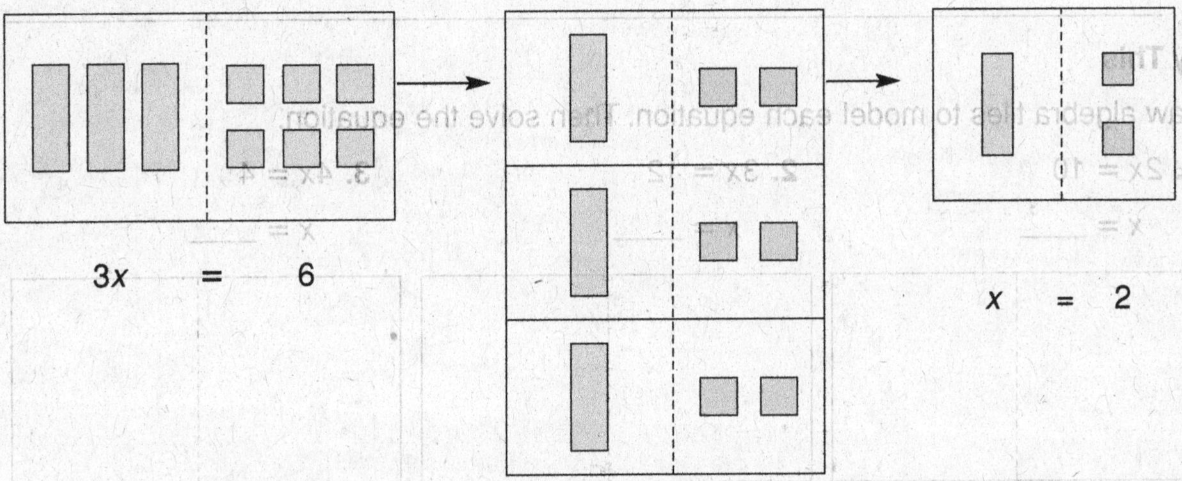

$3x = 6$

$x = 2$

Divide each side into 3 equal parts.

1. Draw the tiles to solve $2x = 8$.

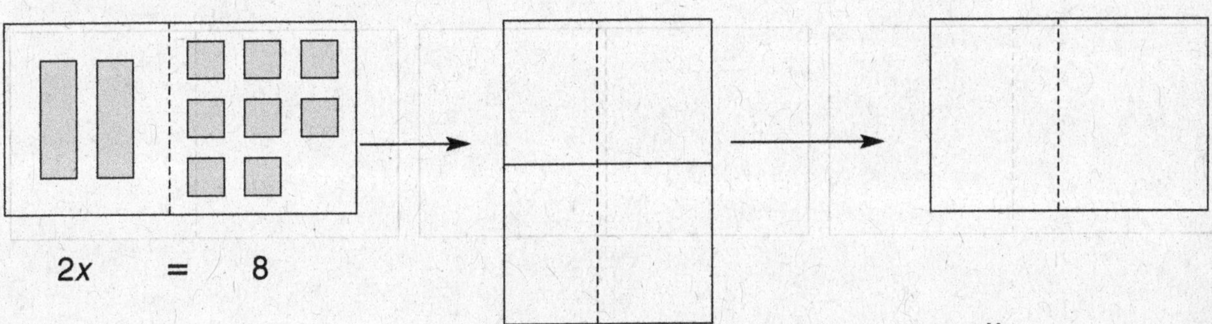

$2x = 8$

$x =$ ___

Divide each side into 2 equal parts.

Name ______________________ Date __________ Class __________

LESSON 2G

Hands-On Lab

Solving Multiplication Equations, continued

Think and Discuss

1. When you put tiles into equal groups, what operation are you representing?

____

2. Give an example of an equation that would require you to divide the tiles into five equal groups to solve.

____

3. How many x blocks and how many unit blocks do you need to represent and solve your equation in Exercise 2?

____

Try This

Draw algebra tiles to model each equation. Then solve the equation.

1. $2x = 10$

   $x =$ ____

2. $3x = 12$

   $x =$ ____

3. $4x = 4$

   $x =$ ____

4. $4x = 8$

   $x =$ ____

5. $3x = 9$

   $x =$ ____

6. $2x = 8$

   $x =$ ____

Name ______________________ Date ____________ Class ________

LESSON **2G**

Review for Mastery

Multiplication Equations

You can use tiles to help you solve multiplication equations.

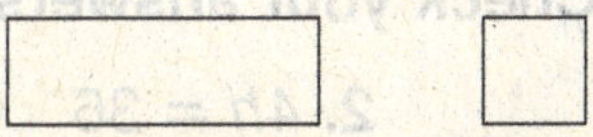

variable add 1

Division undoes multiplication, so you can use division to solve multiplication equations.

To solve $3x = 12$, first use tiles to model the equation.

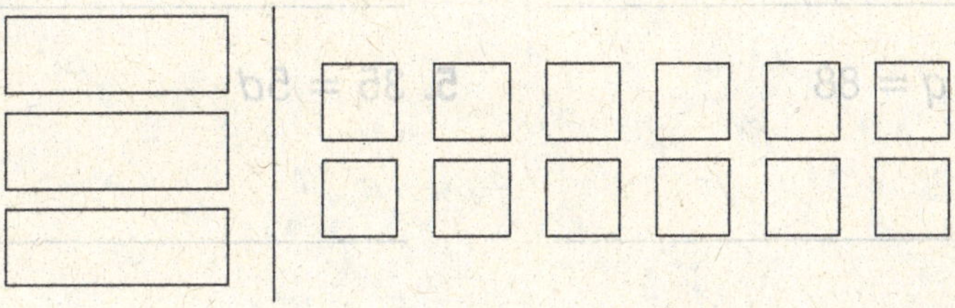

$3x$ = 12

Next, divide each side of the equal sign into 3 equal groups

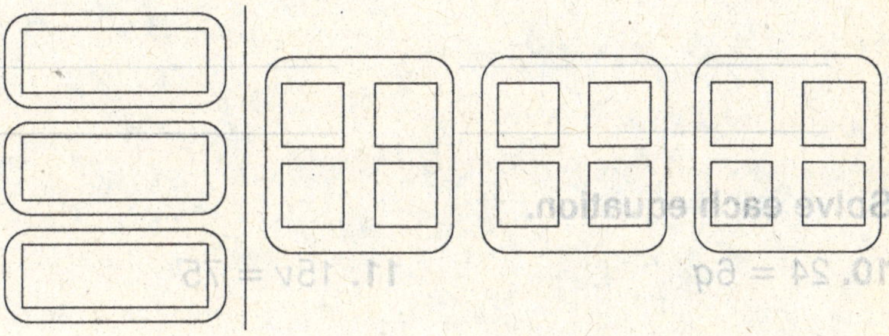

The number of tiles in one group represents the solution.

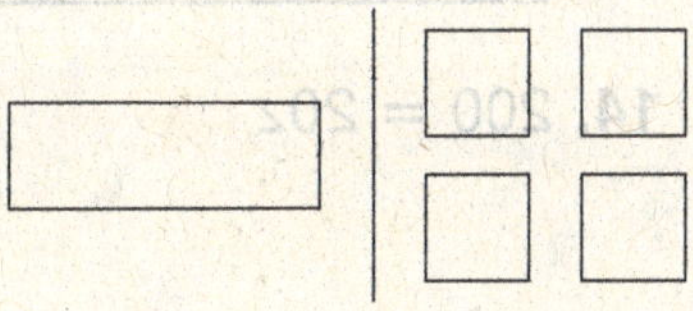

x = 4

Check: $3x = 12$

$3 \cdot 4 \stackrel{?}{=} 12$

$12 \stackrel{?}{=} 12$ ✓

$x = 4$

Use tiles to solve each equation. Then check each answer.

1. $5x = 15$ ____________

2. $2x = 6$ ____________

3. $4x = 16$ ____________

4. $8x = 24$ ____________

5. $3x = 18$ ____________

6. $6x = 12$ ____________

7. $7x = 21$ ____________

8. $9x = 9$ ____________

9. $4x = 24$ ____________

10. $3x = 9$ ____________

11. $8x = 16$ ____________

12. $5x = 25$ ____________

Name ______________________ Date ___________ Class ___________

LESSON 2G

Homework and Practice

Solving Multiplication Equations

Solve each equation. Check your answers.

1. $7w = 42$

2. $4h = 36$

3. $40 = 8f$

4. $11q = 88$

5. $35 = 5d$

6. $9p = 54$

7. $12x = 72$

8. $13r = 143$

9. $51 = 3m$

Solve each equation.

10. $24 = 6q$

11. $15v = 75$

12. $100 = 2s$

13. $48 = 16a$

14. $200 = 20z$

15. $18t = 108$

16. $30c = 90$

17. $24g = 168$

18. $126 = 14k$

19. A hot air balloon flew at 8 miles per hour. Write and solve a multiplication equation to find out how many hours the balloon traveled if it covered a distance of 72 miles.

20. Trevor makes $11 an hour. His last paycheck was for $385. Write and solve a multiplication equation to find out how many hours he worked.

EXPLORATION

Program of Studies and Core Content

MA-6-NPO-S-NO3, MA-6-AT-S-VE01, MA-6-AT-S-EI1, MA-6-AT-S-EI3, MA-06-1.3.2, MA-06-5.3.1

2H Division Equations

Four friends decided to share the cost of a gift for their dance teacher. After dividing the cost by 4, each friend's share is \$25. What was the cost of the gift?

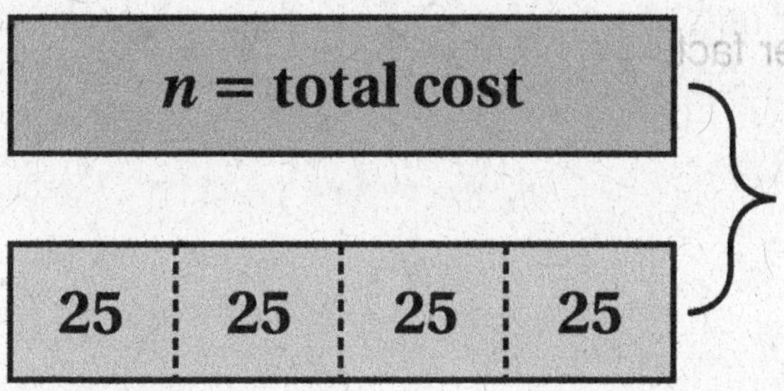

$$\frac{n}{4} = 25 \longrightarrow \frac{100}{4} = 25$$

$$n = 100$$

The gift cost \$100.

Find the value of *n* in each equation.

1. $\frac{n}{2} = 50$

   $n =$ ________

2. $\frac{n}{10} = 2$

   $n =$ ________

3. $20 = \frac{n}{3}$

   $n =$ ________

4. $\frac{n}{7} = 5$

   $n =$ ________

Think and Discuss

5. **Discuss** your strategies for solving the equations.
6. **Explain** how you can find the solution to $\frac{n}{2} = 26$.

Name ______________________ Date __________ Class __________

LESSON 2H

Review for Mastery

Division Equations

You can use multiplication and division to write related number facts.

$3 \cdot 4 = 12 \qquad 12 \div 4 = 3$

Division and multiplication are inverse operations. They undo each other. So you can use multiplication to solve division equations.

To solve $\frac{x}{2} = 3$, think of a related number fact.

If $\frac{x}{2} = 3$, then $3 \cdot 2 = x$.

$3 \cdot 2 = x$
$x = 6$

Check: $\frac{x}{2} = 3$

$\frac{6}{2} \stackrel{?}{=} 3$ *substitute*

$3 \stackrel{?}{=} 3$ ✔

$x = 6$ is the solution for $\frac{x}{2} = 3$.

Use a related number fact to solve each equation. Then check each answer.

1. $\frac{x}{2} = 4$ ____________

2. $\frac{x}{8} = 2$ ____________

3. $\frac{x}{3} = 5$ ____________

4. $\frac{x}{5} = 1$ ____________

5. $\frac{x}{9} = 3$ ____________

6. $\frac{x}{6} = 3$ ____________

7. $\frac{x}{8} = 4$ ____________

8. $\frac{x}{2} = 9$ ____________

9. $\frac{x}{4} = 4$ ____________

10. $\frac{x}{5} = 4$ ____________

11. $\frac{x}{6} = 2$ ____________

12. $\frac{x}{9} = 4$ ____________

Name ______________________ Date __________ Class __________

LESSON 2H

Homework and Practice

Solving Division Equations

Solve each equation. Check your answers.

1. $\frac{a}{2} = 4$ ____
2. $\frac{z}{3} = 6$ ____
3. $5 = \frac{a}{5}$ ____
4. $\frac{k}{4} = 7$ ____
5. $7 = \frac{b}{6}$ ____
6. $\frac{m}{7} = 14$ ____
7. $\frac{r}{9} = 11$ ____
8. $\frac{c}{8} = 8$ ____
9. $12 = \frac{s}{12}$ ____

Solve each equation.

10. $13 = \frac{b}{10}$ ____
11. $\frac{x}{7} = 9$ ____
12. $13 = \frac{v}{13}$ ____
13. $9 = \frac{n}{6}$ ____
14. $20 = \frac{f}{8}$ ____
15. $\frac{d}{4} = 14$ ____
16. $\frac{s}{3} = 70$ ____
17. $\frac{t}{5} = 30$ ____
18. $95 = \frac{u}{2}$ ____

19. All the seats in the theater are divided into 8 groups. There are 40 seats in each group. Write and solve a division equation to find how many seats there are in the theater.

____

____

20. There are 12 inches in one foot. A table is 5 feet long. Write and solve a division equation to find how long the table is in inches.

____

____

EXPLORATION

Program of Studies and Core Content
MA-6-AT-S-EI3, MA-06-5.3.1

2I Introduction to Inequalities

Sally gift wraps identically sized boxes. For each box she will use 2.5 feet of wrapping paper. Sally has one 75-foot roll of wrapper paper to use.

25 boxes
30 boxes
40 boxes
50 boxes

1. Which of the numbers of boxes shown requires Sally to have the following amounts of wrapping paper?

a. Less than one roll

b. Exactly one roll

c. More than one roll

2. Name another number of boxes that Sally could wrap that would use less than one roll of wrapping paper.

3. At most how many boxes could Sally gift wrap if she had 2 rolls of wrapping paper?

Think and Discuss

4. Describe the strategies you used to solve Number 1.

5. Explain how you found numbers of boxes in order to solve Numbers 2 and 3.

Name ______________________________ Date ____________ Class ____________

LESSON 21

Review for Mastery

Introduction to Inequalities

A **solution of an inequality** is a number that makes the inequality true. An inequality usually has more than one solution. All the solutions are contained in the **solution set.**

As with equations, solve a simple inequality by using inverse operations to isolate the variable.

Solve and graph $x + 1 > 6$.

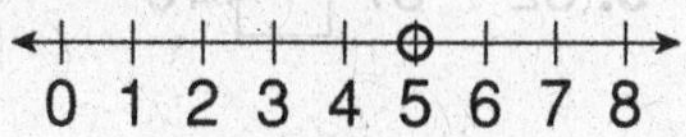

Draw an open circle at 5 to show that 5 is not included in the solution set.

$$\begin{array}{rcl} x + 1 & > & 6 \\ \underline{-1} & & \underline{-1} \\ x & > & 5 \end{array}$$

Subtract 1.

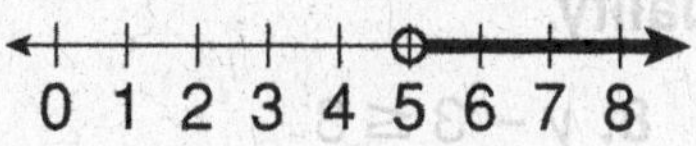

Draw an arrow to the right of 5 to show that all numbers greater than 5 are included in the solutions.

According to the graph, 6 should be a solution and 4 should not be a solution.

Check: $x + 1 > 6$ $x + 1 > 6$

$6 + 1 \overset{?}{>} 6$ $4 + 1 \overset{?}{>} 6$

$7 > 6$ $5 \not> 6$

So, 6 *is* in the solution set and 4 *is not* in the solution set. Thus, the solution set for the inequality $x + 1 > 6$ is $x > 5$.

Write *true* or *false*.

1. $0 < 4$ ____________

2. $-10 \leq 9$ ____________

3. $-3 > 2$ ____________

Using the variable *n*, write the inequality shown by each graph.

4.

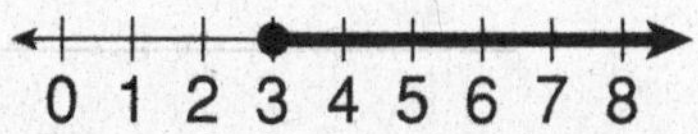

____

5.

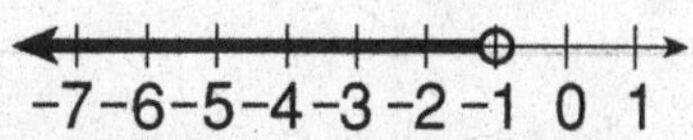

____

Complete. Is the given value in the solution set? Answer *is* or *is not*.

6. 4 ______ in the solution set of $x - 1 > 5$.

$x - 1 > 5$

____ $- 1 \overset{?}{>} 5$

________ > 5

7. 0 ______ in the solution set of $z + 2 \geq 2$.

$z + 2 \geq 2$

____ $+ 2 \geq 2$

____ ≥ 2

8. 14 ______ in the solution set of $w - 10 \leq 25$.

$w - 10 \leq 25$

____ $- 10 \leq 25$

____ ≤ 25

Name ______________________ Date ____________ Class ____________

LESSON 21

Homework and Practice

Solving Simple Inequalities

Use <, >, or = to compare each inequality.

1. $8 + 13 \square 5$ **2.** $23 \square 3(7)$ **3.** $28 - 9 \square 18$

4. $72 \square 9(8)$ **5.** $52 - 37 \square 15$ **6.** $4(15) \square 12(6)$

Solve and graph each inequality.

7. $x + 8 > 12$ **8.** $y - 3 \leq 3$ **9.** $5n \geq 25$

10. $h - 4 \geq 3$ **11.** $\frac{x}{5} < 3$ **12.** $\frac{t}{2} \geq 6$

13. $7d > 35$ **14.** $4m < 24$ **15.** $x - 3 \leq 5$

16. Luis wants to drink at least 72 ounces of water every day while at work. He works 8 hours a day. How many ounces of water must Luis drink each hour? Write and solve an inequality to answer the question.

____

17. Theo's car has an 18-gallon gas tank. He wants to travel 468 miles on one full tank of gas. What is the least miles per gallon Theo's car must get on this trip? Write and solve an inequality to answer the question.

____

LESSON 2

Chapter Summary

Answer these question to summarize the important concepts from Chapter 2 in your own words.

1. Explain what it means to evaluate an expression with values of a variable.

2. When translating from words to math, there are key words that represent each operation. List two key words that represent addition. List two keys words for subtraction. List two key words for multiplication. List two key words for division.

3. What does it mean to be a solution of an equation?

4. Explain how to solve a whole-number equation.

For more review of Chapter 2:

- Complete the Chapter 2 Study Guide and Review on pages 94–96 of your textbook.
- Complete the Ready to Go On quizzes on pages 68 and 88 of your textbook.

Name ______________________ Date __________ Class __________

CHAPTER 2

KCCT Prep

Multiple Choice

1. Martin is 13 years younger than Amy. Amy is 27 years old. Which equation can you solve to find Martin's age?

Ⓐ $m - 13 = 27$ Ⓒ $m = 27 + 13$
Ⓑ $m + 13 = 27$ Ⓓ $m + 27 = 13$

2. Lara added parentheses to the expression 6 + 10 ÷ 2 + 2, but she did not change the value of the expression. Which of the following shows how she may have placed the parentheses?

Ⓐ (6 + 10) ÷ 2 + 2
Ⓑ 6 + (10 ÷ 2) + 2
Ⓒ (6 + 10) ÷ (2 + 2)
Ⓓ 6 + 10 ÷ (2 + 2)

3. Which of these is NOT a term of the sequence shown in the table?

Position	1	2	3	4
Value of term	3	6	9	12

Ⓐ 300 Ⓒ 315
Ⓑ 301 Ⓓ 324

4. Clover Street is twice as long as Grant Road. Grant Road is 16 miles long. Which equation can you solve to find the length of Clover Street?

Ⓐ $2c = 16$ Ⓒ $\frac{c}{2} = 16$
Ⓑ $c + 2 = 16$ Ⓓ $16 \div c = 2$

5. Which of these is an example of the Commutative Property?

Ⓐ 5 + (11 + 2) = (5 + 11) + 2
Ⓑ 17 + 8 = 8 + 17
Ⓒ 13 × (9 × 2) = (13 × 9) × 2
Ⓓ 250 × 1 = 250

6. The number of points Jamal scores on a game show is 5 times the number of questions he answers correctly minus 3 times the number of questions he answers incorrectly. Jamal answers 12 questions correctly and 7 questions incorrectly. Which expression represents his score?

Ⓐ 5 × 12 − 3 × 7
Ⓑ 5 × (12 − 3) × 7
Ⓒ 5 × (12 − 3 × 7)
Ⓓ (5 × 12 − 3) × 7

7. A shirt is on sale for $15 off the regular price. The sale price is $14. Which equation can you solve to find the original price of the shirt?

Ⓐ $15 - n = 14$ Ⓒ $15 = n + 14$
Ⓑ $n - 15 = 14$ Ⓓ $15 + n = 14$

8. Which pattern has 200 as one of its terms?

Ⓐ 1, 3, 9, 27, ...
Ⓑ 5, 2, 7, 2, 9, 2, 11, ...
Ⓒ 10, 7, 20, 17, 30, 27, ...
Ⓓ 15, 13, 25, 22, 35, 32, ...

9. Mari's aquarium has 18 more fish than Alex's aquarium. Alex's aquarium has 34 fish. Which equation can you solve to find the number of fish in Mari's aquarium?

Ⓐ $y + 18 = 34$ Ⓒ $18 - y = 34$
Ⓑ $y = 34 - 18$ Ⓓ $34 = y - 18$

10. The temperature of an oven in degrees Fahrenheit is given by the expression $50x - 3^2$, where x is the time in minutes. At which of these times is the temperature closest to 300°F?

Ⓐ 4 minutes Ⓒ 6 minutes
Ⓑ 5 minutes Ⓓ 7 minutes

11. Which is the best estimate of the solution of $15{,}023 = x - 8{,}957$?

Ⓐ 6,000 Ⓒ 23,000
Ⓑ 7,000 Ⓓ 24,000

12. After Steven grew 2 inches, his height was greater than 50 inches. Which inequality represents his original height?

Ⓐ $h < 48$ Ⓒ $h > 48$
Ⓑ $h \leq 48$ Ⓓ $h \geq 48$

13. Sandra creates a sequence by starting with 15 and adding 6 to each term to get the next term. Her teacher asks her to make a conjecture about the terms of the sequence. Which of these is the best conjecture?

Ⓐ All of the terms are less than 80.
Ⓑ All of the terms are even.
Ⓒ All of the terms are odd.
Ⓓ All of the terms are a multiple of 15.

14. A group of students goes on a trip on 5 buses. Each bus carries 20 students. Which equation can you solve to find the total number of students who go on the trip?

Ⓐ $20 = \frac{x}{5}$ Ⓒ $5 \div x = 20$
Ⓑ $20 = \frac{5}{x}$ Ⓓ $x = 20 \div 5$

15. After Tanesha gave away 6 of her CDs, she had at least 20 CDs in her collection. Which inequallity represents her original number of CDs?

Ⓐ $x > 26$ Ⓒ $x > 14$
Ⓑ $x \geq 26$ Ⓓ $x \leq 14$

Open Response

16. Admission to a chili festival costs $5. Each sample of chili costs $2.

a. Find the total cost of admission and 5 samples of chili.

b. Write an expression that gives the total cost of admission and n samples of chili.

c. At a barbecue festival, the expression $2 + 3n$ gives the total cost of admission plus n barbecue samples. Rosa bought the same number of samples at the chili festival and at the barbecue festival, but she spent more money at the barbecue festival. What can you say about the number of samples she bought? Explain.

17. The table gives the heights of some famous skyscrapers.

Famous Skyscrapers	
Name, Location	**Height (m)**
Sears Tower, Chicago	442
Chrysler Building, New York	319
Transamerica Pyramid, San Francisco	260

The Sears Tower in Chicago is 442 m tall, the Chrysler Building in New York is 319 m tall, and the Transamerica Pryamid in San Francisco is 260 m tall.

a. The Empire State Building is m meters taller than the Chrysler Building. Write an expression for the height of the Empire State Building.

b. The west antenna on the Sears Tower is 356 m shorter than the building itself. The west antenna is 9 m taller than the east antenna. Write and solve one or more equations to find the height of the east antenna.

CHAPTER 3

Project
Decimals: Go for the Gold

Career: Sports Statistician

Are people breaking records by running faster and jumping farther and higher? Records are kept for both professional and amateur sports. Many schools keep records of their individual athletes' and teams' performances. Keeping track of sports records is the job of sports historians. One of the most complete records is that of the Olympic games.

Before you start the project, work through the Olympic History worksheet on the next page.

In this chapter project . . .

You compare athletes' performances from the past and present.

What to do:

- Choose an event from the Olympic History worksheet that your school's track team participates in.
- Research the winning times (or distances) from several years. If the records exist, try to find data from your school for the same years that the Olympics were held. Use data for the men's event, the women's event, or both.

What to present:

- A table that shows the Olympic data, the data from your school, and the difference between the two.
- A line graph that shows how the data for the Olympics and your school have changed over time.

Where's the math?

- Use decimals to calculate the difference between the Olympic and school data.
- Use a line graph to compare the results.

go.hrw.com
Chapter Project Online
KEYWORD: MR7 Ch3

Name ______________________ Date __________ Class __________

CHAPTER 3

Project Recording Sheet

Decimals

Olympic History

Use the table below to help you answer the following questions:

Winning Olympic Performances

Year	Women's 100 Meters (s)	Women's High Jump (m)	Women's Discus (m)	Men's 100 Meters (s)	Men's High Jump (m)	Men's Discus (m)
1900				12.0	1.90	36.04
1928	12.2	1.59	39.62	10.8	1.94	47.32
1952	11.5	1.67	51.4	10.4	2.05	55.02
1988	10.54	1.82	72.3	9.92	2.38	68.81
2000	10.75	2.01	68.4	9.87	2.39	69.29
2004	10.93	2.06	67.02	9.85	2.36	69.89

Tip: A decimal fraction such as 11.5 can also be written as 11.50.

1. How much higher did men jump in 2004 than in 1900? __________
2. How much higher did women jump in 2004 than in 1928? __________
3. How much faster was the men's winning 100-meter time in 2004 compared to 1900? __________
4. How much faster was the women's winning 100-meter time in 2004 compared to 1928? __________
5. How much farther did women throw the discus in 2004 than in 1928? __________
6. How much farther did men throw the discus in 2004 than in 1900? __________
7. Estimate the results in these events for the 2012 Olympic games in London, England.

____

____

____

EXPLORATION

Program of Studies and Core Content

MA-6-NPO-S-NS1, MA-6-NPO-S-NS3, MA-6-NPO-S-NS5, MA-06-1.1.1, MA-06-1.1.2, MA-06-1.1.3

3A Representing, Comparing, and Ordering Decimals

To model a decimal,

- color a 10-by-10-square grid for each whole in the decimal,
- color one 10-by-1-square strip for each tenth in the decimal, and color a small square for each hundredth in the decimal.

For example, the graph paper models the decimal 1.62.

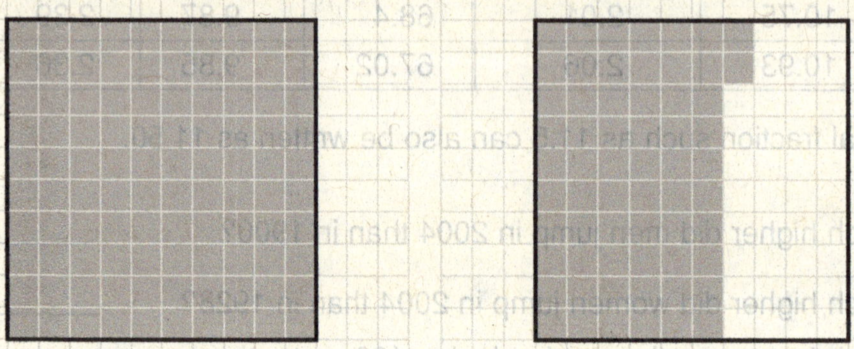

1. Draw a model for each decimal.

a. 1.25 **b.** 2.13 **c.** 1.70 **d.** 1.7

2. Compare the models for **1c** and **1d.** What do you notice about these two decimals?

3. Order the decimals in numbers **1a–1d** from least to greatest. Explain your reasoning.

Think and Discuss

4. Explain why 0.4 = 0.40.

5. Explain why 0.5 is greater than 0.10 even though 10 is greater than 5.

Name ______________________ Date __________ Class __________

LESSON **3A**

Review for Mastery

Representing, Comparing, and Ordering Decimals

You can use place value to write decimals in standard form, expanded form, and word form.

To write 2.14 in expanded form, write the decimal as an addition expression using the place value of each digit.

Ones	Tenths	Hundredths	Thousandths	Ten Thousandths
2.	1	4		

2.14 can be written as 2 + 0.1 + 0.04.

When you write a decimal in word form, the number before the decimal point tells you how many wholes there are. The decimal point stands for the word "and."

Notice that the place value names to the right of the decimal begin with tenths, hundredths, and then thousandths. The "ths" ending indicates a decimal.

2.14 can also be written as *two and fourteen hundredths.*

1. How would you read a number with 4 decimal places?

____

Write each decimal in standard form, expanded form, and word form.

2.

Ones	Tenths	Hundredths	Thousandths	Ten Thousandths
5.	6	9	8	

____

____

____

3.

Ones	Tenths	Hundredths	Thousandths	Ten Thousandths
0.	0	9	4	

____

____

____

4. 7 + 0.8

____

____

5. twelve-hundredths

____

____

Name ______________________ Date ____________ Class ____________

LESSON 3A

Homework and Practice

Representing, Comparing, and Ordering Decimals

Write each in standard form, expanded form, and words.

1. 8 + 0.3

2. one and fourteen thousandths

3. 12.13

4. 11.04

5. 4.23

6. 7 + 0.4 + 0.03

7. eight and twenty-nine thousandths

8. 0.4 + 0.02

9. nine and twenty-nine hundredths

Order the decimals from least to greatest.

10. 2.982, 2.289, 2.829

11. 0.482, 0.4082, 0.8042

12. 0.101, 1.01, 1.001

13. 2.321, 3.12, 2.301, 2.31

14. A hamburger factory makes two types of large hamburgers, the Ultra and the Supreme. The Ultra weighs 0.81 pounds and the Supreme weighs eight tenths of a pound. Which hamburger weighs more?

15. Lauren can jump two and thirteen-hundredths feet high and Terry can jump two and twenty-one hundredths feet high. Which of the two can jump higher?

Program of Studies and Core Content
MA-6-NPO-S-E1, MA-6-NPO-S-E2, MA-06-1.2.1

3B Estimating Decimals

For each problem, estimate a solution. Then compute with a calculator to see how close your estimated solutions are to the actual solutions.

		Estimate	Actual
1.	Seven people want to share the cost of a $33.75 boat rental. How much will each person pay?		
2.	You purchase items that cost the following amounts: $4.95, $1.29, $6.67, $4.19, and $10.39. What is the total cost?		
3.	How much change is there from $200.00 for an item that costs $157.98?		
4.	Ron's gas tank holds 21 gallons and is empty. If gas costs $2.499 per gallon, how much will it cost to fill Ron's tank?		

Think and Discuss

5. Discuss the estimation strategies you used.

6. Describe a situation in which all you need is an estimate and a situation in which you need an exact value.

Name ______________________________ Date ____________ Class ____________

LESSON **3B**

Review for Mastery

Estimating Decimals

You can use rounding to estimate. Round to the indicated place value. Then add or subtract.

A. 3.478 + 7.136; tenths

3.478	7 ≥ 5, so round up	3.5
7.136	3 < 5, so round down	+7.1
		10.6

3.478 + 7.136 is about 10.6.

B. 12.848 − 6.124; hundredths

12.848	8 ≥ 5, so round up	12.85
6.124	4 < 5, so round down	−6.12
		6.73

12.848 − 6.124 is about 6.73.

Estimate by rounding to the indicated place value.

1. 1.04 + 9.37; tenths

1.04 rounds to ______

9.37 rounds to ______

estimate ______

2. 2.17 + 3.56; tenths

2.17 rounds to ______

3.56 rounds to ______

estimate ______

3. 6.753 − 4.245; hundredths

6.753 rounds to ______

4.255 rounds to ______

estimate ______

You can use compatible numbers to estimate. Pick numbers that are close to the actual numbers that are easy to multiply or divide. Then multiply or divide.

A. 4.6 • 3.2
5 and 3 are compatible numbers.
5 • 3 = 15, so 4.6 • 3.2 is about 15.

B. 48.3 ÷ 13.2
48 and 12 are compatible numbers.
48 ÷ 12 = 4, so 48.3 ÷ 13.2 is about 4.

Use compatible numbers to estimate each product or quotient.

4. 9.4 • 5.6 ____________

5. 7.25 • 10.84 ____________

6. 84.8 ÷ 3.9 ____________

7. 21.9 ÷ 3.1 ____________

8. 8.3 • 7.6 ____________

9. 55.7 ÷ 6.9 ____________

10. 5.57 ÷ 2.7 ____________

11. 6.729 • 9.8 ____________

Name ______________________ Date __________ Class __________

LESSON 3B

Homework and Practice

Estimating Decimals

Estimate. Round to the indicated place value.

1. 2.931 + 4.45; tenths

____

2. 8.64202 + 3.328; hundredths

____

3. 3.04891 − 3.03843; thousandths

____

4. 11.49 − 8.5; nearest whole number

____

Estimate each product or quotient.

5. 62.6 ÷ 3.25

____

6. 24.02 ÷ 1.56

____

7. 3.485 × 9.201

____

8. 48.3102 ÷ 7.6871

____

9. 19.872 × 3.9701

____

10. 12.5402 × 4.4134

____

Estimate a range for the sum.

11. 9.4 + 2.7 + 12.7

____

12. 5.82 + 3.14 + 6.29

____

13. Jonathan is painting his house. On Monday he spent 3.3 hours painting and on Tuesday he spent 5.9 hours painting. About how many hours has he painted? If it takes 10.5 hours to paint the entire house, about how many more hours does Jonathan have to spend painting?

____

14. A new space ship, the X-9000, can orbit Earth in 79.62 minutes. If the X-9000 has been flying for 4 hours, about how many times has it orbited Earth? If the pilot wants to circle the Earth 7.3 times, about how many minutes should she fly?

____

EXPLORATION

3C Adding and Subtracting Decimals

Program of Studies and Core Content
MA-6-NPO-S-NS2, MA-6-NPO-S-E1, MA-6-NPO-S-NO1, MA-6-NPO-S-NO2, MA-06-1.1.2, MA-06-1.2.1, MA-06-1.3.1

You have 19¢. How much more do you need to have $1.00? From 19¢ to 20¢, add 1¢. From 20¢ to $1.00, add 80¢. The answer is 1¢ + 80¢, or 81¢.

1. Draw arrows to connect each pair of amounts that would give you a sum of $1.00.

Amount 1	Amount 2
$0.19	$0.63
$0.25	$0.55
$0.76	$0.93
$0.07	$0.75
$0.65	$0.24
$0.37	$0.35
$0.45	$0.81

2. Compute the change from $10.00 on a purchase of each amount. Example: From $1.25 to $2.00, add $0.75. From $2.00 to $10.00, add $8.00. So the change on a purchase of $1.25 is $0.75 + $8.00 = $8.75.

Amount	Change from $10.00
$1.25	$8.75
$2.76	
$3.07	
$4.65	
$5.37	
$6.45	
$7.59	

Think and Discuss

3. Name five different pairs of numbers that each have a sum of $1.00.

4. Describe how you can use the strategy of "adding up" to find 200 − 176.25.

Name ______________________ Date ____________ Class ____________

LESSON 3C

Review for Mastery

Adding and Subtracting Decimals

You can use a place-value chart to help you add and subtract decimals.

Add 1.4 and 0.9.

	Tens	Ones	Tenths	Hundredths	Thousandths
		1.	4		
+		0.	9		

So, 1.4 + 0.9 = 2.3.

Subtract 2.4 from 3.1.

	Tens	Ones	Tenths	Hundredths	Thousandths
		3.	1		
–		2.	4		

So, 3.1 – 2.4 = 0.7.

Find each sum or difference.

1.

	Tens	Ones	Tenths	Hundredths	Thousandths
		2.	6		
+		1.	1	5	

____

2.

	Tens	Ones	Tenths	Hundredths	Thousandths
		2.	5	3	
–		1.	7		

____

3. 4.3 + 1.4

Tens	Ones	Tenths	Hundredths	Thousandths

____

4. 14.4 – 3.8

Tens	Ones	Tenths	Hundredths	Thousandths

____

5. 7.3 + 8.5

Tens	Ones	Tenths	Hundredths	Thousandths

____

6. 12.34 – 6.9

Tens	Ones	Tenths	Hundredths	Thousandths

____

7. 6.3 – 2.5

____

8. 20.65 + 13.24

____

9. 8.9 – 1.95

____

10. 3.42 + 5.25

____

Name ____________________ Date __________ Class __________

LESSON 3C

Homework and Practice

Adding and Subtracting Decimals

Find each sum or difference.

1. 8.9 + 2.4 ____________

2. 12.7 − 9.6 ____________

3. 18.35 − 4.16 ____________

4. 7.21 + 11.6 ____________

5. 0.975 + 3.8 ____________

6. 20.66 − 9.1 ____________

7. Tiffany's job requires a lot of driving. How many miles did she travel during the month of February? ____________

Miles Tiffany Traveled

Week	1	2	3	4
Miles	210.05	195.18	150.25	165.30

8. Shelly babysits after school and on the weekends. How much did she earn in all for the month of April? ____________

Shelly's Earnings for April

Week	1	2	3	4
Earnings	$120.50	$180.75	$205.25	$215.50

Evaluate 5.6 – *a* for each value of *a*.

9. $a = 3.7$ ____________

10. $a = 0.5$ ____________

11. $a = 2.8$ ____________

12. $a = 1.42$ ____________

13. $a = 0.16$ ____________

14. $a = 3.75$ ____________

15. Allen bought a box of envelopes for $2.79 and a pack of paper for $4.50. He paid with a $10 bill. How much change should be receive?

____

16. From a bolt of cloth measuring 25.60 yards, Tina cut a 6.8-yard piece and an 11.9-yard piece. How much material is left on the bolt?

____

EXPLORATION

Program of Studies and Core Content

MA-6-NPO-S-NS2, MA-6-NPO-S-E1, MA-6-NPO-S-NO1, MA-6-NPO-S-NO2, MA-06-1.1.2, MA-06-1.2.1, MA-06-1.3.1

3D Multiplying Decimals

When you multiply decimals, you can use estimation to help you determine the position of the decimal point in the product.

Estimate each product. Then use a calculator to see how reasonable your estimate is.

		Estimate	Actual
1.	4.235 × 16.9		
2.	0.78 × 568		
3.	56.1 × 23		
4.	15.6 × 2.15		

Estimate each product. Use this estimate to decide where to place a decimal point in the answer. Check with your calculator.

5. 70.5 × 4.4 = 3 1 0 2

6. 0.75 × 692 = 5 1 9 0

7. 56 × 3.125 = 1 7 5 0

8. 45.6 × 2.15 = 9 8 0 4

9. 4.17 × 1.2 = 5 0 0 4

10. 125.2 × 7.4 = 9 2 6 4 8

Think and Discuss

11. Discuss your strategies for estimating in numbers 1–4.

12. Explain how you know where to place the decimal point in a product.

Name ______________________ Date ____________ Class ____________

LESSON 3D

Review for Mastery

Multiplying Decimals

You can use a model to help you multiply a decimal by a whole number.

Find the product of 0.12 and 4, using a 10 by 10 grid.

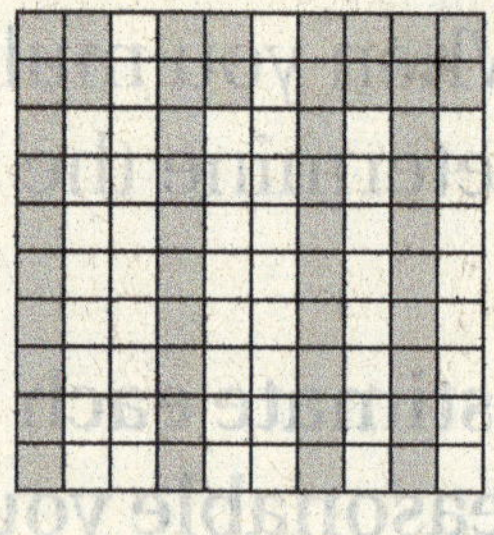

Shade 4 groups of 12 squares. Count the number of shaded squares. Since you have shaded 48 of the 100 squares, 0.12 • 4 = 0.48.

Find each product.

1. 0.23 • 3 ________ **2.** 0.41 • 2 ________ **3.** 0.011 • 5 ________ **4.** 0.32 • 2 ________

5. 0.15 • 3 ________ **6.** 0.42 • 2 ________ **7.** 0.04 • 8 ________ **8.** 0.22 • 4 ________

You can also use a model to help you multiply a decimal by a decimal.

Find the product of 0.4 and 0.6.
0.4 • 0.6 = 0.24

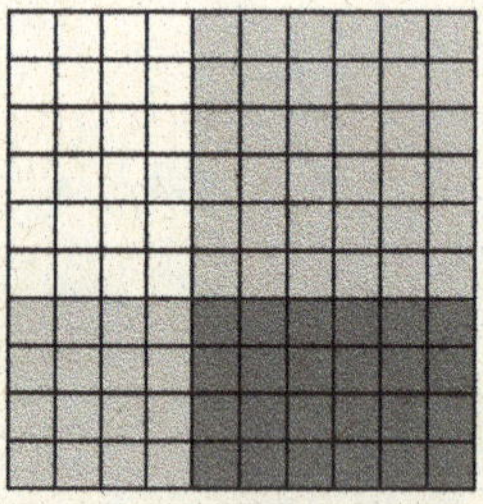

Find each product.

9. 0.2 • 0.8 ________ **10.** 0.7 • 0.9 ________ **11.** 0.5 • 0.5 ________ **12.** 0.3 • 0.6 ________

13. 0.5 • 0.2 ________ **14.** 0.4 • 0.4 ________ **15.** 0.1 • 0.9 ________ **16.** 0.4 • 0.7 ________

Name ______________________ Date __________ Class __________

LESSON 3D

Homework and Practice

Multiplying Decimals

Find each product.

1. 0.6×0.6 ________

2. 0.7×0.8 ________

3. 0.8×0.9 ________

4. 2.5×0.2 ________

5. 0.9×0.27 ________

6. 3.0×0.03 ________

7. 0.9 × 0.41 ________

8. 3.86 × 4.3 ________

9. 8.41 × 0.5 ________

10. 0.7 × 2.32 ________

11. 1.75 × 2.6 ________

12. 3.5 × 2.46 ________

Evaluate 5*x* for each value of *x*.

13. $x = 0.3$ ________

14. $x = 4.8$ ________

15. $x = 0.81$ ________

16. $x = 6.15$ ________

17. $x = 0.672$ ________

18. $x = 35.78$ ________

19. Dimitri normally blinks an average of 21.3 times per minute. How many times will he blink in 2.8 minutes? If he blinks 78.5 times per minute when he is nervous, how many times will he blink in 1.38 minutes while he is nervous?

____

20. A landscaping company charges $2.25 per square foot of flowers added to a yard. If a family wants 12.4 square feet of flowers added to a garden, how much will they have to pay?

____

EXPLORATION

Program of Studies and Core Content
MA-6-NPO-S-NS2, MA-6-NPO-S-NO1, MA-6-NPO-S-NO2, MA-06-1.3.1

3E Dividing Decimals by Whole Numbers

1. Four friends go on a vacation together. They decide to share all expenses evenly. Estimate the cost of each item per person, and then compute the actual cost with a calculator.

Item	Total Cost	Estimated Cost per Person	Actual Cost per Person
Cab fare	$50.00		
Pizza	$13.92		
Movie rental	$10.00		
Dinner	$76.20		
Boat ride	$35.96		

Estimate each quotient. Use this estimate to decide where to place a decimal point in the answer. Check with your calculator.

2. 125.2 ÷ 25 = 5 0 0 8

3. 40 ÷ 16 = 2 5

4. 7.5 ÷ 5 = 1 5

5. 75 ÷ 12 = 6 2 5

Think and Discuss

6. Discuss your strategies for estimating in number **1.**

7. Explain how you know where to place the decimal point in a quotient.

Name ______________________ Date __________ Class __________

LESSON **3E**

Review for Mastery

Dividing Decimals by Whole Numbers

You can use decimal grids to help you divide decimals by whole numbers.

To divide 0.35 by 7, first shade in a decimal grid to show thirty-five hundredths.

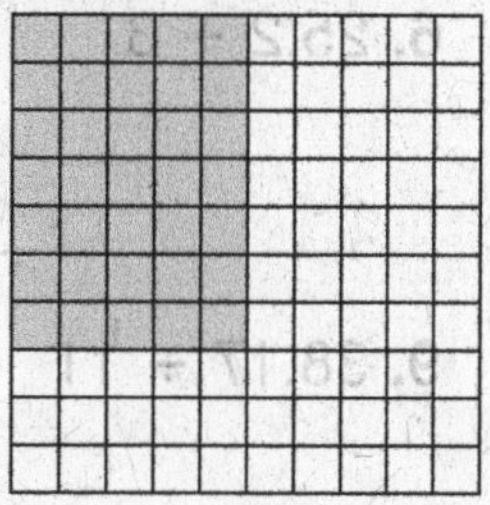

0.35 ÷ 7 means "divide 0.35 into 7 equal groups." Show this on the decimal grid.

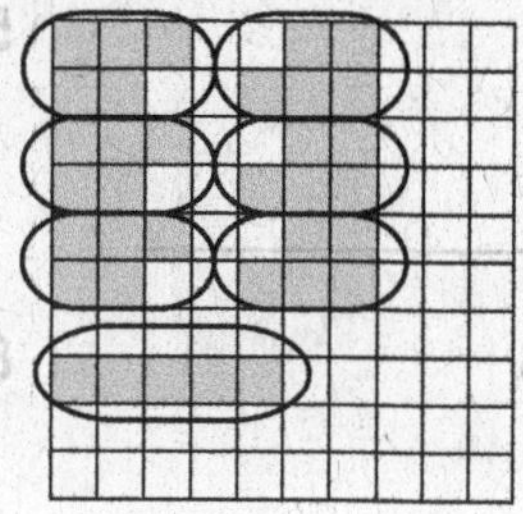

The number of units in each group is the quotient.

So, 0.35 ÷ 7 = 0.05.

Use decimal grids to find each quotient.

1. 0.24 ÷ 4

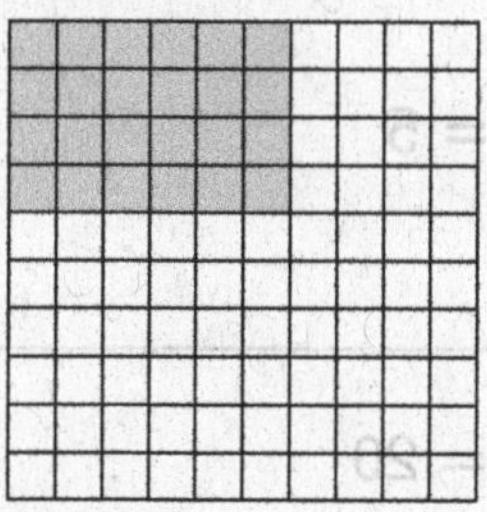

____

2. 0.48 ÷ 12

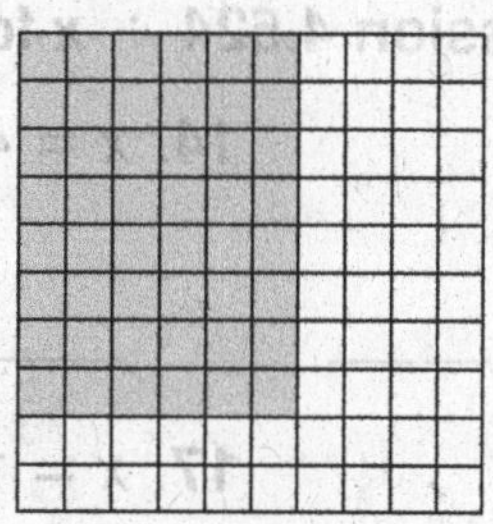

____

3. 0.50 ÷ 10

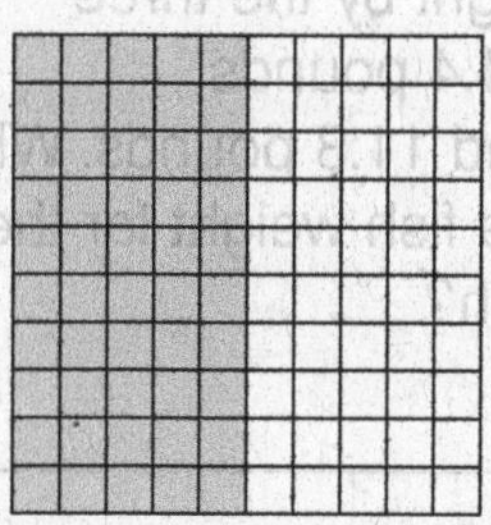

____

4. 0.98 ÷ 7

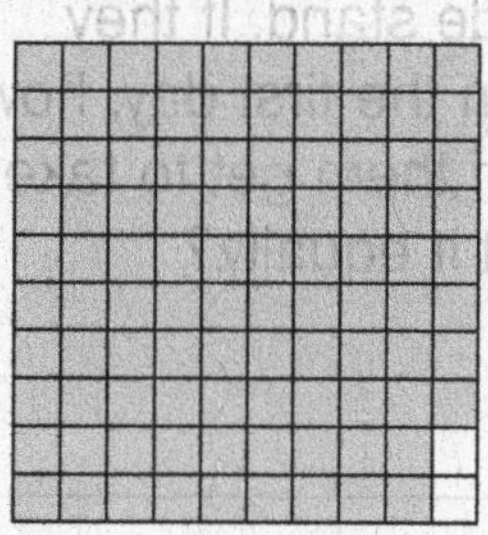

____

5. 0.6 ÷ 5

____

6. 0.78 ÷ 6

____

7. 0.99 ÷ 11

____

8. 0.32 ÷ 4

____

Name ____________________ Date __________ Class __________

LESSON 3E

Homework and Practice

Dividing Decimals by Whole Numbers

Find each quotient.

1. 0.57 ÷ 3 ________

2. 2.73 ÷ 7 ________

3. 5.4 ÷ 5 ________

4. 8.1 ÷ 4 ________

5. 48.24 ÷ 8 ________

6. 25.2 ÷ 3 ________

7. 5.16 ÷ 6 ________

8. 0.387 ÷ 9 ________

9. 38.17 ÷ 11 ________

10. 0.48 ÷ 4 ________

11. 0.0468 ÷ 5 ________

12. 61.5 ÷ 3 ________

Evaluate the expression 4.524 ÷ *x* for the given value of *x*.

13. $x = 3$ ________

14. $x = 4$ ________

15. $x = 5$ ________

16. $x = 6$ ________

17. $x = 12$ ________

18. $x = 20$ ________

19. William, Robert, Wendi, and Robin set up a lemonade stand. If they earned $15.68 on the first day, how much will each of them get to take home if they split it equally?

____

20. During a fishing tournament, the biggest fish caught by the three winners were 14.4 pounds, 13.6 pounds, and 11.3 pounds. What was the average fish weight for the three biggest fish?

____

EXPLORATION

Program of Studies and Core Content
MA-6-NPO-S-NS2, MA-6-NPO-S-NO1, MA-6-NPO-S-NO2, MA-06-1.3.1

3F Dividing by Decimals

A CD store carries the packages of recordable CDs listed in the table. The third column shows the cost of one CD for different packages. This is called the ***unit cost.*** The fourth column shows how many CDs one dollar can buy. For example, if you buy the 5-pack at $4.95, one dollar buys 1.01 CD (a little more than one CD). This is called the ***purchasing power*** of $1.00.

Use a calculator to find the cost of 1 CD and the purchasing power of $1.00 for each package.

	Item	Cost	Cost of 1 CD	Purchasing Power of $1.00
1.	Single CD	$1.19		
2.	5-pack	$4.95	$4.95 \div 5 = 0.99$	$5 \div 4.95 = 1.01$
3.	10-pack	$8.95		
4.	20-pack	$16.95		
5.	50-pack	$35.95		

6. Which package gives you the highest unit cost?

7. Which package gives you the lowest unit cost?

8. Which package gives you the greatest purchasing power per dollar? the least purchasing power per dollar?

Think and Discuss

9. Describe the relationship between unit cost and purchasing power by looking at the numbers in the third and fourth columns above.

Name ______________________________ Date __________ Class __________

LESSON 3F

Technology Lab

Dividing by Decimals

Use your calculator to divide whole numbers and decimals without having to change the divisor or the dividend.

Activity

Your class has decided to raise money for the local food pantry by selling home made ice cream cones at this year's school basketball tournament.

How much will a customer pay for each ounce of ice cream if the typical serving size is 5.57 ounces (oz), and each cone costs $3.50?

Estimate the cost of just one ounce of ice cream.

Enter the total cost: 3.50

Press ÷

Enter the total size in ounces: 5.57

Press ENTER

Compare the actual result to your estimate. How close were you to the $0.628 price?

If your cost to make the ice cream cone is $1.25, how much does it cost you to produce each ounce of ice cream?
Enter the total cost: $1.25

Press ÷

Enter the total size in ounces: 5.57

Press ENTER

Your cost is $0.224 per oz.

Name ______________________ Date __________ Class __________

LESSON 3F

Technology Lab

Dividing by Decimals, continued

Think and Discuss

1. If a gallon of ice cream is 128 oz, how many 5.57 oz cones can you make from each gallon? Divide to find the solution.

____

2. To increase your profits a bit, you decide to sell smaller portions, but charge the same price. How many cones per gallon of ice cream can you serve if the new portion size is 4.25 oz?

____

Try This

1. Calculate the price per ounce of the following foods.

a. 12 oz bag of organic raisins costs $1.91

____

b. 6 oz cup of yogurt cost $0.79

____

c. 5 pounds (lbs) of cashews cost $22.50

____

____

Name ______________________ Date __________ Class __________

LESSON 3F

Review for Mastery

Dividing by Decimals

You can use powers of ten to help you divide a decimal by a decimal.

To divide 0.048 by 0.12, first multiply each number by the least power of ten that makes the divisor a whole number.

$0.048 \div 0.12$

$0.12 \cdot 10^2 = 12$ Move the decimal point 2 places to the right.

$0.048 \cdot 10^2 = 4.8$ Move the decimal point 2 places to the right.

Then divide.

$4.8 \div 12$ **Step 1:** Divide as you would divide a whole number by a whole number.

$$\begin{array}{r} 0.4 \\ 12\overline{)4.8} \\ \underline{4\,8} \\ 0 \end{array}$$

Step 2: Think $48 \div 12 = 4$.

Step 3: Bring the decimal into the quotient and add a zero placeholder if necessary.

So, $0.048 \div 0.12 = 0.4$.

Find each quotient.

1. $0.7\overline{)0.42}$ **2.** $0.08\overline{)0.4}$ **3.** $0.5\overline{)0.125}$ **4.** $0.02\overline{)0.3}$

5. $0.4\overline{)0.08}$ **6.** $0.9\overline{)0.63}$ **7.** $0.008\overline{)0.4}$ **8.** $0.04\overline{)0.032}$

9. $0.3\overline{)0.06}$ **10.** $0.04\overline{)0.2}$ **11.** $0.007\overline{)4.9}$ **12.** $0.6\overline{)0.012}$

Name ______________________ Date __________ Class __________

LESSON 3F

Homework and Practice

Dividing Decimals by Decimals

Find each quotient.

1. 7.7 ÷ 0.7 ____________

2. 14.76 ÷ 4.1 ____________

3. 48.6 ÷ 8.1 ____________

4. 12.06 ÷ 1.8 ____________

5. 27.84 ÷ 5.8 ____________

6. 20.9 ÷ 3.8 ____________

7. 9.6 ÷ 1.2 ____________

8. 12.1 ÷ 1.1 ____________

9. 33.18 ÷ 4.2 ____________

10. 49.68 ÷ 5.4 ____________

11. 34.84 ÷ 6.7 ____________

12. 5.2 ÷ 0.5 ____________

Evaluate the expression $x \div 2.3$ for the given value of x.

13. $x = 43.7$ ____________

14. $x = 6.44$ ____________

15. $x = 13.8$ ____________

16. $x = 51.29$ ____________

17. $x = 21.85$ ____________

18. $x = 1.495$ ____________

19. A potato grinder is grinding potatoes into mashed potatoes at a rate of 5.6 potatoes per minute. How long would it take the grinder to grind 30.8 potatoes?

____

20. An automated call machine can handle 8.86 minutes of calls before it sends a report. How many reports would it send for 53.16 minutes of calls?

____

EXPLORATION

Program of Studies and Core Content

MA-6-NPO-S-NS1, MA-6-NPO-S-NO2, MA-06-1.2.1

3G Interpret the Quotient

For each problem, estimate a solution. Then compute with a calculator.

	Estimate	Actual
1. At Juan's school, each lunch special costs $3.65. How many lunches can Juan buy with $20.00?		
2. Gasoline costs $2.499 per gallon. How many gallons can Sue buy with $25.00?		
3. On Jorge's map, 0.15 cm represents 1 mi. He measures a road which is 7.8 cm. How many mi long is the actual road?		
4. Ofelia makes $6.79 per hour at her summer job. If she wants to make $200 per week, how many hours should she work?		

Think and Discuss

5. Explain the estimation strategies you used.

6. Describe a problem that can be solved by division.

Name ______________________________ Date ___________ Class ___________

LESSON 3G

Review for Mastery

Interpret the Quotient

There are three ways the decimal part of a quotient can be interpreted when you solve a problem.

If the question asks for an exact number, use the entire quotient.
If the question asks how many whole groups are needed to put the dividend into a group, round the quotient up to the next whole number.
If the question asks how many whole groups can be made when you divide, drop the decimal part of the quotient.

To interpret the quotient, decide what the question is asking.

In the school library, there are tables that seat 4 students each. If there are 30 students in a class, how many tables are needed to seat all of the students?

To solve, divide 30 by 4.

$30 \div 4 = 7.5$

The question is asking how many tables (whole groups) are needed to put all of the students in the class (dividend) into a group.

So, round 7.5 up to the next whole number.

8 tables are needed to seat all of the students.

Interpret the quotient to solve each problem.

1. A recipe that serves 6 requires 9 cups of milk. How much milk is needed for each serving?

____

2. A storage case holds 24 model cars. Marla has 84 model cars. How many storage cases does she need to store all of her cars?

____

3. Kenny has $4.25 to spend at the school carnival. If game tickets are $0.50 each, how many games can Kenny play?

____

Name ______________________ Date __________ Class __________

LESSON 3G

Homework and Practice

Interpreting the Quotient

Circle the letter of the correct answer.

1. Isabel's backpack has 3.75 kg of food. How many packs of food does she have if each pack weighs 0.25 kg?

A 66
B 15
C 14
D 9

2. If each cake can serve 6 people, how many cakes do you need to serve 34 people?

A 5
B 6
C 8
D 11

3. How many bags of candy does Mrs. Janet need to buy if each student will eat 0.2 bags and she has 22 students?

A 3
B 4
C 5
D 6

4. At the end of the day, a toy salesman has earned $71.25. How many toys did the salesman sell if each toy earns $2.85?

A 20
B 22
C 23
D 25

Write the correct answer.

5. Anthony and Michelle have walked 92.4 km. How many days have they walked if they can cover 13.2 km per day?

____

6. If you make $8.50 every day in tips, how long do you need to work to earn $110.50?

____

7. Every salad at an upcoming party will have 6 tomatoes. How many salads can be made with a box of 100 tomatoes?

____

8. One hundred and forty-four nails weighs 360 g. How much does each individual nail weigh?

____

EXPLORATION

Program of Studies and Core Content

MA-6-NPO-S-NO1, MA-6-NPO-S-NO3, MA-6-AT-S-VEO1, MA-6-AT-S-EI1, MA-6-AT-S-EI3, MA-06-1.3.2, MA-06-5.2.2, MA-06-5.3.1

3H Solving Decimal Equations

For each equation, estimate the solution. Then use a calculator to solve the equation. Compare the calculated solution with your estimated solution.

		Estimate	Actual
1.	$1.25 + x = 10$		
2.	$20 - x = 1.95$		
3.	$6x = 15$		
4.	$\frac{x}{4.5} = 10$		
5.	$\frac{x}{100} = 1.609$		

6. Write a real-world situation for the equation in Exercise **3.**

Think and Discuss

7. Explain which equations it was easiest to estimate a solution for.

8. Describe a real-world situation that you could model with a decimal equation.

Name ______________________ Date __________ Class __________

LESSON 3H

Hands-On Lab

Using Decimal Models to Solve Decimal Equations

You can solve decimal equations using decimal grids.

KEY

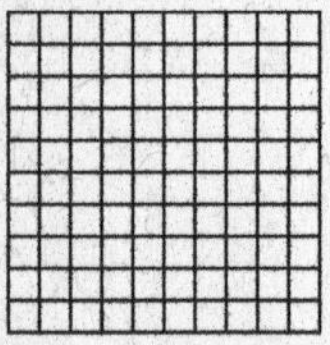

Activity

1. Solve $x + 0.24 = 0.76$ using decimal grids.
Use decimal grids and try to get x by itself.

Draw the equation with decimal grids.

x + 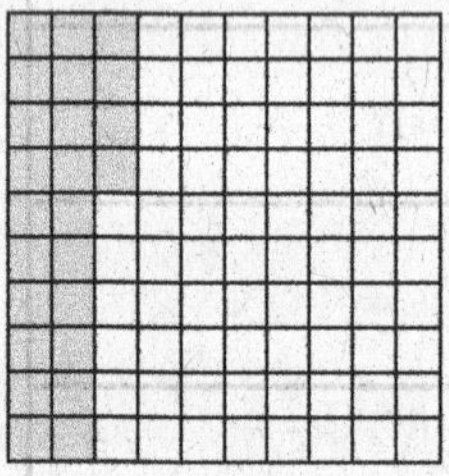 = 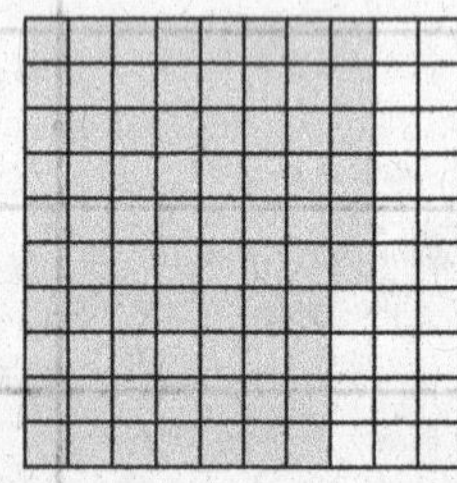

Cross out 24 squares from the decimal grids on each side.

x + 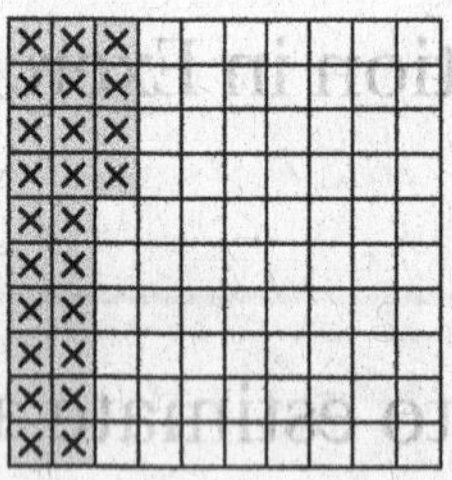 = 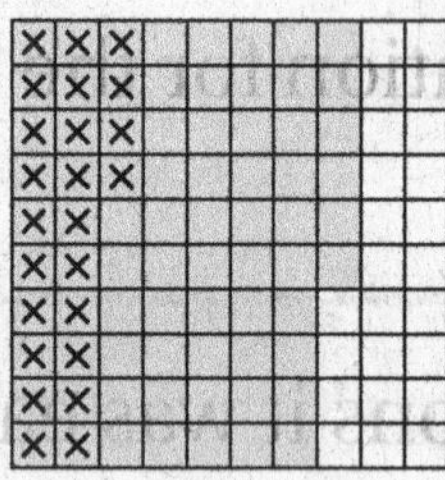

Write the solution.

x = 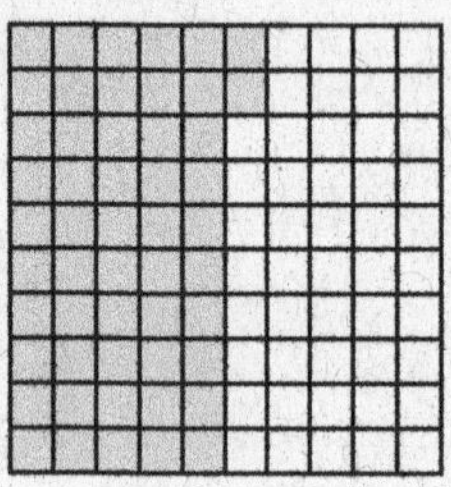 = 0.52

Name ______________________________ Date ___________ Class __________

LESSON 3H

Hands-On Lab

Using Decimal Models to Solve Decimal Equations, continued

2. Solve $x + 0.31 = 0.78$ using decimal grids.

Draw the equation with decimal grids.

x + 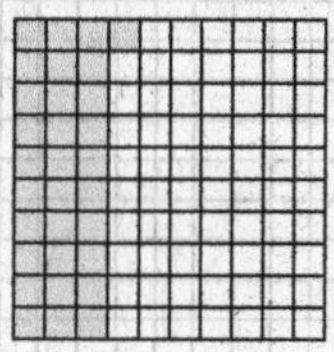 = 

Cross out 31 squares from the decimal grids on both sides.

x + 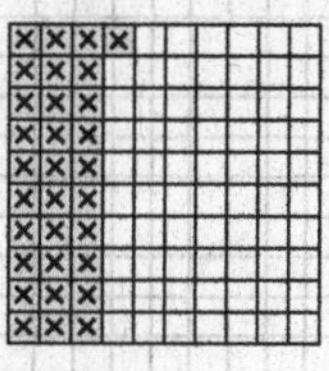 = 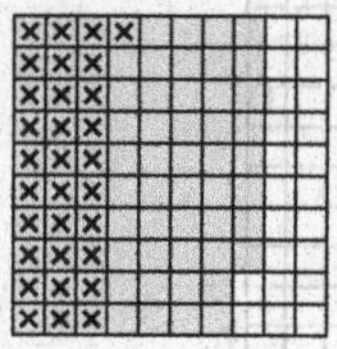

Write the solution.

x = 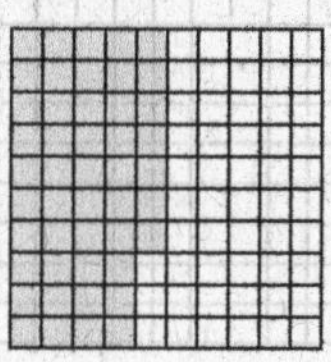 = 0.47

Think and Discuss

1. By crossing out squares on a decimal grid, what operation are you performing?

____

2. Compare solving the equation $x + 0.24 = 0.76$ using decimal grids with solving it algebraically.

____

Name ______________________ Date ___________ Class ___________

LESSON **3H**

Hands-On Lab

Using Decimal Models to Solve Decimal Equations, continued

Try This

Solve each equation using decimal grids.

1. $x + 0.36 = 0.78$

$x =$ ______

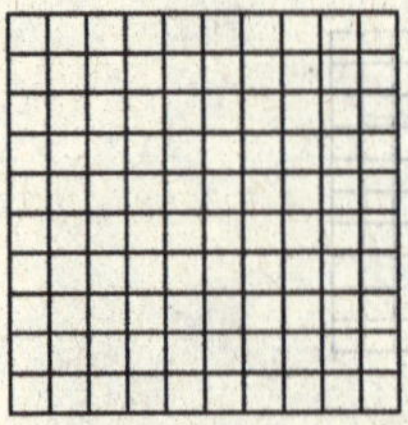
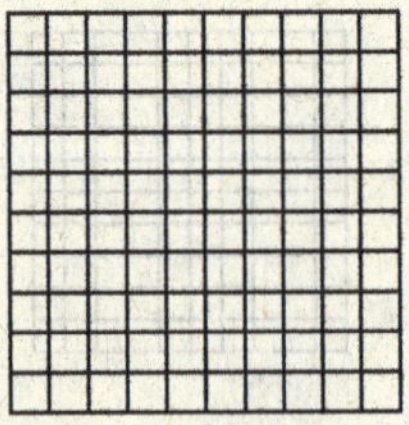

2. $x + 0.13 = 0.67$

$x =$ ______

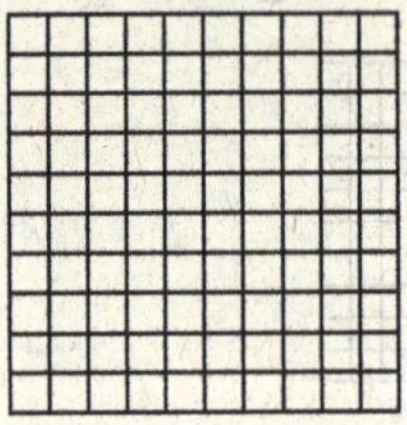
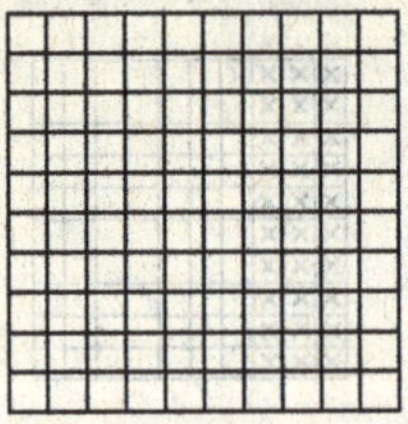

3. $0.83 + x = 0.98$

$x =$ ______

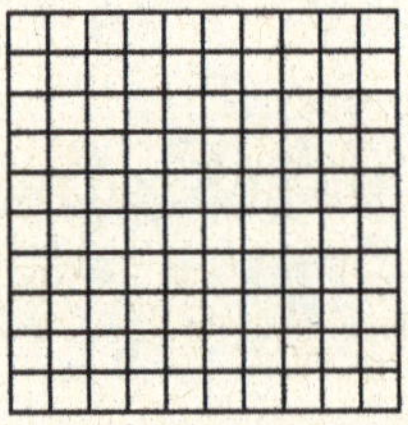
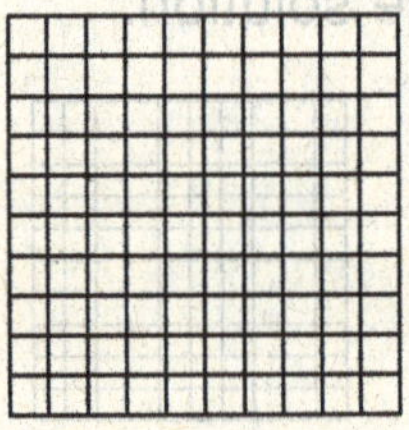

4. $x + 0.61 = 0.75$

$x =$ ______

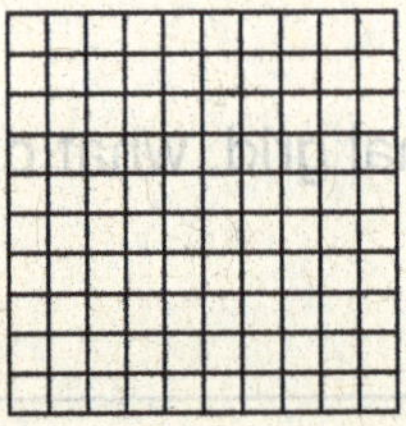
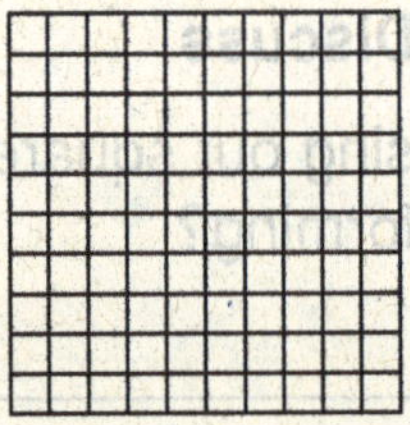

5. $x + 0.49 = 0.68$

$x =$ ______

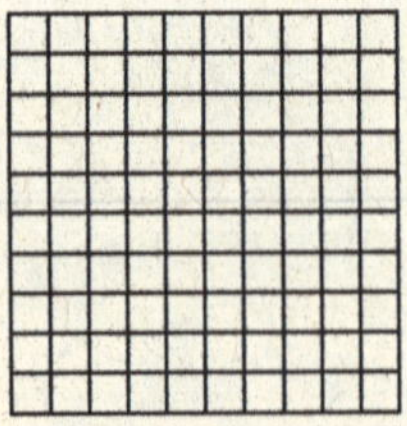
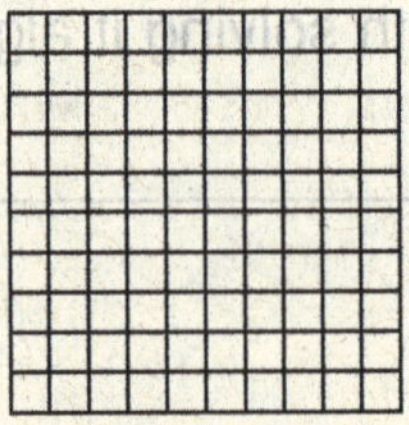

6. $x + 0.15 = 0.56$

$x =$ ______

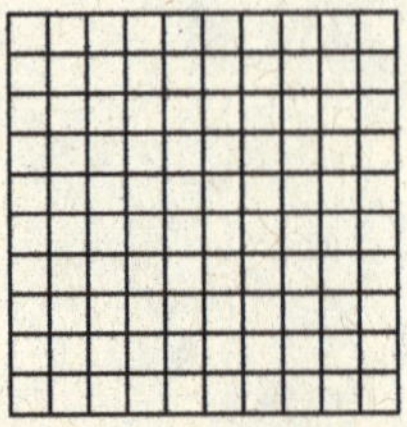
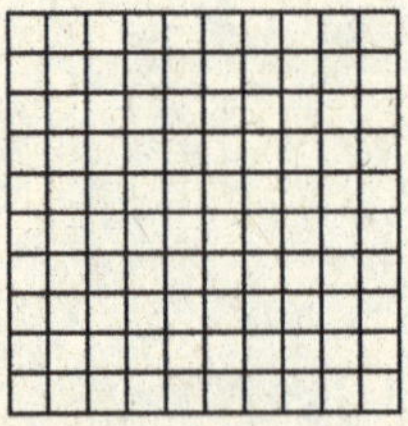

Name ________________________ Date ____________ Class ____________

LESSON **3H**

Review for Mastery

Solving Decimal Equations

You can write related equations for addition and subtraction equations.

$7.4 + 6.2 = 13.6$ $\quad$ $13.6 - 6.2 = 7.4$

Use related equations to solve each of the following.

A. $x + 4.5 = 7.9$

Think: $7.9 - 4.5 = x$

$x = 3.4$

Check $x + 4.5 = 7.9$

$3.4 + 4.5 \stackrel{?}{=} 7.9$ substitute

$7.9 = 7.9$

B. $x - 0.08 = 6.2$

Think: $6.2 + 0.08 = x$

$x = 6.28$

Check $x - 0.08 = 6.2$

$6.28 - 0.08 \stackrel{?}{=} 6.2$ substitute

$6.2 = 6.2$

Use related facts to solve each equation. Then check each answer.

1. $x + 8.7 = 12.9$ **2.** $x + 8.4 = 16.6$ **3.** $x - 2.65 = 7.8$ **4.** $x - 0.8 = 2.3$

________ ________ ________ ________

You can write related equations for multiplication and division equations.

$3.2 \cdot 2.4 = 7.68$ $\quad$ $7.68 \div 2.4 = 3.2$

Use related equations to solve each of the following.

C. $3x = 1.5$

Think: $1.5 \div 3 = x$

$x = 0.5$

Check: $3x = 1.5$

$3 \cdot 0.5 \stackrel{?}{=} 1.5$ substitute

$1.5 = 1.5$

D. $x \div 6 = 1.2$

Think: $1.2 \cdot 6 = x$

$x = 7.2$

Check: $x \div 6 = 1.2$

$7.2 \div 6 \stackrel{?}{=} 1.2$ substitute

$1.2 = 1.2$

Use related facts to solve each equation. Then check each answer.

5. $x \div 3 = 6.3$ **6.** $x \div 0.2 = 3.4$ **7.** $7x = 4.2$ **8.** $5x = 4.5$

________ ________ ________ ________

Name ____________________ Date __________ Class __________

LESSON 3H

Homework and Practice

Decimal Equations

Solve each equation. Check your answer.

1. $e \div 8.2 = 15$

2. $n + 4.1 = 5.9$

3. $8.4z = 4.2$

4. $g - 6.3 = 3.6$

5. $2 + r = 5.6$

6. $c + 6.9 = 7.3$

7. $4.9b = 107.8$

8. $8.94 - p = 4.79$

9. $7.11 \div a = 0.3$

10. $2.1v = 1.68$

11. $i + 3.15 = 12.65$

12. $4x = 4.68$

13. $x \div 1.5 = 4.34$

14. $9.7 + z = 10.79$

15. $n - 0.22 = 7.7$

16. After dinner, six people contributed equally to pay the bill. The total amount collected was $42.90. How much did each person pay?

17. Farmer O'Brien has 48 feet of fence. Each sheep pen needs 12.35 feet of fence to keep a sheep safe. How many feet of fence will farmer O'Brien have left if he makes as many sheep pens as he can?

CHAPTER 3

Chapter Summary

Answer these questions to summarize the important concepts from Chapter 3 in your own words.

1. Explain how to compare two decimals.

2. List methods used to estimate the answer to decimal computation problems.

3. List the steps when adding or subtracting decimals.

4. Explain when you should round a decimal quotient up or down to the nearest whole number.

5. List the steps when dividing by decimals.

For more review of Chapter 3:

- Complete the Chapter 3 Study Guide and Review on pages 152–154.
- Complete the Ready to Go On quizzes on pages 122 and 148.

Name ______________________ Date __________ Class __________

CHAPTER 3

KCCT Prep

Multiple Choice

1. Lashonda bought peaches that cost \$1.45 per pound and a melon that cost \$1.20. She spent a total of \$6.42. How many pounds of peaches did she buy?

   Ⓐ 2.8 pounds　Ⓒ 3.77 pounds
   Ⓑ 3.6 pounds　Ⓓ 4.1 pounds

2. Which is the best estimate for $61.23 \div 14.88$?

   Ⓐ 2　Ⓒ 4
   Ⓑ 3　Ⓓ 5

3. A model train can travel at any of the following speeds. At which speed does the train go once around the track in the least amount of time?

   Ⓐ 3.08 ft/s　Ⓒ 3.5 ft/s
   Ⓑ 3.35 ft/s　Ⓓ 3.30 ft/s

4. The sequence shown below gives the number of songs on Miranda's mp3 player each month. Emily has twice as many songs on her mp3 player each month. Which expression describes the number of songs on Emily's mp3 player?
   10, 15, 20, 25, 30, ...

   Ⓐ $2n$　Ⓒ $10 + 5n$
   Ⓑ $5 + 5n$　Ⓓ $10 + 10n$

5. Which is the best estimate for the value of the expression $6.07^2 - 5.2 \times 3.9$?

   Ⓐ 16　Ⓒ 56
   Ⓑ 29　Ⓓ 124

6. Four friends recorded the distance of their best long jumps at a track meet; Chris: 5.68 m, Dana: 5.09 m, Freddie: 5.8 m, and Lee: 5.77 m. Who came closest to jumping 6 meters?

   Ⓐ Chris　Ⓒ Freddie
   Ⓑ Dana　Ⓓ Lee

7. Which is the best estimate of the value of $6x$ when $x = 3.198$?

   Ⓐ 9　Ⓒ 18
   Ⓑ 10　Ⓓ 24

8. Miguel drives 30.8 miles each day going to and from work. He also drives 1.9 miles each day picking up his son from school. How many miles does he drive in 5 days?

   Ⓐ 6.54 miles　Ⓒ 150.5 miles
   Ⓑ 16.35 miles　Ⓓ 163.5 miles

9. The dance club has 11 more members than the soccer club. There are 34 students in the soccer club. Which equation represents the number of students in the dance club?

   Ⓐ $d + 11 = 34$　Ⓒ $d - 11 = 34$
   Ⓑ $d + 34 = 11$　Ⓓ $d = 34 - 11$

10. A ribbon is 14.9 ft long. Mitch needs pieces of ribbon that are 2.5 ft long. What is the greatest number of such pieces that he can cut from the ribbon?

    Ⓐ 5　Ⓒ 12
    Ⓑ 6　Ⓓ 13

11. Which number is closest to 8?

Ⓐ 7.46 Ⓒ 7.408
Ⓑ 7.009 Ⓓ 7.05

12. Gasoline costs $3.15 per gallon. Anna filled the tank of her car and spent a total of $27.72. Which is the best estimate for the amount of gas she bought?

Ⓐ 8 gallons Ⓒ 11 gallons
Ⓑ 9 gallons Ⓓ 24 gallons

13. Dara has a large sack of flour. She plans to scoop out equal portions of the flour. She can use a scoop with any of the following sizes. Which scoop should she use in order to get the greatest number of portions?

Ⓐ 1.5 cups Ⓒ 1.40 cups
Ⓑ 1.05 cups Ⓓ 1.45 cups

14. Eleven friends go a restaurant. The bill comes to a total of $180.95. They decide to split the bill evenly. Which is the best estimate of the amount each friend owes?

Ⓐ $11 Ⓒ $16
Ⓑ $14 Ⓓ $20

15. What is the standard form of the decimal that has the expanded form 0.01 + 0.004 + 0.0003?

Ⓐ 0.0143 Ⓒ 0.143
Ⓑ 0.08 Ⓓ 0.8

16. The following are the weights of four lion cubs. Each cub is expected to gain 2.9 kg in the next month. At the end of the month, which cub will be closest to 12 kg?

Ⓐ 8.08 kg Ⓒ 8.7 kg
Ⓑ 8.49 kg Ⓓ 8.60 kg

Open Response

17. Ellyn has a catering company. She is preparing rice pilaf for a large party.

a. Ellyn needs to prepare 30 servings. If each serving is 2.4 ounces, how many ounces of rice pilaf should she make?

b. Ellyn is transporting the rice pilaf in 12.5 ounce containers. How many containers will she need? Explain.

18. The table shows the populations of four cities in the United States.

2003 Populations	
City	**Population (millions)**
Dallas	1.208
Phoenix	1.388
San Antonio	1.215
San Diego	1.267

a. List the cities in order from the least to the greatest population.

b. The population of San Antonio is about three times the population of Omaha. Explain how to estimate the population of Omaha.

19. The table shows the cost of ordering tickets to a school play. Each ticket has the same price, and there is a flat service fee for each order.

Number of Tickets	1	2	3	4	5
Total Cost ($)	10	17	24	31	38

a. What is the cost of each ticket?

b. What is the service charge? Explain.

c. Greg spent a total of $87 on tickets. Use one or more equations to find the number of tickets Greg bought.

CHAPTER 4

Project
Fractions: Down the Drain

Career: Plumber

Do you like working with your hands to solve problems? If so, you might want to become a skilled trade worker, such as a master plumber. To calculate the cost of parts and labor, plumbers use basic mathematical formulas. For example, some plumbers might calculate the cost of a new sewer line with a formula like the following:

$$\text{cost of installed line} = \frac{\text{cost of pipe}}{3} \times 49 + \frac{\text{cost of pipe fittings}}{2}$$

Before you start the project, work through the Plumber's Cost/Profit Analysis worksheet on the next page.

In this chapter project . . .

You design a piece of art made of plumbing supplies.

What to do:

- Artists often use unusual materials to create works of art. Design a sculpture made from the plumbing supplies listed on the next page. Use the web or your local hardware store to find out what $\frac{1}{4}$, $\frac{1}{6}$, and $\frac{1}{8}$ bend connections look like.
- List all of the pieces you would need to create your sculpture and the cost of each piece.
- Calculate what the supplies you need would cost. Would there be a large difference in cost between ABS (plastic) pipe and copper pipe?

What to present:

- The sculpture that you created.
- A table that shows how many of each item you would need, the cost of each item, and the total costs.
- A cost estimate based on the same sculpture but made using a different material.

Where's the math?

- Use tables to organize lists of supplies and prices.
- Use operations on decimals and fractions to calculate total costs.

go.hrw.com
Chapter Project Online
KEYWORD: MR7 Ch4

Name ________________________ Date __________ Class __________

CHAPTER 4

Project Recording Sheet

Number Theory and Fractions

Plumber's Cost/Profit Analysis

Plumbers and other trades people figure the cost to the consumer by examining the costs of materials and the amount that they should receive for their time, labor, and overhead. Plumbers use a formula to determine the price of a job.

$$\text{Cost to consumer} = \left(\frac{\text{cost of pipe}}{3}\right) \times 49 + \frac{\text{cost of pipe fittings}}{2}$$

The numbers in the formula may change depending upon the plumber's costs.

Use the formula and the prices given for materials in the table below to calculate the costs of four different jobs.

Installed Drain Lines

Job	Feet of Pipe	Fittings	Cost of Installed Line ($)
1	30	two $\frac{1}{4}$-bends, one $\frac{1}{6}$-bend	
2	26	two $\frac{1}{4}$-bends	
3	75	three $\frac{1}{8}$-bends	
4	150	three $\frac{1}{6}$-bends, one $\frac{1}{8}$-bend	

ABS Plastic Drain Pipe Prices

Component	Cost ($)
Pipe 4 in. x 10 ft	11.99
Pipe 4 in. x 20 ft	22.57
Straight coupling	2.19
$\frac{1}{4}$-bend connection	6.49
$\frac{1}{8}$-bend connection	5.99
$\frac{1}{6}$-bend connection	7.49

What happens to the cost if the numbers in the formula change? Make the following changes in the formula and calculate the new cost for Job 1 from the table: The cost of the pipe doubles; the cost of the fittings doubles; the number 49 that represents time, labor, and overhead is changed to 60. ____________________

Which change affects the cost the most? Why?

____

____

EXPLORATION

Program of Studies and Core Content
MA-6-NPO-S-PNO1, MA-06-1.5.1

4A Divisibility

Some calculators have an INT ÷ key, which returns a quotient and a remainder.

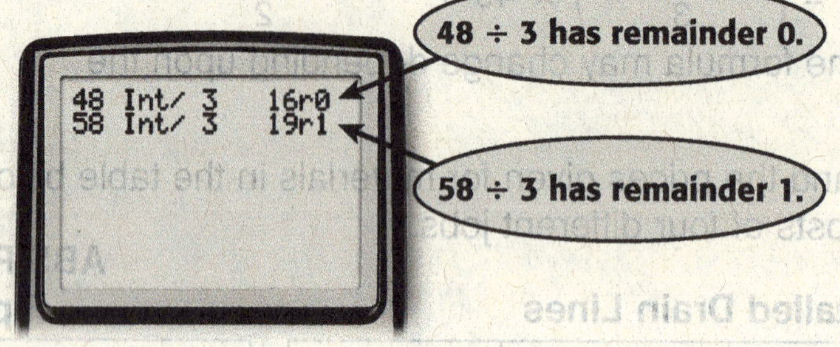

1. Use mental math or a calculator to determine each quotient and remainder. Then add the digits of the dividend.

	Dividend	Divisor	Quotient	Remainder	Sum of Digits
a.	48	3	16	0	$4 + 8 = 12$
b.	58	3	19	1	$5 + 8 = 13$
c.	256	3			
d.	1,011	3			
e.	72	3			
f.	74	3			
g.	129	3			
h.	130	3			

Think and Discuss

2. **Explain** whether 3,129 is divisible by 3.
3. **Describe** the pattern between the remainder and the sum of the digits in the table.

Name ______________________________ Date ___________ Class ___________

LESSON 4A

Review for Mastery

Divisibility

A number is divisible by another number if the quotient is a whole number with no remainder.

15 is divisible by 3 because $15 \div 3 = 5$. There are rules to help you figure out if a number is divisible by another number.

Divisibility Rules

A number is divisible by:

2 if the last digit is even.
3 if the sum of the digits is divisible by 3.
4 if the last two digits form a number that is divisible by 4.
5 if the last digit is 0 or 5.
6 if the number is divisible by both 2 and 3.
9 if the sum of the digits is divisible by 9.
10 if the last digit is 0.

To tell whether 315 is divisible by 2, 3, 4, 5, 6, 9, and 10, you can use the divisibility rules listed above.

5 is not an even number, so 315 **is not divisible** by 2.
$3 + 1 + 5 = 9$. 9 is divisible by 3, so 315 **is divisible** by 3.
15 is not divisible by 4, so 315 **is not divisible** by 4.
The last digit is 0 or 5, so 315 **is divisible** by 5.
315 is divisible by 3, but not 2. So, 315 **is not divisible** by 6.
The sum of the digits is 9 and 9 is divisible by 9. So 315 **is divisible** by 9.
The last digit is not 0, so 315 **is not divisible** by 10.

315 is divisible by 3, 5, and 9.

Use divisibility rules to tell whether each number is divisible by 2, 3, 4, 5, 6, 9, and 10.

1. 120 ____________________

2. 435 ____________________

3. 228 ____________________

4. 540 ____________________

5. 144 ____________________

6. 634 ____________________

7. 402 ____________________

8. 320 ____________________

Name ______________________ Date ___________ Class ___________

LESSON 4A

Review for Mastery

Divisibility, continued

Numbers that are divisible by more than 2 numbers are called composite numbers.

12 is a composite number because it is divisible by 1, 2, 3, 4, 6, and 12.

$12 \div 1 = 12$ $\quad$ $12 \div 2 = 6$ $\quad$ $12 \div 3 = 4$

$12 \div 4 = 3$ $\quad$ $12 \div 6 = 2$ $\quad$ $12 \div 12 = 1$

Numbers that are divisible by exactly 2 numbers, 1 and itself, are called prime numbers.

5 is a prime number because it is divisible by only 1 and 5.

$5 \div 1 = 5$ $\quad$ $5 \div 5 = 1$

Tell whether each number is prime or composite.

9. 7 ____________ **10.** 12 ____________ **11.** 17 ____________ **12.** 28 ____________

13. 37 ____________ **14.** 45 ____________ **15.** 79 ____________ **16.** 88 ____________

17. 72 ____________ **18.** 59 ____________ **19.** 94 ____________ **20.** 21 ____________

21. 47 ____________ **22.** 55 ____________ **23.** 67 ____________ **24.** 81 ____________

Name ______________________ Date __________ Class __________

LESSON **4A**

Homework and Practice

Divisibility

Tell whether each number is divisible by 2, 3, 4, 5, 6, 9, and 10.

1. 10 ______

2. 68 ______

3. 24 ______

4. 370 ______

5. 200 ______

6. 284 ______

7. 354 ______

8. 280 ______

9. 178 ______

Tell whether each number is prime or composite.

10. 29 ______

11. 81 ______

12. 39 ______

13. 37 ______

14. 71 ______

15. 27 ______

16. 91 ______

17. 59 ______

18. 36 ______

19. Carlos is having a pizza party. If 38 pieces of pizza were eaten, did each of his 8 guests have the same number of slices? How can you tell?

______

20. A costume maker is creating monster costumes for a new movie. If she has 49 fake eyes, can she make 7 costumes that all have the same number of eyes? How can you tell?

______

EXPLORATION

Program of Studies and Core Content
MA-6-NPO-S-PN01,
MA-06-1.5.1

4B Factors and Prime Factorization

1. The rectangle measures 4 units by 6 units and has an area of 24 square units. Use graph paper to draw rectangles that have different whole-number dimensions but each has an area of 24 square units. (*Hint:* $4 \times 6 = 24$. What factors other than 1×24 give you 24?)

2. The rectangle measures 3 units by 5 units and has an area of 15 square units. Is it possible to draw a rectangle that has whole-number dimensions other than 3×5 (and 1×15) and still have an area of 15 square units?

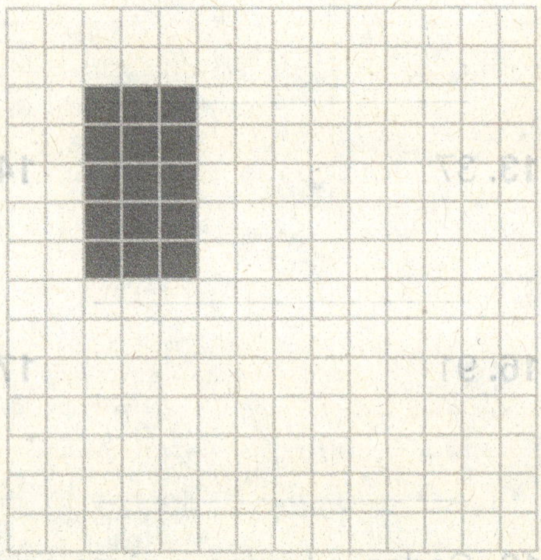

Think and Discuss

3. **Explain** how you can use rectangles to determine factors of numbers.
4. **Explain** why it is possible to draw more than two different rectangles with an area of 24 square units, but it is not possible to draw more than two different rectangles with an area of 15 square units (each with whole-number dimensions).

Name ______________________ Date __________ Class __________

LESSON **4B**

Review for Mastery

Factors and Prime Factorization

Factors of a product are the numbers that are multiplied to find that product. A factor is also a whole number that divides the product with no remainder.

To find all of the factors of 24, make a list of multiplication facts.

$1 \cdot 24 = 24$
$2 \cdot 12 = 24$
$3 \cdot 8 = 24$
$4 \cdot 6 = 24$

The factors of 24 are 1, 2, 3, 4, 6, 8, 12, and 24.

Write multiplication facts to find the factors of each number.

1. 20 ______________________ ______________________

2. 16 ______________________ ______________________

3. 35 ______________________ ______________________

4. 31 ______________________ ______________________

A number written as the product of prime factors is called the prime factorization of the number.

To write the prime factorization of 24, first write it as product of 2 numbers. Then rewrite each factor as the product of 2 numbers until all of the factors are prime numbers.

$24 = \mathbf{4} \cdot 6$	(Write 24 as the product of 2 numbers.)
$= 2 \cdot 2 \cdot \mathbf{6}$	(Rewrite 4 as the product of 2 prime numbers.)
$= 2 \cdot 2 \cdot 2 \cdot 3$	(Rewrite 6 as the product of 2 prime numbers.)

So, the prime factorization of 24 is $2 \cdot 2 \cdot 2 \cdot 3$ or $2^3 \cdot 3$.

Find the prime factorization of each number.

5. 28 __________

6. 45 __________

7. 50 __________

8. 72 __________

Name ______________________ Date __________ Class __________

LESSON **4B**

Homework and Practice

Factors and Prime Factorization

List all the factors for each number.

1. 49

2. 94

3. 48

4. 74

5. 80

6. 53

7. 68

8. 58

9. 55

Write the prime factorization of each number.

10. 34

11. 80

12. 20

13. 76

14. 68

15. 30

16. 91

17. 21

18. 185

19. To win a game show, Theodore must find an odd prime factor of 56. How many odd prime factors of 56 are there? Name the odd prime factors.

20. Ronald writes the prime factorization of 90 as $2 \times 3 \times 3 \times 5$ on the board. Donald writes the prime factorization of 90 as $3^2 \times 5$. Who is correct?

EXPLORATION

Program of Studies and Core Content
MA-6-NPO-S-PN01, MA-06-1.5.1

4C Greatest Common Factor

The sixth-grade band, which has 60 members, and the seventh-grade band, which has 48 members, are getting ready for a parade. How can they march together in blocks with the same number of columns? The model and table below show one possible formation.

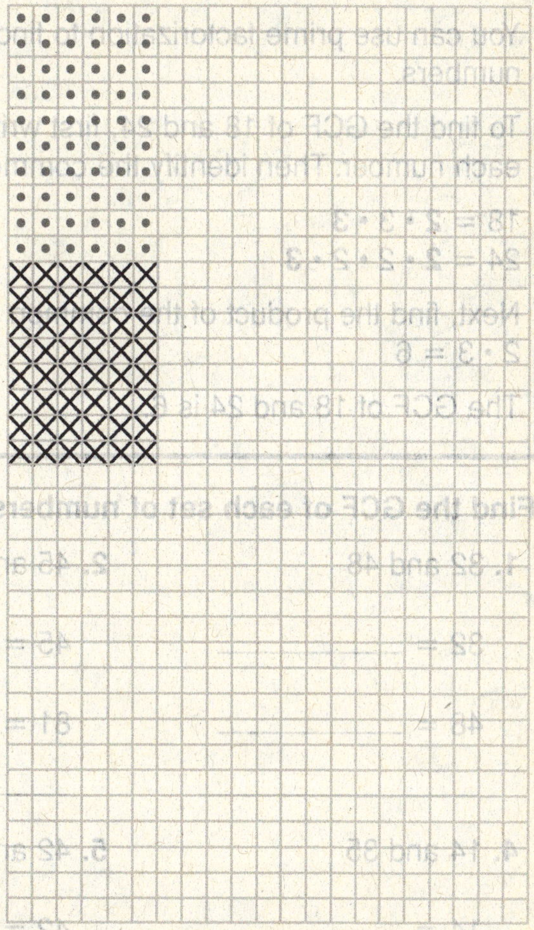

1. Use graph paper to draw a model of two other formations that would work.
2. Complete the table to show the number of rows and columns in the other two formations.

	Formation 1		Formation 2		Formation 3	
	Rows	Columns	Rows	Columns	Rows	Columns
6th Grade	10	6				
7th Grade	8	6				

Think and Discuss

3. **Discuss** which formation the band director should select if she wants the bands to pass through the parade as quickly as possible.
4. **Explain** why both 48 and 60 must be divisible by the number of columns.

Name ______________________ Date __________ Class __________

LESSON 4C

Review for Mastery

Greatest Common Factor

The greatest common factor, or GCF, is the largest number that is the factor of any set of at least two numbers.

You can use prime factorization to find the GCF of two or more numbers.

To find the GCF of 18 and 24, first write the prime factorization of each number. Then identify the common prime factors.

18 = **2** • 3 • **3**
24 = **2** • 2 • 2 • **3**

Next, find the product of the common prime factors.
2 • 3 = 6

The GCF of 18 and 24 is 6.

Find the GCF of each set of numbers.

1. 32 and 48

32 = __________

48 = __________

____

2. 45 and 81

45 = __________

81 = __________

____

3. 18 and 36

18 = __________

36 = __________

____

4. 14 and 35

14 = __________

35 = __________

____

5. 42 and 72

42 = __________

72 = __________

____

6. 56 and 64

56 = __________

64 = __________

____

7. 28, 56, and 84

28 = __________

56 = __________

84 = __________

____

8. 30, 45, and 75

30 = __________

45 = __________

75 = __________

____

9. 36, 45, and 54

36 = __________

45 = __________

54 = __________

____

Name ______________________ Date __________ Class __________

LESSON **4C**

Homework and Practice

Greatest Common Factor

Find the GCF of each set of numbers.

1. 8 and 12 ______

2. 10 and 25 ______

3. 14 and 21 ______

4. 18 and 81 ______

5. 32 and 20 ______

6. 26 and 48 ______

7. 60 and 72 ______

8. 75 and 90 ______

9. 70 and 165 ______

10. 12, 24, and 30 ______

11. 40, 60, and 80 ______

12. 15, 42, and 84 ______

13. 44, 66, and 121 ______

14. 28, 56, and 60 ______

15. 20, 30, and 105 ______

16. 39, 91, and 130 ______

17. 48, 168, and 280 ______

18. 42, 63, 84 ______

19. Jason is trying to make picnic lunches. He has 12 sandwiches, 18 apples, and 30 pieces of candy. How many lunches can he make if he wants each lunch to have the same number of each kind of food and use all of the food?

______

20. Helga is organizing her marble collection into boxes. She has 54 red marbles, 135 yellow marbles, and 189 clear marbles. If she wants every box to have the same number of each color marbles, how many boxes can she make if she uses all of the marbles?

______

EXPLORATION

Program of Studies and Core Content
MA-6-NPO-S-NS1, MA-6-NPO-S-NS5, MA-06-1.1.1, MA-06-1.1.2, MA-06-1.1.3

4D Decimals and Fractions

Use the model to complete the table of equivalent fractions and decimals.

	Fraction	Decimal
1.	$\frac{1}{5}$	
2.		0.6
3.	$\frac{2}{5}$	
4.		0.8

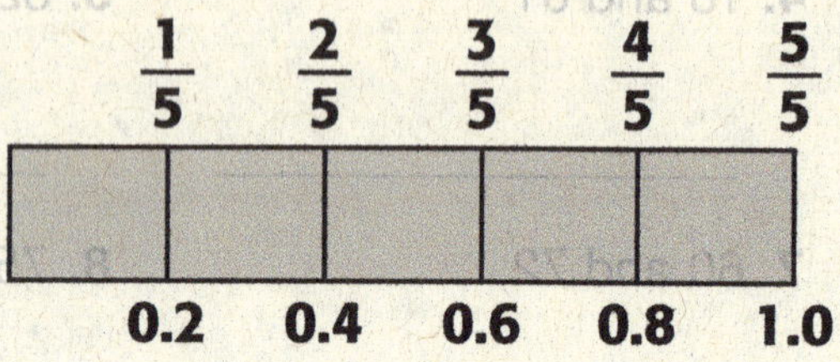

Use the model to complete the table of equivalent fractions and decimals.

	Fraction	Decimal
5.	$\frac{10}{100}$	
6.		0.45
7.	$\frac{40}{100}$	
8.		0.30

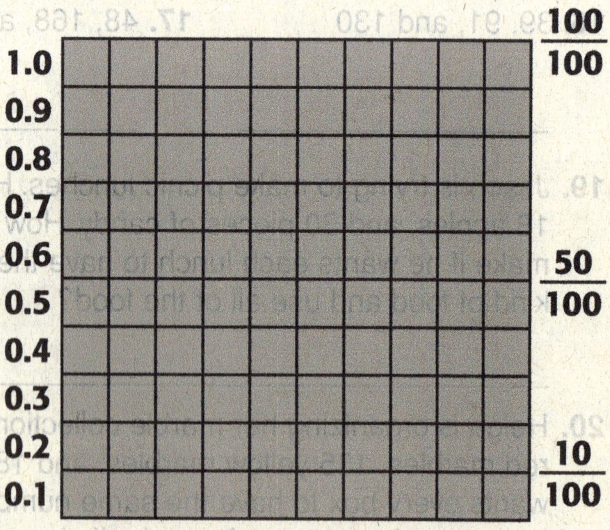

Think and Discuss

9. Describe a situation in which decimals are typically used.

10. Describe a situation in which fractions are typically used.

Name ______________________________ Date ____________ Class __________

LESSON 4D

Review for Mastery

Decimals and Fractions

You can write decimals as fractions or mixed numbers. A place value chart will help you read the decimal. Remember the decimal point is read as the word "and."

To write 0.47 as a fraction, first think about the decimal in words.

Ones	Tenths	Hundredths	Thousandths	Ten Thousandths
0.	4	7		

0.47 is read "forty-seven hundredths." The place value of the decimal tells you the denominator is 100.

$0.47 = \frac{47}{100}$

To write 8.3 as a mixed number, first think about the decimal in words.

Ones	Tenths	Hundredths	Thousandths	Ten Thousandths
8.	3			

8.3 is read "eight and three tenths." The place value of the decimal tells you the denominator is 10. The decimal point is read as the word "and."

$8.3 = 8\frac{3}{10}$

Write each decimal as a fraction or mixed number.

1. 0.61 ____________ **2.** 3.43 ____________ **3.** 0.009 ____________ **4.** 4.7 ____________

5. 1.5 ____________ **6.** 0.13 ____________ **7.** 5.002 ____________ **8.** 0.021 ____________

Name ______________________ Date __________ Class __________

LESSON 4D

Review for Mastery

Decimals and Fractions, continued

Fractions and mixed numbers can be written as decimals.

To write $\frac{1}{4}$ as a decimal, first think about the expression in words.

$\frac{1}{4}$ means "1 divided by 4."

Then do the division.

$$\begin{array}{r} 0.25 \\ 4\overline{)1.00} \\ \underline{-8} \\ 20 \\ \underline{-20} \\ 0 \end{array}$$

0.25 is a terminating decimal because it has an end.

$\frac{1}{4} = 0.25$

A number that contains a whole number and a fraction is called a mixed number. $2\frac{1}{3}$ is an example of a mixed number. To write $2\frac{1}{3}$ as a decimal, first think about the expression in words.

$2\frac{1}{3}$ means "2 and 1 divided by 3." Keep 2 as the whole number.

Then do the division.

$$\begin{array}{r} 0.33 \\ 3\overline{)1.00} \\ \underline{-9} \\ 10 \\ \underline{-09} \\ 1 \end{array}$$

$1 \div 3 = 0.33\ldots$, or $0.\overline{3}$.

$0.\overline{3}$ is a repeating decimal because it does not end.

So, $2\frac{1}{3} = 2.\overline{3}$.

Write each fraction or mixed number as a decimal.

9. $\frac{3}{5}$ ____________

10. $3\frac{3}{4}$ ____________

11. $\frac{2}{3}$ ____________

12. $1\frac{2}{9}$ ____________

13. $\frac{1}{6}$ ____________

14. $2\frac{1}{8}$ ____________

15. $\frac{5}{6}$ ____________

16. $8\frac{1}{9}$ ____________

Name ______________________ Date __________ Class __________

LESSON 4D

Homework and Practice

Decimals and Fractions

Write each decimal as a fraction or mixed number.

1. 0.4 ________ **2.** 1.5 ________ **3.** 0.75 ________

4. $2.\overline{1}$ ________ **5.** 2.8 ________ **6.** 1.1 ________

Write each fraction or mixed number as a decimal.

7. $\frac{1}{2}$ ________ **8.** $1\frac{1}{3}$ ________ **9.** $2\frac{9}{10}$ ________

10. $\frac{1}{8}$ ________ **11.** $4\frac{2}{9}$ ________ **12.** $\frac{7}{8}$ ________

Order the fractions and decimals from least to greatest.

13. $\frac{5}{6}$, 0.8, 0.75 ________ **14.** 0.5, $\frac{4}{5}$, $\frac{3}{8}$ ________ **15.** $\frac{1}{3}$, 0.2, $\frac{2}{5}$ ________

Order the fractions and decimals from greatest to least.

16. 0.75, $\frac{7}{10}$, $\frac{1}{7}$ ________ **17.** $\frac{2}{3}$, 0.5, $\frac{7}{8}$ ________ **18.** 0.8, $\frac{5}{9}$, $\frac{1}{2}$ ________

19. Daniel, Natalie, and Sean each ordered a small pizza pie for lunch. If Daniel ate $\frac{3}{8}$ of his pie, Natalie had 0.5 of her pie, and Sean ate $\frac{2}{5}$ of his pie, who ate the most? Who ate the least?

____

20. Box A contains 3 quarters, Box B contains 14 nickels, and Box C contains 5 dimes. Which box has the most money?

____

Program of Studies and Core Content
MA-6-NPO-S-PNO2, MA-6-NPO-S-NS1, MA-06-1.1.2

4E Equivalent Fractions

Equivalent fractions are fractions that have the same value.

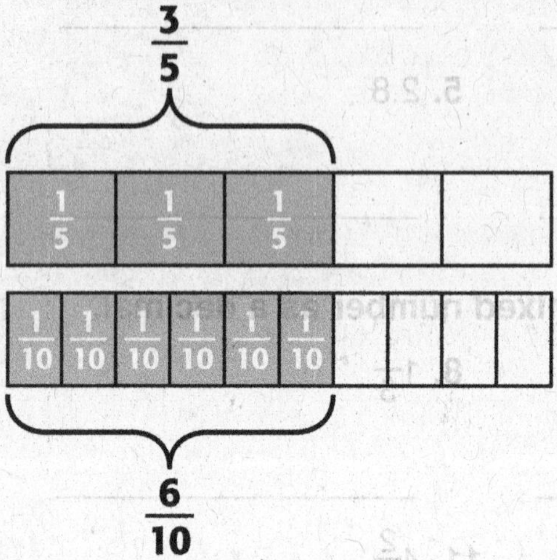

Use the models below to find equivalent fractions.

1. $\frac{1}{2} = \square = \square = \square$

2. $\frac{2}{3} = \square = \square$

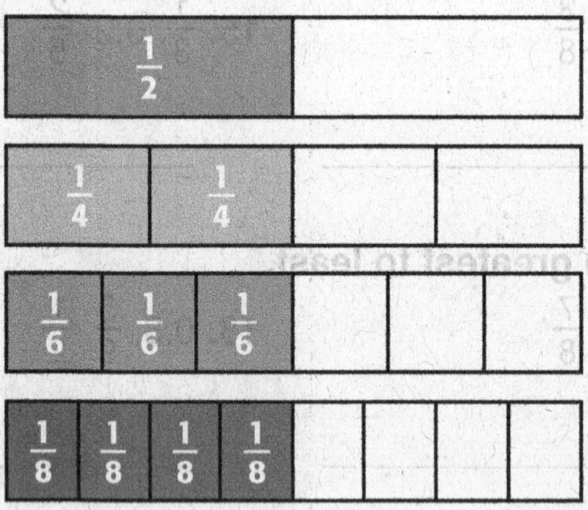

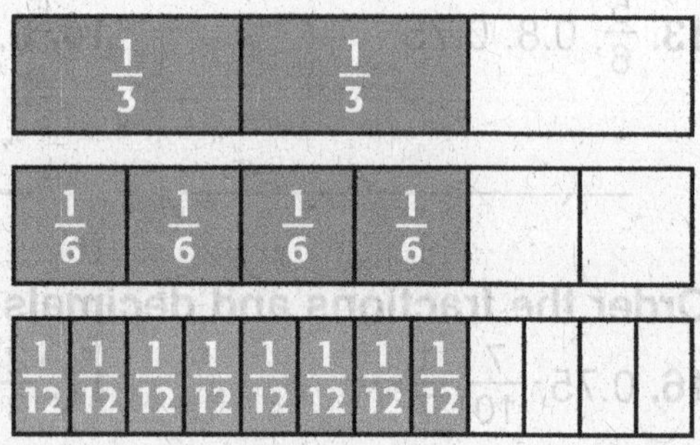

Think and Discuss

3. Describe how you could write $\frac{1}{2}$ as an equivalent fraction with a denominator of 24.

4. Discuss whether $\frac{6}{12}$ and $\frac{9}{18}$ are equivalent fractions.

Name ______________________ Date __________ Class __________

LESSON **4E**

Review for Mastery

Equivalent Fractions

Fractions that have the same value are equivalent fractions.

$\frac{3}{4} = \frac{6}{8}$

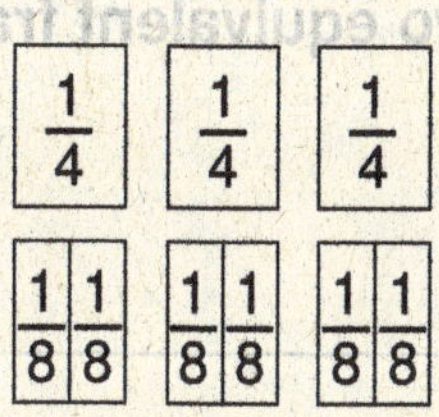

$\frac{3}{4}$ and $\frac{6}{8}$ are equivalent fractions.

You can use fraction strips to help you find equivalent fractions.

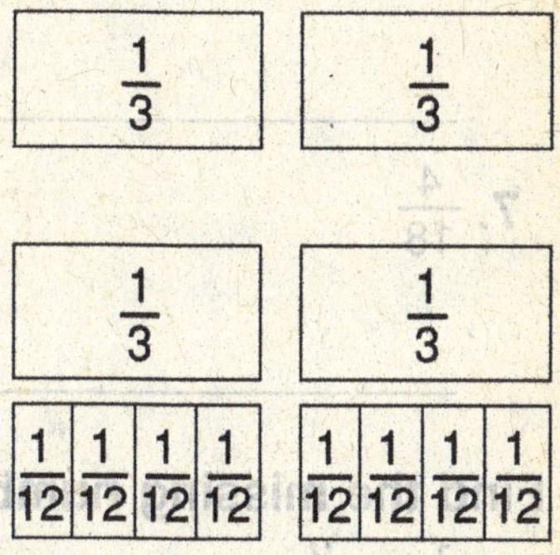

To solve $\frac{2}{3} = \frac{?}{12}$, first use fraction strips to model the first fraction.

Then use $\frac{1}{12}$ fraction pieces to make a length as long as the $\frac{2}{3}$ strip.

You need eight $\frac{1}{12}$ pieces, so $\frac{2}{3} = \frac{8}{12}$.

Find the missing number that makes the fractions equivalent.

1. $\frac{3}{4} = \frac{?}{12}$ ________

2. $\frac{8}{10} = \frac{?}{5}$ ________

3. $\frac{6}{9} = \frac{?}{3}$ ________

4. $\frac{5}{6} = \frac{?}{12}$ ________

A fraction is in simplest form when the GCF of the numerator and the denominator is 1.

To write $\frac{4}{6}$ in simplest form, divide the numerator and the denominator by their GCF, 2.

$4 \div 2 = 2$

$6 \div 2 = 3$

So, $\frac{4}{6}$ in simplest form is $\frac{2}{3}$.

Write each fraction in simplest form.

5. $\frac{3}{9}$ ________

6. $\frac{12}{16}$ ________

7. $\frac{14}{18}$ ________

8. $\frac{8}{20}$ ________

Name ____________________ Date __________ Class __________

LESSON 4E

Homework and Practice

Equivalent Fractions and Mixed Numbers

Find two equivalent fractions for the given fraction.

1. $\frac{4}{6}$ ____________

2. $\frac{3}{9}$ ____________

3. $\frac{1}{5}$ ____________

4. $\frac{4}{7}$ ____________

5. $\frac{4}{9}$ ____________

6. $\frac{12}{26}$ ____________

7. $\frac{4}{18}$ ____________

8. $\frac{9}{12}$ ____________

9. $\frac{20}{50}$ ____________

Find the missing numbers that make the fractions equivalent.

10. $\frac{1}{4} = \frac{?}{32}$ ____________

11. $\frac{2}{9} = \frac{?}{27}$ ____________

12. $\frac{30}{6} = \frac{?}{4}$ ____________

13. $\frac{16}{20} = \frac{?}{5}$ ____________

14. $1\frac{3}{4} = \frac{?}{100}$ ____________

15. $\frac{3}{9} = \frac{1}{?}$ ____________

Write each fraction in simplest form.

16. $\frac{14}{16}$ ____________

17. $\frac{20}{25}$ ____________

18. $\frac{9}{24}$ ____________

19. $\frac{12}{24}$ ____________

20. Jeffrey has 30 coins in his collection. He sold 8 and gave 4 to a friend. Write the fraction of the original set that he has left in simplest form. If someone later gave him 15 coins, what fraction of the original set would he have? Write this as a mixed number.

____

EXPLORATION

4F Mixed Numbers and Improper Fractions

Program of Studies and Core Content
MA-6-NPO-S-NS1, MA-06-1.1.1, MA-06-1.1.2

An *improper fraction* is a fraction in which the numerator is greater than or equal to the denominator. The model shows that $\frac{7}{4} = \frac{4}{4} + \frac{3}{4} = 1 + \frac{3}{4}$.

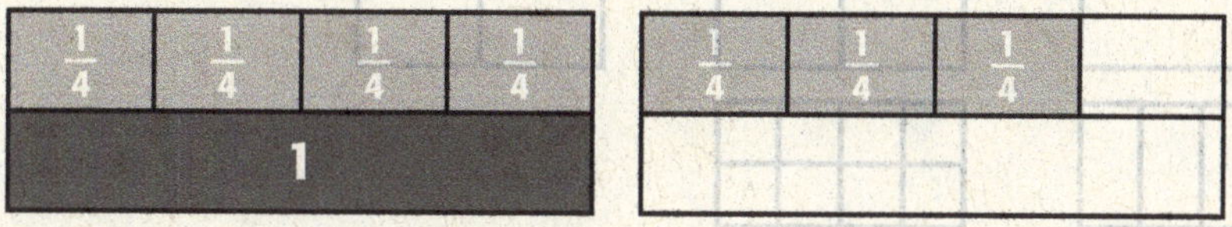

The mixed number $1\frac{3}{4} = 1 + \frac{3}{4}$.

For each improper fraction, complete the model and write the improper fraction as a mixed number.

1. $\frac{5}{3}$

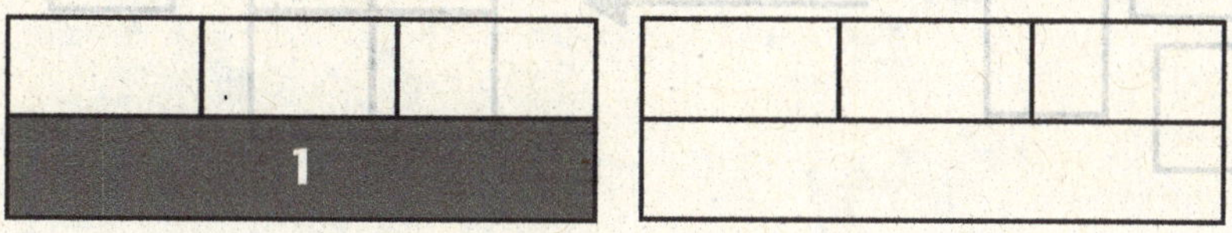

______

2. $\frac{7}{6}$

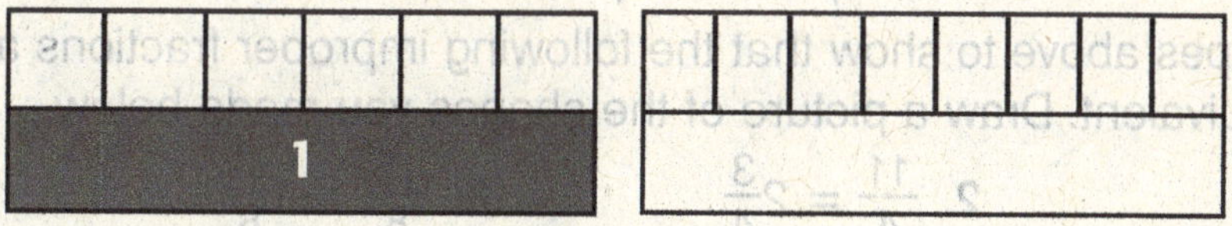

______

3. $\frac{6}{4}$

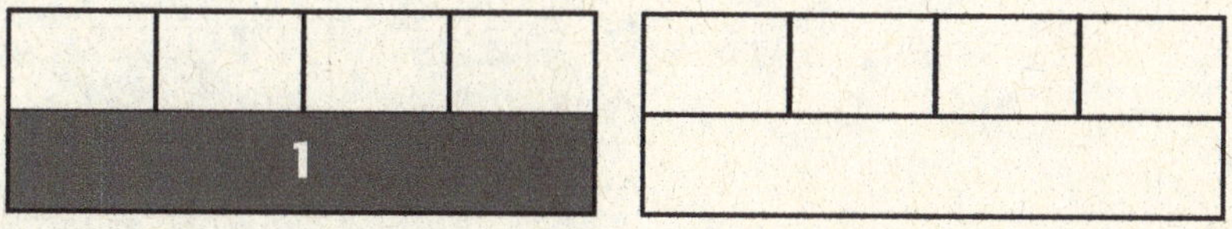

______

Think and Discuss

4. Explain why $\frac{6}{4} = 1\frac{1}{2}$.

5. Discuss a situation in which mixed numbers are used.

Name ______________________ Date __________ Class __________

LESSON 4F

Hands-On Lab

Convert Improper Fractions to Mixed Numbers

Activity

Make 3 copies of each shape below on a separate piece of paper and cut the shapes out. Cut each piece up into its smaller pieces.

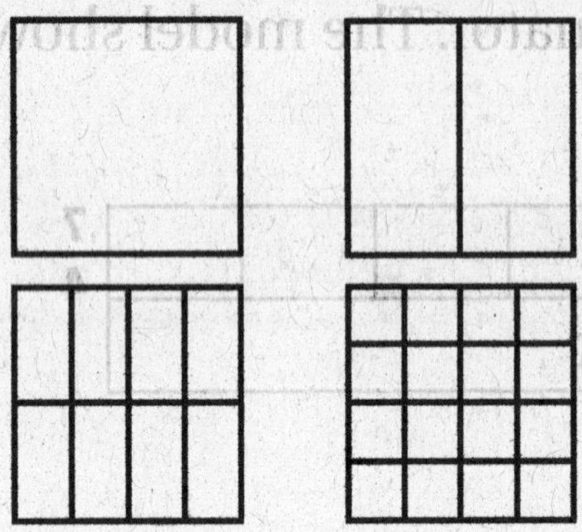

You can assemble pieces of different sizes to show that an improper fraction is equivalent to a mixed number.

$$\frac{5}{4} = 1\frac{1}{4}$$

Use the shapes above to show that the following improper fractions and mixed numbers are equivalent. Draw a picture of the shapes you made below.

1. $\frac{5}{2} = 2\frac{1}{2}$ **2.** $\frac{11}{4} = 2\frac{3}{4}$ **3.** $\frac{15}{8} = 1\frac{7}{8}$ **4.** $\frac{33}{16} = 2\frac{1}{16}$

Name ______________________________ Date ___________ Class ___________

LESSON 4F

Hands-On Lab

Convert Improper Fractions to Mixed Numbers, continued

Think and Discuss

1. Is there more than one way to convert $\frac{20}{8}$ to a mixed number? Explain.

____

____

____

2. How could you use the squares you cut out to show this?

____

____

____

Try This

Use a ruler, paper, and scissors to make 6 more big squares, but cut up 3 of the big squares into thirds. Now cut the other 3 big squares into sixths.

1. Use both this new set and the original set of fractional shapes to find 3 improper fractions equivalent to $2\frac{1}{2}$. Write the fractions below.

____

2. Use the shapes to find 2 improper fractions equivalent to $1\frac{2}{3}$. Write the fractions below.

____

3. Why can't you use any of the first set of fractional shapes (halves, fourths, eighths, sixteenths) to find improper fractions equivalent to $1\frac{2}{3}$? Explain.

____

____

____

Name ______________________ Date ___________ Class ___________

LESSON 4F

Review for Mastery

Mixed Numbers and Improper Fractions

A proper fraction is a fraction whose numerator is less than its denominator.
$\frac{2}{3}$, $\frac{1}{4}$, and $\frac{2}{7}$ are examples of proper fractions.

An improper fraction is a fraction whose numerator is greater than or equal to its denominator.
$\frac{3}{2}$, $\frac{8}{3}$, and $\frac{5}{5}$ are examples of improper fractions.

Some improper fractions can be written as mixed numbers.
To write $\frac{7}{4}$ as a mixed number, draw circles divided into $\frac{1}{4}$ sections.
Then shade in 7 of the $\frac{1}{4}$ sections.

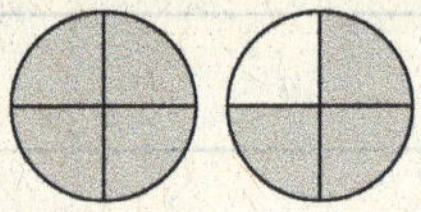

There is one circle and $\frac{3}{4}$ of a circle shaded.
So, $\frac{7}{4} = 1\frac{3}{4}$.

Write each improper fraction as a mixed number. Draw models to help you.

1. $\frac{14}{3}$ ____________ **2.** $\frac{11}{2}$ ____________ **3.** $\frac{15}{4}$ ____________ **4.** $\frac{19}{6}$ ____________

Mixed numbers can be written as improper fractions.
To write $2\frac{1}{3}$ as an improper fraction, draw 3 circles. Divide each circle into $\frac{1}{3}$ sections. Next, shade in 2 whole circles and one $\frac{1}{3}$ section of the last circle.

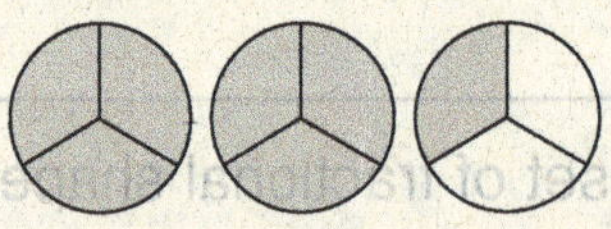

Then find the total number of $\frac{1}{3}$ sections that are shaded.
Seven $\frac{1}{3}$ sections are shaded, so $2\frac{1}{3} = \frac{7}{3}$.

Write each mixed number as an improper fraction. Draw models to help you.

5. $3\frac{1}{4}$ ____________ **6.** $5\frac{2}{3}$ ____________ **7.** $4\frac{1}{2}$ ____________ **8.** $1\frac{5}{6}$ ____________

Name ______________________ Date __________ Class __________

LESSON 4F

Homework and Practice

Mixed Numbers and Improper Fractions

Draw a model of each mixed number. Then write each mixed number as an improper fraction.

1. $2\frac{2}{3}$ **2.** $1\frac{7}{8}$ **3.** $6\frac{3}{8}$

______ ______ ______

4. $1\frac{3}{4}$ **5.** $8\frac{3}{16}$ **6.** $7\frac{1}{4}$

______ ______ ______

7. $5\frac{1}{8}$ **8.** $2\frac{1}{16}$ **9.** $5\frac{3}{5}$

______ ______ ______

Draw a model of each improper fraction. Then write each improper fraction as a mixed number or whole number. Tell whether your answer is a mixed number or whole number.

10. $\frac{6}{2}$ **11.** $\frac{13}{4}$ **12.** $\frac{16}{3}$

______ ______ ______

13. $\frac{40}{8}$ **14.** $\frac{7}{4}$ **15.** $\frac{5}{4}$

______ ______ ______

16. $\frac{130}{12}$ **17.** $\frac{150}{8}$ **18.** $\frac{133}{10}$

______ ______ ______

19. A javelin thrower threw a javelin $\frac{148}{3}$ yards. Write this length as a mixed number.

______

20. Phillip's ghost costume needed $\frac{183}{4}$ inches of piping for a hem and Shandra's needed 45 inches. Whose needed more?

______

EXPLORATION

Program of Studies and Core Content

MA-6-NPO-S-NS1, MA-6-NPO-S-NS5, MA-06-1.1.2, MA-06-1.1.3

4G Comparing and Ordering Fractions

Use the model to decide whether the fraction on the left is greater than (>), less than (<), or equal to (=) the fraction on the right.

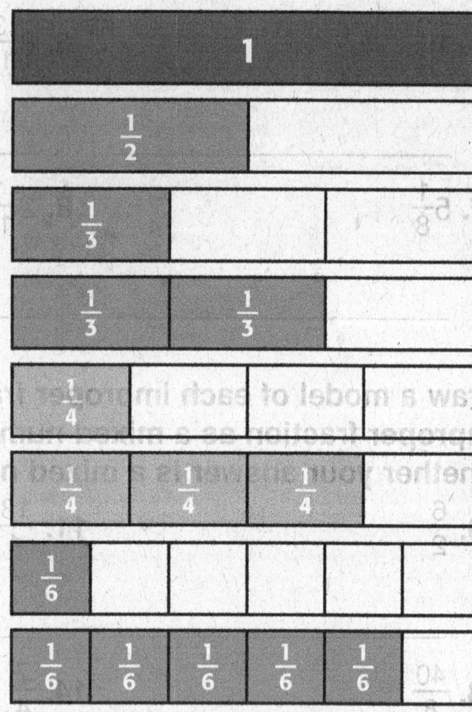

1. $\frac{1}{2} \square \frac{1}{3}$

2. $\frac{1}{2} \square \frac{1}{4}$

3. $\frac{1}{2} \square \frac{2}{3}$

4. $\frac{1}{3} \square \frac{1}{4}$

5. $\frac{3}{4} \square \frac{2}{3}$

6. $\frac{3}{4} \square \frac{5}{6}$

7. $\frac{1}{6} \square \frac{1}{4}$

8. Look at the calculator screen to compare $\frac{1}{2}$ and $\frac{2}{3}$. Which fraction is greater? How do you know?

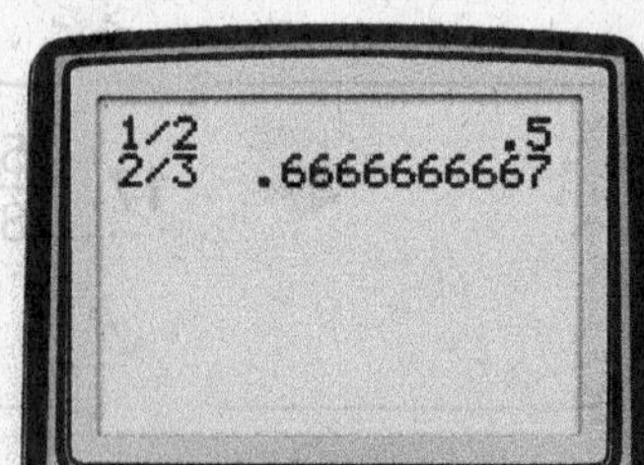

Think and Discuss

9. Explain how you could compare a fraction and a decimal.

10. Explain why $\frac{1}{17}$ is greater than $\frac{1}{18}$.

Name ______________________________ Date ____________ Class ____________

LESSON **4G**

Hands-On Lab

Compare and Order Fractions

Activity

Make 3 copies of each square below on a separate piece of paper and cut the shapes out. Cut each square up into its smaller pieces.

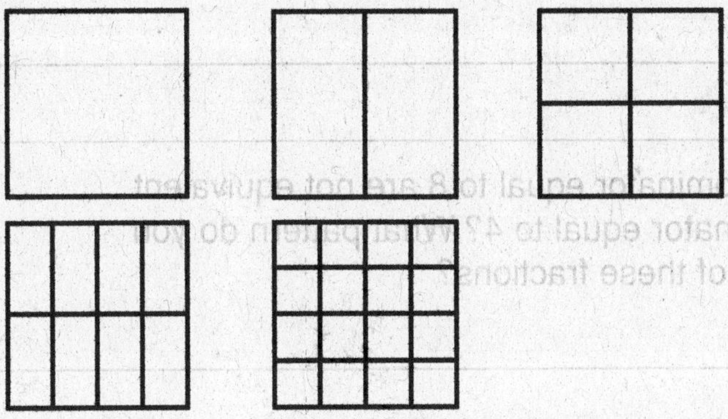

The area of a shape you make represents a fraction. You can use these shapes to compare fractions.

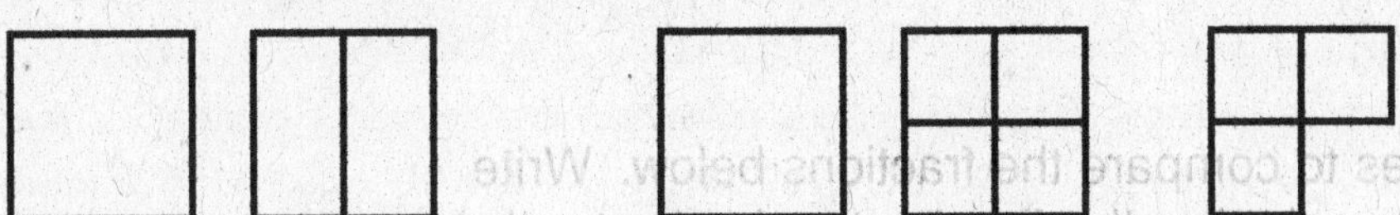

$$\frac{1}{4} = \frac{2}{8} \qquad \frac{1}{4} = \frac{4}{16} > \frac{3}{16}$$

Use the cut up shapes to make each of the shapes below, then fill in the blanks to complete the expressions.

1.

$\_\_\_\_ = \frac{4}{8} < \_\_\_\_$

2.

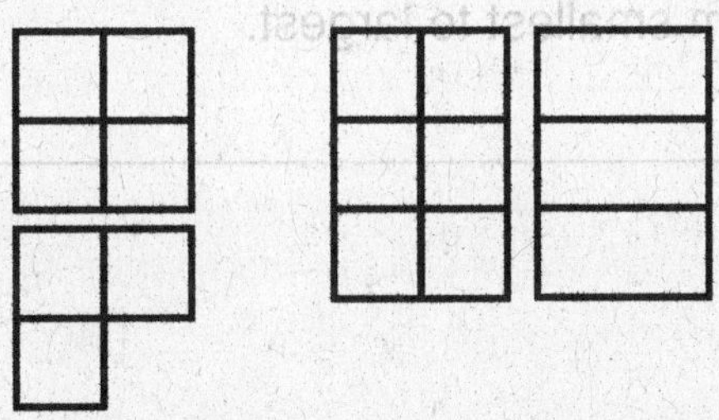

$\frac{7}{16} > \_\_\_\_ = \_\_\_\_$

Name ______________________________ Date __________ Class __________

LESSON 4G

Hands-On Lab

Compare and Order Fractions, continued

Think and Discuss

1. How many fractions equivalent to $\frac{1}{2}$ can you make using the fractional shapes? List them.

____

____

2. Which fractions with a denominator equal to 8 are not equivalent to a fraction with a denominator equal to 4? What pattern do you observe in the numerators of these fractions?

____

____

____

Try This

Use the fractional shapes to compare the fractions below. Write either <, >, or = to show whether the first fraction is greater than, less than, or equal to the second fraction.

1. $\frac{3}{8}$ ___ $\frac{1}{4}$ 2. $\frac{7}{8}$ ___ $\frac{13}{16}$ 3. $\frac{5}{16}$ ___ $\frac{1}{2}$

4. Use the fractional shapes to order the fractions $\frac{3}{16}$, $\frac{3}{4}$, $\frac{5}{8}$, $\frac{1}{2}$, and $\frac{13}{16}$ from smallest to largest.

____

Name ______________________ Date __________ Class __________

LESSON **4G**

Technology Lab

Comparing and Ordering Fractions

> **Remember**
>
> To compare like fractions, compare their numerators. To compare unlike fractions, first rename the fractions so they have the same denominator, then compare their numerators.

Activity

The graph shown represents the amount of sandwiches, salads, beverages and desserts served in the school cafeteria at lunch each day. Compare and order the fractions.

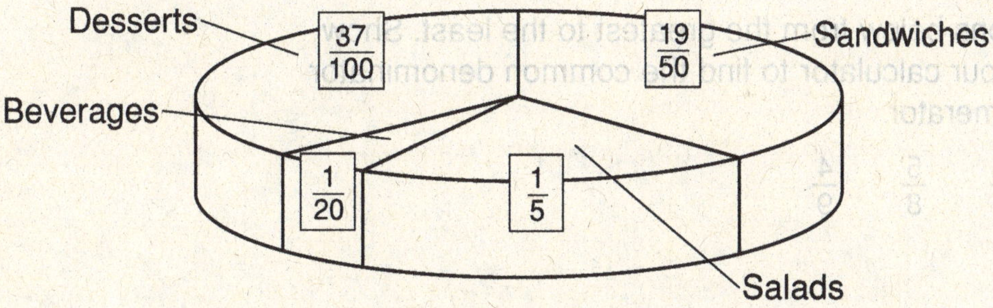

To compare the fractions, rewrite each fraction with a common denominator. The easiest common denominator to use is 100. Multiply the numerator of each fraction by the same number as you used to get 100 in the denominator:

37 × 1 = 37 **19 × 2 = 38** **1 × 5 = 5** **1 × 20 = 20**

Arrange the numerators above in order from least to greatest:

5 < 20 < 37 < 38

Now arrange the original fractions in the same order.

$\frac{1}{20} < \frac{1}{5} < \frac{37}{100} < \frac{19}{50}$

Name ______________________________ Date ____________ Class ____________

LESSON 4G

Technology Lab

Comparing and Ordering Fractions, continued

Think and Discuss

1. Why was it easy to find a common denominator for the fractions in the graph?

____

____

____

____

____

Try This

1. Order the fractions below from the greatest to the least. Show how you used your calculator to find the common denominator and the new numerator.

$\frac{1}{4}$ $\frac{3}{16}$ $\frac{7}{32}$ $\frac{5}{8}$ $\frac{4}{9}$

____

____

____

____

Name ______________________________ Date ____________ Class ____________

LESSON 4G

Review for Mastery

Comparing and Ordering Fractions

$\frac{3}{4}$ and $\frac{2}{3}$ are unlike fractions because they have different denominators. To compare these fractions, graph the fractions on the same number line.

The fraction that is farther to the right on the number line is greater in value.

$\frac{9}{12} > \frac{8}{12}$, so $\frac{3}{4} > \frac{2}{3}$.

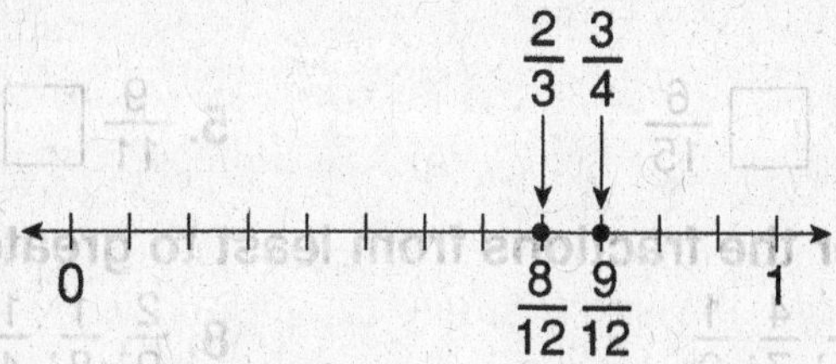

Use the number line to compare the fractions. Write < or >.

1.

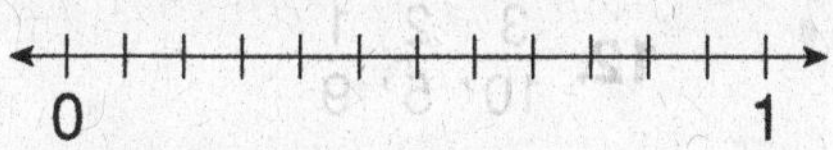

$\frac{1}{3}$ ☐ $\frac{5}{6}$

2.

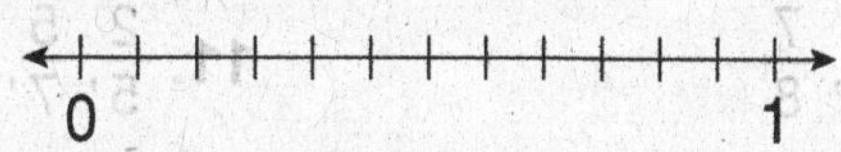

$\frac{5}{10}$ ☐ $\frac{1}{4}$

3.

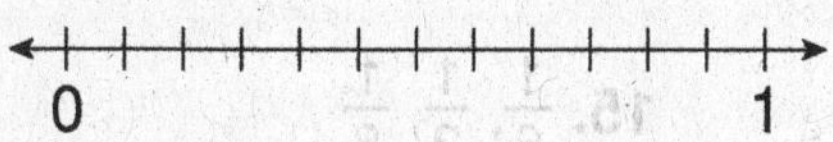

$\frac{11}{12}$ ☐ $\frac{4}{6}$

4.

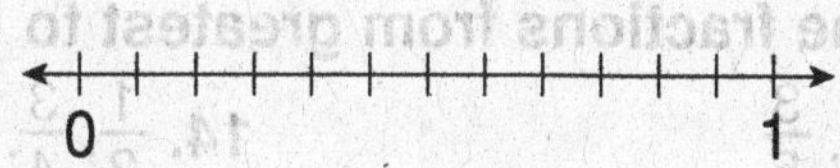

$\frac{6}{9}$ ☐ $\frac{1}{6}$

You can use number lines to order fractions from least to greatest.

To order $\frac{1}{2}$, $\frac{5}{6}$, and $\frac{3}{4}$, graph the values on a number line.

Then list the fractions as they appear from left to right. From least to greatest the fractions are $\frac{1}{2}$, $\frac{3}{4}$, and $\frac{5}{6}$.

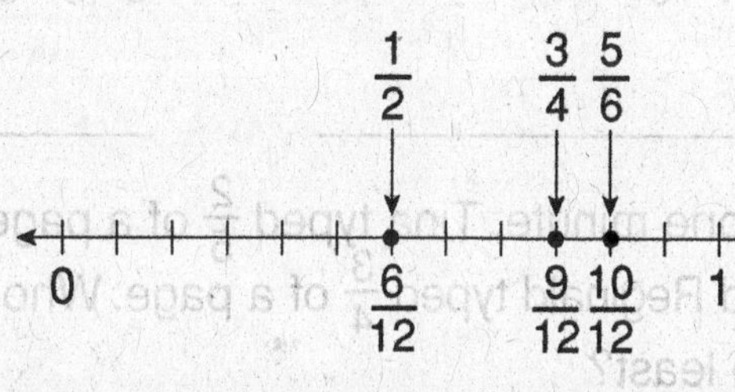

Use the number line to order the fractions from least to greatest.

5.

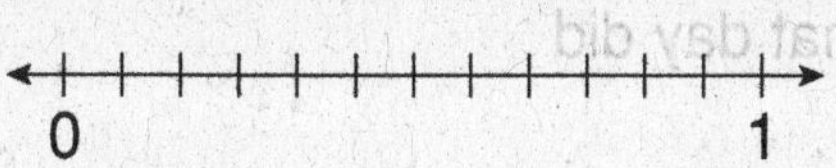

$\frac{7}{12}, \frac{1}{4}, \frac{4}{8}$ ____________________

6.

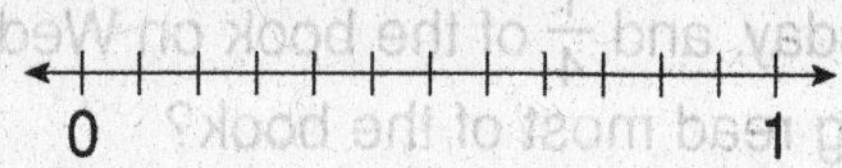

$\frac{2}{3}, \frac{7}{12}, \frac{2}{8}$ ____________________

7.

$\frac{1}{2}, \frac{3}{4}, \frac{3}{9}$ ____________________

8.

$\frac{5}{12}, \frac{1}{3}, \frac{5}{6}$ ____________________

Name ________________________________ Date __________ Class ________

LESSON 4G

Homework and Practice

Comparing and Ordering Fractions

Compare. Write <, >, or =.

1. $\frac{3}{8}$ ☐ $\frac{4}{5}$

2. $\frac{3}{8}$ ☐ $\frac{1}{3}$

3. $\frac{4}{9}$ ☐ $\frac{4}{5}$

4. $\frac{2}{5}$ ☐ $\frac{6}{15}$

5. $\frac{9}{11}$ ☐ $\frac{8}{13}$

6. $\frac{2}{5}$ ☐ $\frac{3}{6}$

Order the fractions from least to greatest.

7. $\frac{1}{4}, \frac{4}{7}, \frac{1}{2}$ ____________

8. $\frac{2}{9}, \frac{1}{8}, \frac{1}{4}$ ____________

9. $\frac{4}{5}, \frac{3}{10}, \frac{1}{6}$ ____________

10. $\frac{7}{9}, \frac{2}{3}, \frac{7}{8}$ ____________

11. $\frac{2}{5}, \frac{5}{7}, \frac{1}{3}$ ____________

12. $\frac{3}{10}, \frac{2}{5}, \frac{1}{9}$ ____________

Order the fractions from greatest to least.

13. $\frac{1}{3}, \frac{1}{9}, \frac{3}{5}$ ____________

14. $\frac{1}{2}, \frac{3}{4}, \frac{1}{3}$ ____________

15. $\frac{1}{6}, \frac{1}{2}, \frac{1}{8}$ ____________

16. $\frac{5}{8}, \frac{4}{7}, \frac{2}{3}$ ____________

17. $\frac{3}{5}, \frac{1}{5}, \frac{3}{8}$ ____________

18. $\frac{7}{8}, \frac{4}{5}, \frac{2}{3}$ ____________

19. In one minute, Tina typed $\frac{2}{5}$ of a page, Troy typed $\frac{9}{16}$ of a page, and Reginald typed $\frac{3}{4}$ of a page. Who typed neither the most, nor the least?

____

20. Trung read $\frac{2}{3}$ of the book on Monday, $\frac{1}{12}$ of the book on Tuesday, and $\frac{1}{4}$ of the book on Wednesday. On what day did Trung read most of the book?

____

EXPLORATION

4H Adding and Subtracting with Like Denominators

Program of Studies and Core Content
MA-6-NPO-S-NS2, MA-6-NPO-S-NO1, MA-6-NPO-S-NO2, MA-06-1.1.2, MA-06-1.3.1

Suzanne runs on a $\frac{3}{4}$-mile track. She ran one lap and then decided to run one more lap.

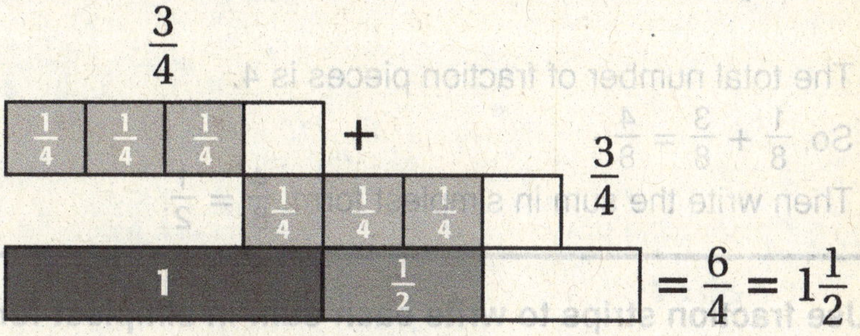

Susan ran $1\frac{1}{2}$ miles.

Draw a model to solve each addition problem.

1. $\frac{3}{5} + \frac{4}{5}$ **2.** $\frac{3}{10} + \frac{7}{10}$

Paul is recording a CD and has $\frac{1}{6}$ of the work completed. How much recording is left?

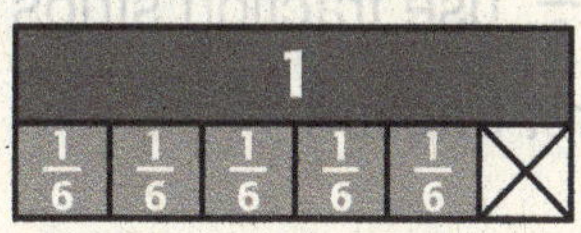

Paul needs to record $\frac{5}{6}$ of the CD. $1 - \frac{1}{6} = \frac{6}{6} - \frac{1}{6} = \frac{5}{6}$

Draw a model to solve each subtraction problem.

3. $\frac{4}{5} - \frac{3}{5}$ **4.** $1 - \frac{3}{10}$

Think and Discuss

5. Explain how to add and subtract fractions with like denominators.

6. Explain how to subtract a fraction from 1.

Name ______________________ Date __________ Class __________

LESSON 4H

Review for Mastery

Adding and Subtracting with Like Denominators

You can use fraction strips to add like fractions.

To find $\frac{1}{8} + \frac{3}{8}$, first model the expression using $\frac{1}{8}$ fraction pieces.

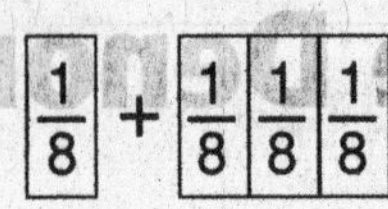

The total number of fraction pieces is 4.

So, $\frac{1}{8} + \frac{3}{8} = \frac{4}{8}$.

Then write the sum in simplest form. $\frac{4}{8} = \frac{1}{2}$.

Use fraction strips to write each sum in simplest form.

1. $\frac{1}{4} + \frac{1}{4}$ ______
2. $\frac{1}{5} + \frac{3}{5}$ ______
3. $\frac{1}{6} + \frac{5}{6}$ ______
4. $\frac{5}{12} + \frac{1}{12}$ ______

You can use fraction strips to subtract like fractions.

To find $\frac{5}{6} - \frac{1}{6}$, use fraction strips to model the first fraction.

Next, take away fraction strips that represent the second fraction.

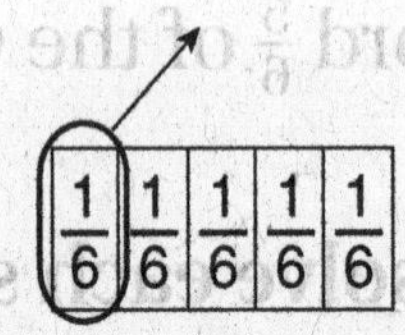

The remaining fraction strips represent the difference.

$\frac{5}{6} - \frac{1}{6} = \frac{4}{6}$

$\frac{4}{6} = \frac{2}{3}$ in simplest form.

Use fraction strips to write each difference in simplest form.

5. $\frac{5}{8} - \frac{3}{8}$ ______
6. $\frac{7}{12} - \frac{5}{12}$ ______
7. $\frac{3}{4} - \frac{1}{4}$ ______
8. $\frac{9}{10} - \frac{3}{10}$ ______

Name ______________________ Date __________ Class __________

LESSON **4H**

Homework and Practice

Adding and Subtracting Like Fractions

Subtract. Write your answers in simplest form.

1. $1 - \frac{3}{8}$

2. $\frac{7}{11} - \frac{3}{11}$

3. $3\frac{4}{5} - 2\frac{1}{5}$

4. $11\frac{3}{4} - 3\frac{1}{4}$

5. $6\frac{1}{2} - 5$

6. $9\frac{3}{10} - 4\frac{1}{10}$

7. $2\frac{5}{6} - 1\frac{1}{6}$

8. $9\frac{4}{13} - 6\frac{2}{13}$

9. $2 - 1\frac{3}{16}$

Evaluate $\frac{11}{13} - x$ for each value of x. Write your answers in simplest form.

10. $x = \frac{2}{13}$

11. $x = \frac{8}{13}$

12. $x = \frac{1}{13}$

13. $x = \frac{10}{13}$

Write the sum or difference in simplest form.

14. $\frac{3}{5} + \frac{2}{5}$

15. $\frac{5}{9} + \frac{2}{9}$

16. $\frac{3}{4} - \frac{1}{4}$

17. $5\frac{22}{66} + 2\frac{19}{66}$

18. $\frac{8}{25} - \frac{5}{25}$

19. $\frac{50}{73} + \frac{12}{73} + \frac{1}{73}$

20. Richard has $3\frac{1}{5}$ dollars and Marc has $2\frac{3}{5}$ dollars. How much money do they have together?

21. George, Michael, and Edith are playing a game of cards. If George wins $\frac{5}{24}$ of the games and Michael wins $\frac{7}{24}$ of the games, what fraction of the games did Edith win?

EXPLORATION

41 Estimating Fraction Sums and Differences

Program of Studies and Core Content
MA-6-NPO-S-E1, MA-6-NPO-S-E2, MA-06-1.1.2, MA-06-1.2.1

Out of 80 students, 49 are in athletics. As the number lines show, approximately half the students are in athletics. In other words, $\frac{49}{80}$ is close to $\frac{1}{2}$.

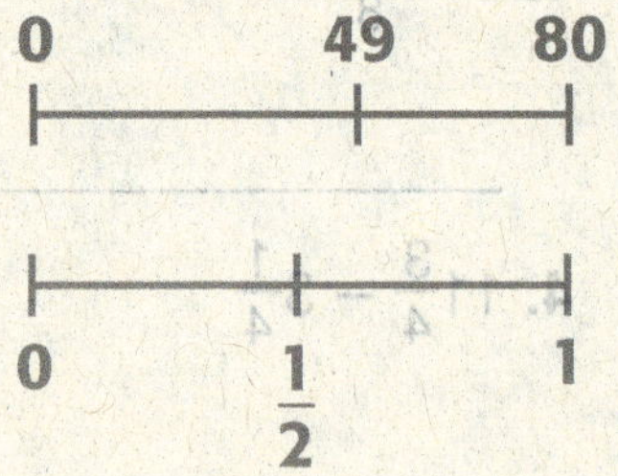

Use a number line to determine whether each fraction is closest to 0, $\frac{1}{2}$, or 1.

1. $\frac{79}{99}$

2. $\frac{22}{213}$

3. $\frac{15}{27}$

4. $\frac{22}{45}$

5. $\frac{300}{475}$

6. $\frac{400}{475}$

Use the estimates you found in numbers 1–6 to estimate each sum or difference.

7. $\frac{79}{99} + \frac{15}{27}$

8. $\frac{22}{213} + \frac{22}{45}$

9. $\frac{300}{475} - \frac{22}{45}$

10. $\frac{15}{27} - \frac{22}{45}$

Think and Discuss

11. Discuss your strategies for determining whether the fractions were closest to 0, $\frac{1}{2}$, or 1.

12. Explain how you know $\frac{237}{475}$ is less than $\frac{1}{2}$.

Name ______________________ Date ____________ Class ____________

LESSON **4I**

Review for Mastery

Estimating Fraction Sums and Differences

You can use number lines to help you estimate fraction sums and differences.

To estimate the sum of $\frac{5}{6}$ and $\frac{1}{3}$, locate each fraction on a number line. Then round each fraction to 0, $\frac{1}{2}$, or 1.

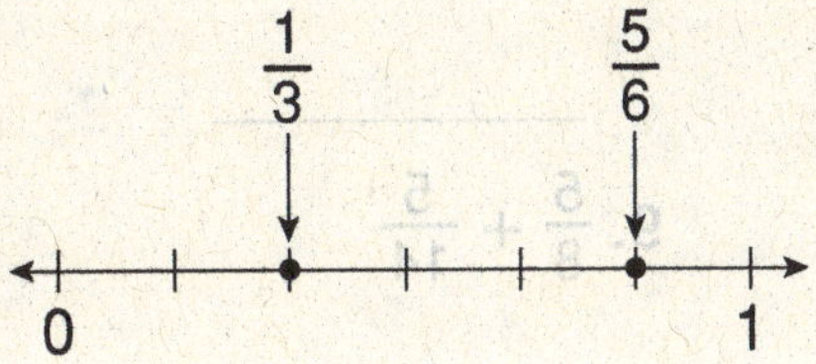

$\frac{5}{6} + \frac{1}{3} \approx$

$1 + \frac{1}{2} = 1\frac{1}{2}$

So, $\frac{5}{6} + \frac{1}{3}$ is about $1\frac{1}{2}$.

To estimate the difference between $\frac{7}{8}$ and $\frac{2}{4}$, locate each fraction on a number line. Then round each fraction to 0, $\frac{1}{2}$, or 1.

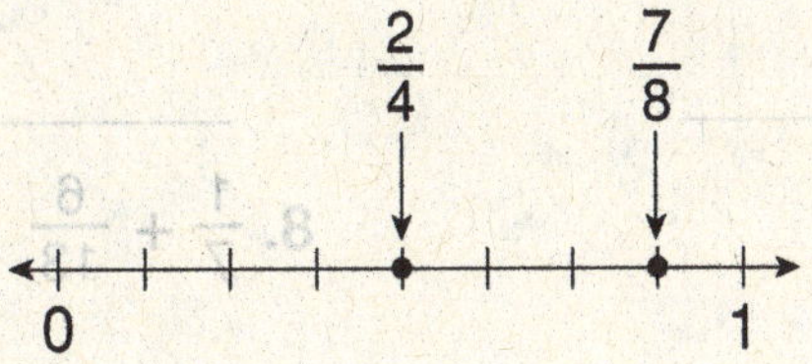

$\frac{7}{8} - \frac{2}{4} \approx$

$1 - \frac{1}{2} = \frac{1}{2}$

So, $\frac{7}{8} - \frac{2}{4}$ is about $\frac{1}{2}$.

Use the number line to round each fraction to 0, $\frac{1}{2}$, or 1 to estimate each sum or difference.

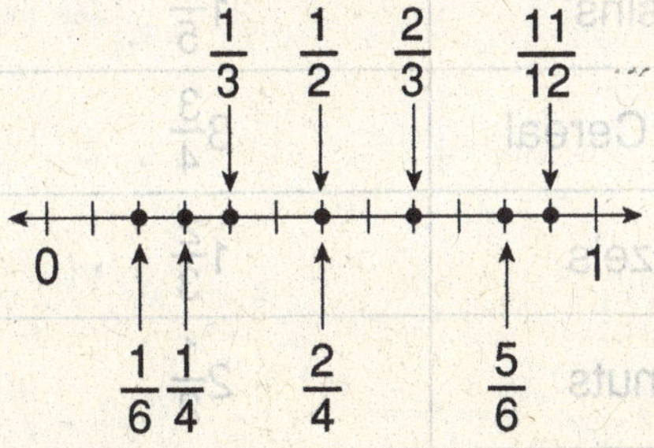

1. $\frac{5}{6} + \frac{1}{6}$ ____________

2. $\frac{11}{12} - \frac{1}{2}$ ____________

3. $\frac{2}{3} + \frac{2}{4}$ ____________

4. $\frac{1}{4} - \frac{1}{3}$ ____________

Use the number line to round each fraction to 0, $\frac{1}{2}$, or 1 to estimate each sum or difference.

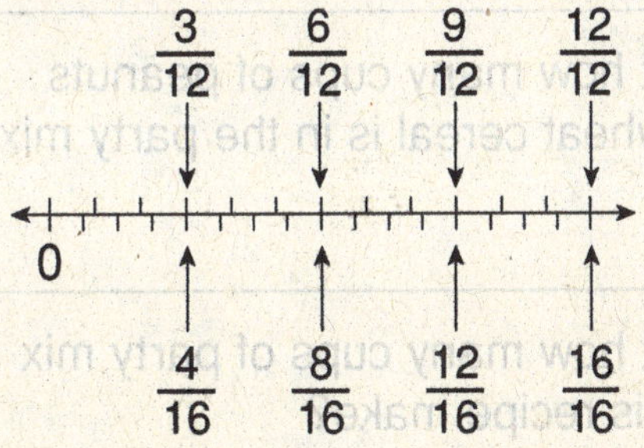

5. $\frac{7}{12} + \frac{2}{6}$ ____________

6. $\frac{5}{6} - \frac{3}{8}$ ____________

7. $\frac{1}{4} + \frac{2}{6}$ ____________

8. $\frac{7}{8} + \frac{14}{16}$ ____________

Name ____________________ Date __________ Class __________

LESSON 4I

Homework and Practice

Estimating Fraction Sums and Differences

Estimate each sum or difference by rounding to 0, $\frac{1}{2}$, or 1.

1. $\frac{3}{7} - \frac{2}{5}$ ________
2. $\frac{4}{7} + \frac{3}{5}$ ________
3. $\frac{7}{9} - \frac{3}{4}$ ________
4. $\frac{13}{14} + \frac{1}{6}$ ________
5. $\frac{4}{5} - \frac{6}{13}$ ________
6. $\frac{8}{11} + \frac{1}{3}$ ________
7. $\frac{3}{8} - \frac{2}{11}$ ________
8. $\frac{1}{7} + \frac{6}{13}$ ________
9. $\frac{5}{8} + \frac{5}{14}$ ________
10. $2\frac{1}{8} + 1\frac{2}{15}$ ________
11. $1\frac{1}{3} - \frac{5}{6}$ ________
12. $5\frac{1}{2} + 2\frac{5}{6}$ ________

Use the table for exercises 13–15.

13. About how many more cups are there of pretzels than raisins?

____

14. About how many cups of peanuts and wheat cereal is in the party mix?

____

15. About how many cups of party mix will this recipe make?

____

Party Mix

Ingredient	Amount (cups)
Raisins	$1\frac{1}{5}$
Wheat Cereal	$3\frac{3}{4}$
Pretzels	$1\frac{2}{3}$
Peanuts	$2\frac{1}{2}$

16. Harvey drove $2\frac{1}{6}$ hours to a wedding, then $1\frac{5}{9}$ hours to the reception. About how long did Harvey drive?

____

CHAPTER 4

Chapter Summary

Answer these questions to summarize the important concepts from Chapter 4 in your own words.

1. Explain the difference between a prime number and a composite number.

2. List two methods that can be used to find the greatest common factor of a set of numbers.

3. Explain how to write a fraction as a decimal. Explain how to write a decimal as a fraction.

4. Explain how to change a mixed number to an improper fraction and how to change an improper fraction to a mixed number.

For more review of Chapter 4:

- Complete the Chapter 4 Study Guide and Review on pages 216–218 of your textbook.
- Complete the Ready to Go On quizzes on pages 178, 196, and 210 of your textbook.

Name ______________________ Date __________ Class __________

CHAPTER 4

KCCT Prep

Multiple Choice

1. The prime factorization of a number is $2^{\square}3^2$, but the exponent of the 2 is missing. Which of these could be the number?

Ⓐ 24 Ⓒ 72
Ⓑ 54 Ⓓ 75

2. Rosa runs $3\frac{5}{8}$ miles per day. How should she write this distance as a decimal?

Ⓐ 3.29 miles Ⓒ 3.625 miles
Ⓑ 3.58 miles Ⓓ 3.875 miles

3. James has a piece of string that is $4\frac{7}{10}$ meters long. He cuts off a piece that is $1\frac{3}{10}$ meters long. What is the length of the remaining piece of string?

Ⓐ $3\frac{2}{5}$ meters Ⓒ $5\frac{2}{5}$ meters
Ⓑ $3\frac{7}{10}$ meters Ⓓ 6 meters

4. The equation $15p = 960$ gives the number of pages p in a magazine. How many pages are there altogether in 6 copies of the magazine?

Ⓐ 90 Ⓒ 384
Ⓑ 160 Ⓓ 2400

5. Which sequence includes the term 444?

Ⓐ 5, 9, 13, 17, 21, ...
Ⓑ 22, 18, 14, 10, 6, ...
Ⓒ 9, 2, 11, 2, 13, 2, ...
Ⓓ 12, 16, 20, 24, 28, ...

6. The solution of the equation $30 = y \square 6$ is $y = 24$. What is the missing operation in the equation?

Ⓐ ÷ Ⓒ +
Ⓑ × Ⓓ −

7. The table shows the amount of water needed to cook equal amounts of several different grains. Which list shows the grains in order, from the one requiring the least amount of water to the one requiring the most?

Cooking Instructions	
Grain	**Amount of Water**
Buckwheat	$\frac{1}{3}$ cup
Millet	$\frac{5}{12}$ cup
Oats	$\frac{2}{3}$ cup
Wild Rice	$\frac{1}{2}$ cup

Ⓐ Wild Rice, Millet, Buckwheat, Oats
Ⓑ Buckwheat, Wild Rice, Millet, Oats
Ⓒ Wild Rice, Buckwheat, Oats, Millet
Ⓓ Buckwheat, Millet, Wild Rice, Oats

8. Which fraction is equal to 0.35?

Ⓐ $\frac{7}{20}$ Ⓒ $\frac{3}{5}$
Ⓑ $\frac{3}{8}$ Ⓓ $\frac{11}{15}$

9. Jermaine is hiking at a national park. The length of the Waterfall Trail is $\frac{5}{8}$ mile. The length of the Mountain Trail is $1\frac{5}{8}$ miles. Jermaine hikes both trails. Find the total distance hiked.

Ⓐ $1\frac{1}{4}$ miles Ⓒ $2\frac{1}{4}$ miles
Ⓑ $1\frac{4}{5}$ miles Ⓓ $2\frac{5}{8}$ miles

10. Which number is closest to 3.7 on a number line?

Ⓐ 3.08 Ⓒ 3.56
Ⓑ 3.8 Ⓓ 3.65

11. What is the greatest common factor of 24, 12, and 30?

Ⓐ 3 Ⓒ 6
Ⓑ 4 Ⓓ 12

12. Which expression describes the sequence in the table?

Position	1	2	3	4	*n*
Value of term	9	19	29	39	■

Ⓐ $n + 8$ Ⓒ $9n$
Ⓑ $10n - 1$ Ⓓ $9n + 1$

13. Dee has $\frac{1}{2}$ gallon of blue paint, $\frac{3}{5}$ gallon of green paint, $\frac{5}{8}$ gallon of red paint, and $\frac{3}{4}$ gallon of yellow paint. Which list shows the colors in increasing order of amount of paint?

Ⓐ Blue, Green, Red, Yellow
Ⓑ Green, Blue, Red, Yellow
Ⓒ Blue, Red, Yellow, Green
Ⓓ Green, Red, Blue, Yellow

14. Which decimal is farthest from 1 on a number line?

Ⓐ 0.8 Ⓒ 0.89
Ⓑ $0.\overline{8}$ Ⓓ 0.09

15. The fraction $\frac{n}{6}$ is greater than $\frac{7}{12}$ and less than 1. Which of these could be the value of the numerator n?

Ⓐ 3 Ⓒ 6
Ⓑ 5 Ⓓ 7

Open Response

16. The table shows the fraction of the total U.S. population that lives in Florida, Texas, and Iowa.

U.S. Population Data	
State	**Fraction of Total Population**
Florida	$\frac{3}{50}$
Texas	$\frac{2}{25}$
Iowa	$\frac{1}{100}$

a. List the states in order from the state with the greatest population to the state with the least population. Explain your answer.

b. Indiana has $\frac{1}{50}$ of the U.S. population. What fraction of the population lives in Florida or Indiana?

17. Yuri is ordering a window shade. He measures the window and finds that its length is $3\frac{7}{8}$ ft.

a. Yuri wants the shade to extend $\frac{3}{8}$ ft beyond the bottom of the window. What will be the shade's total length?

b. To order online, Yuri needs to enter the length as a decimal. What decimal should he enter?

c. The shades cost $6.24 per foot. Will Yuri be able to purchase his window shade for less than $25? Explain.

CHAPTER 5

Project

Fraction Operations: Paint the Town

Career: Painter

Have you ever wondered how painters estimate how much to charge for a job? Professional painters might paint houses, schools, office buildings, music halls, or even sports stadiums. To estimate how much to charge, many painters use a table that lists the average time it should take to prepare and paint certain objects.

Before you start the project, work through the Painter's Estimates worksheet on the next page.

In this chapter project . . .

You design a mural for your bedroom wall.

What to do:

- Learn about the time it takes to prepare and paint various surfaces.
- Choose a wall in your bedroom, measure it, and then design, draw, and color a mock up for a mural you'd like to paint.
- Estimate the time it would take to paint the mural, and go to a local hardware store to research the amount and cost of the paint you would need. (Remember, more colors mean more money!)

What to present:

- The mock up for your mural. Show how the mural will fit on the wall, and indicate the distance each side of the mural is from the ends of the wall.
- An estimate of the amount of time and money required to make the mural.

Where's the math?

- Use operations with fractions, such as addition and subtraction, to find the lengths and distances of the mural and the wall.
- Calculate the time and cost associated with making the mural, and compare these to the estimates you made.

go.hrw.com
Chapter Project Online
KEYWORD: MR7 Ch5

Name ______________________ Date __________ Class __________

CHAPTER 5

Project Recording Sheet

Fraction Operations

Painter's Estimates

Use the table to calculate the time required for each painting job below.

Painting Times

Task	Time (hr)	Task	Time (hr)
Wall preparation (100 ft^2)		**Painting wall** (100 ft^2)	
Plaster	$\frac{1}{2}$	Painting	$\frac{3}{10}$
Wallboard	$\frac{2}{5}$	**Floor** (100 ft^2)	
Chair rail (10 ft)		Sanding	$\frac{3}{4}$
Preparing	$\frac{4}{5}$	Preparing	$\frac{3}{5}$
Painting	$\frac{3}{5}$	Varnishing	$\frac{1}{2}$
		Polishing	$\frac{2}{5}$

Job One

You are refinishing the school basketball court. The total floor space is 110 ft by 70 ft. You will sand, prepare, varnish, and polish the floor. Complete the table.

Basketball Court Floor

Task	Time Needed (hr)
Sanding	
Preparing	
Varnishing	
Polishing	
Total time for job	

Job Two

You are painting a room. The old wall paper has been removed, but the plaster walls must be prepared and painted. The total area of the walls is 100 ft^2. The room contains a chair rail that is 10 ft long that must be prepared and painted. How long will it take to complete each task? How long will the entire job take?

preparing walls: ______ hr; preparing chair rail: ______ hr; painting chair rail: ______ hr;

painting walls: ______ hr; total job: ______ hr

EXPLORATION

Program of Studies and Core Content
MA-6-NPO-S-PNO1, MA-6-NPO-S-PNO3, MA-06-1.5.1

5A Least Common Multiple

1. Sarah and Jane enter a walkathon for charity. They start together, but Sarah completes one lap every **6** minutes while Jane completes one lap every **8** minutes.

Number of Laps Completed	Sarah's Time (min)	Jane's Time (min)
1	$6 \cdot 1 = 6$	$8 \cdot 1 = 8$
2	$6 \cdot 2 = 12$	$8 \cdot 2 = 16$
3	$6 \cdot 3 = 18$	$8 \cdot 3 = 24$
4	$6 \cdot 4 = 24$	$8 \cdot 4 = 32$
5	$6 \cdot 5 = 30$	$8 \cdot 5 = 40$
6	$6 \cdot 6 = 36$	$8 \cdot 6 = 48$
7	$6 \cdot 7 = 42$	$8 \cdot 7 = 56$
8	$6 \cdot 8 = 48$	$8 \cdot 8 = 64$

a. After how many minutes will Sarah and Jane meet at the start again?

b. When will they meet the next time?

Think and Discuss

2. Discuss the solution to number **1a** using the term *common multiple.*

3. Compare the solution to number **1a** with the solution to number **1b**, and describe these solutions using the terms ***common multiple*** and ***least common multiple.***

Name ______________________________ Date ____________ Class ____________

LESSON 5A

Technology Lab

Least Common Multiple

Spreadsheet software such as Excel lets you find the least common multiple of two or more numbers easily. Once you've entered two simple formulae into a worksheet you can find the LCM for any numbers simply by entering them.

Activity

Follow the instructions below to find the LCM of 6 and 15 using an Excel spreadsheet.

Open an Excel spreadsheet and name it "lcm.xls".

1. Starting in Cell B1, enter the numbers 1 through 15 in the first row.

2. Enter the number 6 in Cell A2.

3. Enter the number 15 in Cell A3.

4. Select Cell B2 and write the following formula: =B1*6.

5. Select Cell B3 and and write the following formula: =B1*15.

	A	B	C	D	E	F	G	H
1		1	2	3	4	5	6	7
2	6	=B1*6						
3	15	=B1*15						

6. Then select the cells B2 through P2. Next, choose "Fill Right" from the Edit menu, or press Control-R. This will extend the formula to each cell you selected.

7. Then select the cells B3 through P3. Next, choose "Fill Right" from the Edit menu, or press Control-R. This will extend the formula to each cell you selected.

Name ______________________________ Date ___________ Class ___________

LESSON 5A

Technology Lab

Least Common Multiple, continued

8. Spreadsheet "lcm.xls" should now look like the sample below. Look for the number that appears in both Row 2 and Row 3. In this case, the LCM is 30.

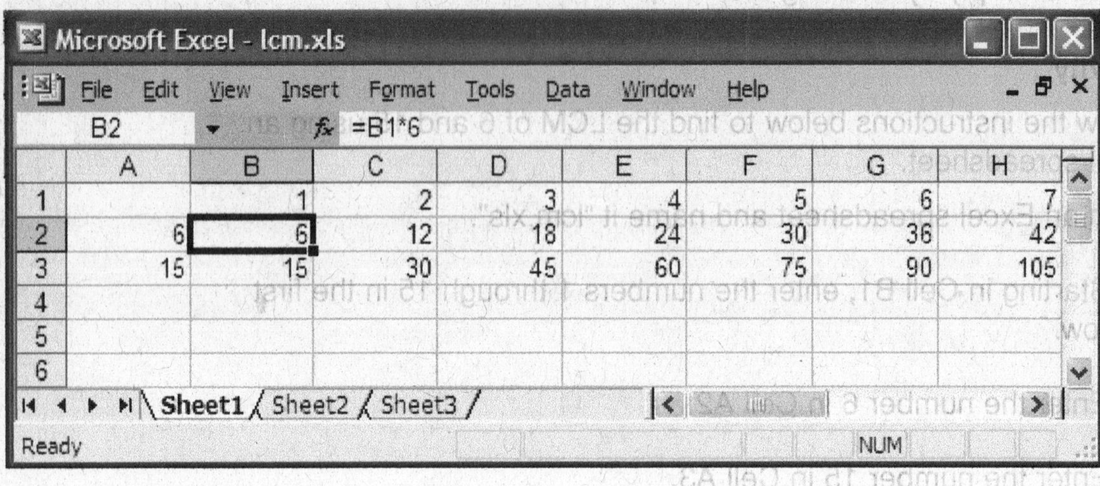

	A	B	C	D	E	F	G	H
1		1	2	3	4	5	6	7
2	6	6	12	18	24	30	36	42
3	15	15	30	45	60	75	90	105
4								
5								
6								

Think and Discuss

1. Using the instructions above, make a spreadsheet to find the LCM of 176 and 325. Can you find the LCM in any of the first 15 cells? What does this suggest about the limitations of the spreadsheet method?

____

____

____

____

Try This

Make a spreadsheet using 1 through 25 as the factors in Row 1. Then use the spreadsheet method to find the LCM of each pair of numbers.

1. 14, 4 **2.** 15, 90 **3.** 60, 250

__________ __________ __________

Name ______________________________ Date __________ Class __________

LESSON **5A**

Review for Mastery

Least Common Multiple

The smallest number that is a multiple of two or more numbers is called the least common multiple (LCM).

To find the least common multiple of 3, 6, and 8, list the multiples for each number and put a circle around the LCM in the three lists.

Multiples of 3: 3, 6, 9, 12, 15, 18, 21, 24

Multiples of 6: 6, 12, 18, 24, 30, 36, 42

Multiples of 8: 8, 16, 24, 32, 40, 48, 56

So 24 is the LCM of 3, 6, and 8.

List the multiples of each number to help you find the least common multiple of each group.

1. 3 and 4

Multiples of 3: ______

Multiples of 4: ______

LCM: ______

2. 5 and 7

Multiples of 5: ______

Multiples of 7: ______

LCM: ______

3. 8 and 12

Multiples of 8: ______

Multiples of 12: ______

LCM: ______

4. 2 and 9

Multiples of 2: ______

Multiples of 9: ______

LCM: ______

5. 4 and 6

Multiples of 4: ______

Multiples of 6: ______

LCM: ______

6. 4 and 10

Multiples of 4: ______

Multiples of 10: ______

LCM: ______

7. 2, 5, and 6

Multiples of 2: ______

Multiples of 5: ______

Multiples of 6: ______

LCM: ______

8. 3, 4, and 9

Multiples of 3: ______

Multiples of 4: ______

Multiples of 9: ______

LCM: ______

9. 8, 10, and 12

Multiples of 8: ______

Multiples of 10: ______

Multiples of 12: ______

LCM: ______

Name ______________________ Date __________ Class __________

LESSON 5A

Homework and Practice

Least Common Multiple

Find the least common multiple (LCM).

1. 3 and 5	**2.** 2 and 9	**3.** 5 and 6
________	________	________
4. 7 and 3	**5.** 6 and 7	**6.** 2 and 8
________	________	________
7. 12 and 18	**8.** 6 and 15	**9.** 6 and 9
________	________	________
10. 3, 4, and 6	**11.** 2, 5, and 7	**12.** 6, 8, and 12
________	________	________
13. 2, 4, and 11	**14.** 4, 9, and 12	**15.** 2, 3, and 7
________	________	________

16. Coach Wilson is ordering shirts and shorts for his track team. There are 30 members on his track team. Shorts come in packs of 3, and shirts come in packs of 6. What is the least number of packs of each he should order so that each member will have 1 pair of shorts and 1 shirt, and none will be left over?

____

17. Sally is selling breakfast bags. She needs to sell 24 bags today. Muffins come 3 to a pack, bagels come in packs of 4, and fruit bars come in packs of 6. How many packs of each item will she need to buy to make the 24 bags with no leftovers?

____

18. A restaurant needs to make 60 lunches. Cartons of juice come in packs of 4, bags of chips come in packs of 6, and toys come in packs of 15. How many of each pack will the restaurant need to buy with no leftovers?

____

EXPLORATION

Program of Studies and Core Content

MA-6-NPO-S-NS2, MA-6-NPO-S-NO1, MA-6-NPO-S-NO2, MA-06-1.1.2, MA-06-1.3.1

5B Adding and Subtracting with Unlike Denominators

Fractions are pieces of a whole. When you add or subtract fractions with unlike denominators, you are usually adding or subtracting pieces of different sizes. Look at the model used to solve the problem below.

Phil combines $\frac{1}{4}$ gallon of paint with $\frac{1}{2}$ gallon of paint. How much paint does he have now?

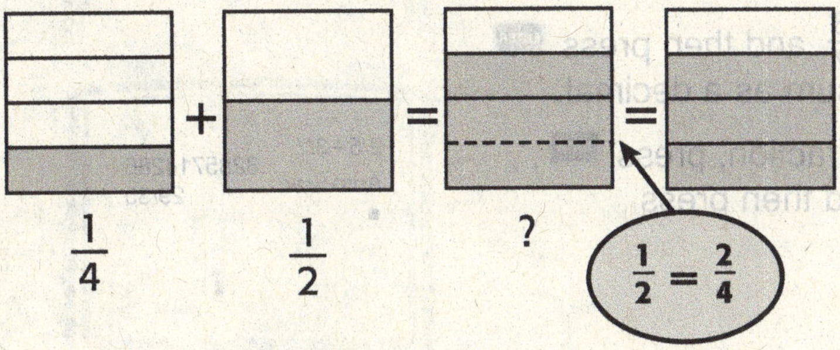

$\rightarrow \frac{1}{4} + \frac{1}{2} = \frac{1}{4} + \frac{2}{4} = \frac{3}{4}$

Phil has $\frac{3}{4}$ gallon of paint.

1. Draw a model to show that $1 - \frac{1}{4} = \frac{3}{4}$.

Draw a model to solve each problem. Simplify your answers.

2. $\frac{1}{2} + \frac{1}{3}$

3. $\frac{4}{5} - \frac{1}{2}$

Think and Discuss

4. Explain how to add and subtract fractions with unlike denominators.

5. Draw a model to show $\frac{1}{2} + \frac{2}{3} = 1\frac{1}{6}$.

Name ______________________ Date __________ Class __________

LESSON 5B

Technology Lab

Fraction Operations

To enter a fraction, like $\frac{1}{8}$, into a graphing calculator, use [÷]. When you press [ENTER], the fraction is converted to a decimal. To convert a decimal number, such as 0.625, to a fraction, enter the number on your calculator. Press [MATH], and then press [ENTER] to select Frac.

When you press [ENTER] again, the calculator displays the fraction equivalent.

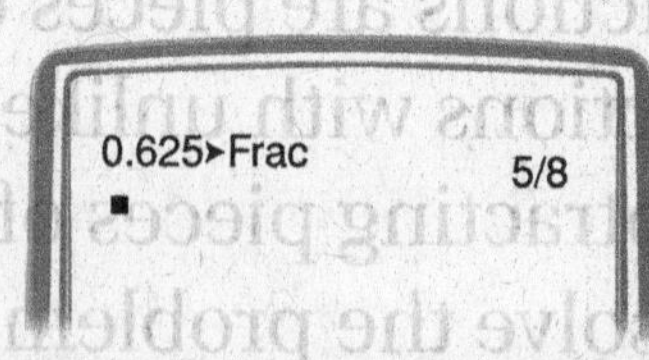

The decimal 0.625 is equivalent to $\frac{5}{8}$.

Activity

Use your calculator to find the value of $\frac{2}{5} + \frac{3}{7}$. Write the sum as a fraction.

Step 1: To find the sum, enter $\frac{2}{5} + \frac{3}{7}$, and then press [ENTER]. The calculator displays the sum as a decimal.

Step 2: To convert the decimal to a fraction, press [MATH], press [ENTER] to select Frac, and then press [ENTER] again.

$$\frac{2}{5} + \frac{3}{7} = \frac{29}{35}$$

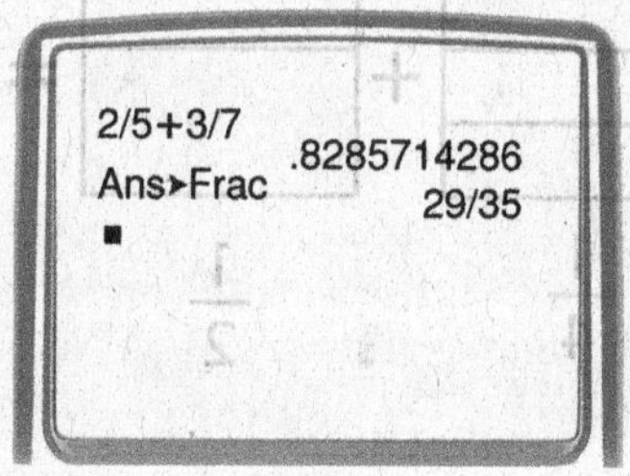

Think and Discuss

1. Without using a calculator, write 0.10001 as a fraction. Then use a graphing calculator to convert 0.10001 to a fraction. What happens? Why do you think this happened?

____

____

Name ______________________ Date ___________ Class ___________

LESSON 5B

Technology Lab

Fraction Operations, continued

Try This

Add or subtract. Write the answer as a fraction or a mixed number and as a decimal rounded to the nearest hundredth.

1. $\frac{1}{4} + \frac{2}{3}$ ______________

2. $\frac{2}{3} - \frac{1}{4}$ ______________

3. $\frac{3}{5} + \frac{5}{3}$ ______________

4. $\frac{4}{3} - \frac{3}{4}$ ______________

Name ______________________________ Date ____________ Class ____________

LESSON 5B

Review for Mastery

Adding and Subtracting with Unlike Denominators

Unlike fractions have different denominators. To add and subtract fractions, you must have a common denominator. The least common denominator (LCD) is the least common multiple of the denominators.

To add or subtract unlike fractions, first find the LCD of the fractions.

$\frac{2}{3} + \frac{1}{4}$

Multiples of 4: 4, 8, **12,...**

Multiples of 3: 3, 6, 9, **12,...**

The LCD is 12.

Next, use fraction strips to find equivalent fractions.

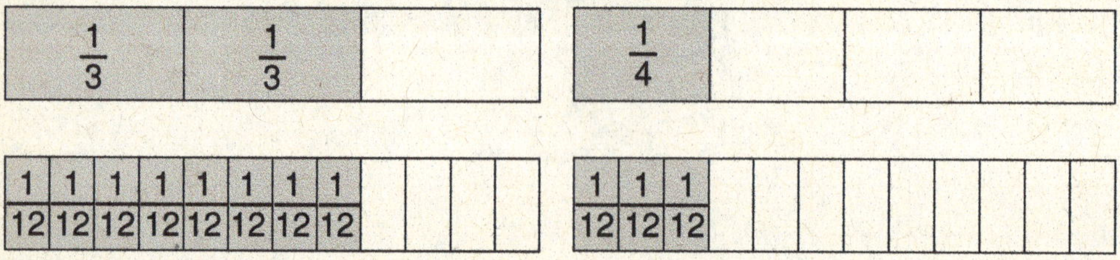

Then use fraction strips to find the sum or difference.

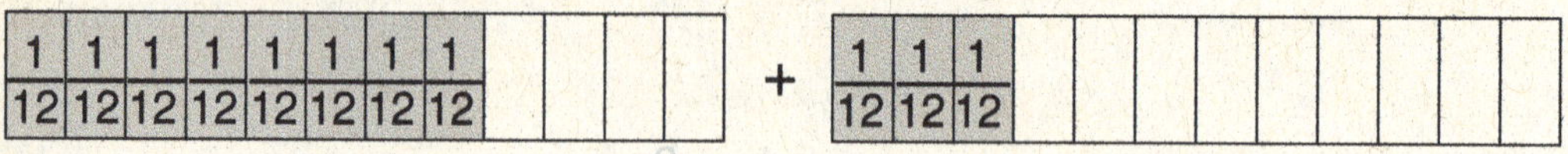

$\frac{8}{12} + \frac{3}{12} = \frac{11}{12}$

So, $\frac{2}{3} + \frac{1}{4} = \frac{11}{12}$.

Use fraction strips to find each sum or difference. Write your answer in simplest form.

1. $\frac{1}{4} + \frac{1}{8}$ ____________

2. $\frac{5}{6} - \frac{2}{3}$ ____________

3. $\frac{3}{4} - \frac{1}{3}$ ____________

4. $\frac{3}{5} + \frac{3}{10}$ ____________

5. $\frac{3}{4} + \frac{1}{6}$ ____________

6. $\frac{1}{2} + \frac{3}{8}$ ____________

7. $\frac{2}{3} - \frac{1}{6}$ ____________

8. $\frac{1}{3} - \frac{1}{4}$ ____________

Name ______________________ Date __________ Class __________

LESSON 5B

Homework and Practice

Adding and Subtracting Unlike Fractions

Add or subtract. Write your answers in simplest form.

1. $\frac{4}{5} + \frac{2}{3}$
2. $\frac{5}{12} - \frac{1}{7}$
3. $\frac{3}{4} + \frac{5}{6}$
4. $\frac{7}{12} - \frac{1}{10}$
5. $\frac{4}{5} + \frac{5}{11}$
6. $\frac{2}{5} - \frac{2}{15}$
7. $\frac{7}{8} - \frac{8}{15}$
8. $\frac{8}{9} + \frac{2}{5}$
9. $\frac{11}{15} - \frac{4}{13}$
10. $\frac{1}{2} - \frac{3}{8}$
11. $\frac{4}{7} + \frac{1}{12}$
12. $\frac{3}{7} + \frac{11}{16}$

Evaluate each expression for $a = \frac{2}{5}$. Write your answers in simplest form.

13. $a - \frac{1}{4}$
14. $\frac{9}{14} - a$
15. $\frac{1}{9} + a$
16. $\frac{1}{3} + a$
17. $\frac{5}{16} + a$
18. $\frac{7}{8} - a$

19. Five-eighths of the marching band members are seniors and $\frac{1}{5}$ of the marching band members are juniors. What fraction of the marching band members are freshmen and sophomores?

20. Fred leaves his house for work, which is exactly two miles away. After Fred has traveled $\frac{2}{3}$ of a mile, he must turn around to pick up some papers he forgot. After he travels $\frac{1}{6}$ of a mile back home, how far away from work is he?

EXPLORATION

Program of Studies and Core Content

MA-6-NPO-S-NS2, MA-6-NPO-S-NO1, MA-6-NPO-S-NO2, MA-06-1.3.1

5C Adding and Subtracting Mixed Numbers

The graph shows typical rainfall levels for a city in the Southwest for the first 5 months of the year. What is the approximate total rainfall from January through May?

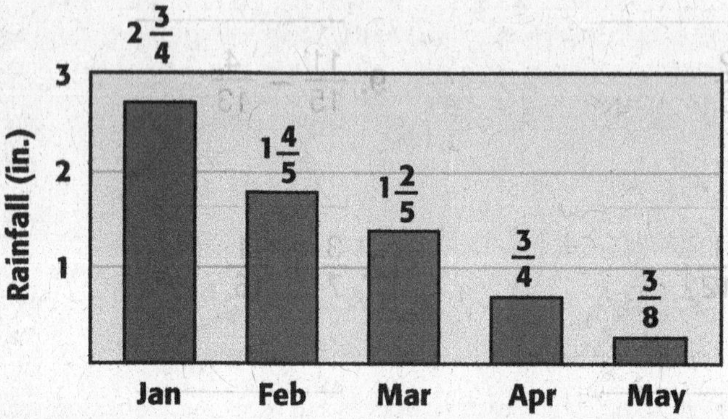

A mixed number contains a whole number and a fraction. To estimate with mixed numbers, round each mixed number to the nearest whole number.

Actual $2\frac{3}{4} + 1\frac{4}{5} + 1\frac{2}{5} + \frac{3}{4} + \frac{3}{8}$

Estimated $3 + 2 + 1 + 1 + 0 = 7$ in.

The total rainfall is about 7 inches.

Estimate each sum or difference.

1. $13\frac{1}{2} - 9\frac{27}{32}$

2. $1\frac{1}{2} + 9\frac{3}{8} - 2\frac{1}{4}$

3. $17\frac{7}{8} + 19\frac{1}{10}$

4. $4\frac{1}{8} - 1\frac{3}{10} + 3\frac{1}{4}$

Think and Discuss

5. Discuss the estimation strategies you used.

6. Describe a real-world situation in which mixed numbers are added or subtracted.

Name ______________________________ Date __________ Class __________

LESSON **5C**

Technology Lab

Adding and Subtracting Mixed Numbers

You can evaluate expressions with mixed numbers using a calculator by learning to use the machine's Memory function. Instead of stopping to write calculations as you do them, the calculator lets you store numbers and recall them later. Once a number is stored, you can make other calculations without affecting the number in memory.

Activity 1

Evaluate the following expression using a calculator.

$3\frac{3}{4} + 2\frac{5}{9}$

1. Determine the common denominator for 4 and 9, which is 36. Then, convert the first mixed number to an improper fraction by entering the following keystrokes:
   $4 \times 3 + 3 = 15$

   The improper fraction is written as $\frac{15}{4}$.

2. Since the common denominator is 36, and 36/4 = 9, multiply 15 by 9. When you see the product, 135, press STO> ALPHA A ENTER. This enters the number into memory for later recall.

3. Convert the second mixed number to an improper fraction by entering the following keystrokes: $9 \times 2 + 5 = 23$

   The improper fraction is written as $\frac{23}{9}$.

4. Since the common denominator is 36, and 36/9 = 4, multiply 23 by 4. When you see the product, 92, press + ALPHA A ENTER.
   Store the result as A. The sum of 135 and 92 is 227.
   So the sum of $3\frac{3}{4} + 2\frac{5}{9} = \frac{135}{36} + \frac{92}{36} = \frac{227}{36}$.

Name ______________________ Date ____________ Class ____________

LESSON 5C

Technology Lab

Adding and Subtracting Mixed Numbers, continued

5. You can also use a calculator to reduce a fraction to a mixed number.

Change $\frac{227}{36}$ to a mixed number.

Divide 227 by 36. The quotient is 6.3055, meaning that 227 divided by 36 equals 6 plus a remainder. Clear the quotient from the calculator and multiply 36 times 6. When you see the product, 216, press STO> ALPHA B ENTER. Then press ALPHA A − ALPHA B ENTER. This subtracts 216 from 227, which are stored in the calculator's memory.
227 − 216 = 11

Write the final sum as $6\frac{11}{36}$.

Think and Discuss

1. How can you use mental math to check your answer?

____

____

____

____

____

2. Do you think this process would work with subtraction?

____

____

Name ______________________________ Date ___________ Class ___________

LESSON 5C

Technology Lab

Adding and Subtracting Mixed Numbers, continued

Try This

Use your calculator to evaluate the following expressions with mixed numbers.

1. $2\frac{7}{8} + 2\frac{1}{3}$

____

2. $5\frac{1}{5} + 2\frac{3}{8}$

____

3. $7\frac{13}{40} + 3\frac{2}{5}$

____

Name ______________________ Date ____________ Class ____________

LESSON 5C

Review for Mastery

Adding and Subtracting Mixed Numbers

You can use what you know about improper fractions to add and subtract mixed numbers.

To find the sum or difference of mixed numbers, first write the mixed numbers as improper fractions.

A. $3\frac{1}{4} + 2\frac{1}{3}$	**B.** $4\frac{1}{2} - 2\frac{2}{3}$
$= \frac{13}{4} + \frac{7}{3}$	$= \frac{9}{2} - \frac{8}{3}$

Next, find equivalent fractions with a least common denominator.

$\frac{13}{4} + \frac{7}{3}$	$\frac{9}{2} - \frac{8}{3}$
$= \frac{39}{12} + \frac{28}{12}$	$= \frac{27}{6} - \frac{16}{6}$

Then add or subtract the like fractions.

$\frac{39}{12} + \frac{28}{12}$	$\frac{27}{6} - \frac{16}{6}$
$= \frac{67}{12}$	$= \frac{11}{6}$

Write the answer as a mixed number in simplest form.

$\frac{67}{12}$	$\frac{11}{6}$
$= 5\frac{7}{12}$	$= 1\frac{5}{6}$
So, $3\frac{1}{4} + 2\frac{1}{3} = 5\frac{7}{12}$.	So, $4\frac{1}{2} - 2\frac{2}{3} = 1\frac{5}{6}$.

Find each sum or difference. Write your answer in simplest form.

1. $1\frac{1}{4} + 1\frac{1}{2}$

$= \frac{\quad}{4} + \frac{\quad}{2}$

$= \frac{\quad}{4} + \frac{\quad}{4}$

____

2. $3\frac{1}{6} + 1\frac{2}{3}$

$= \frac{\quad}{6} + \frac{\quad}{3}$

$= \frac{\quad}{6} + \frac{\quad}{6}$

____

3. $2\frac{1}{8} + 4\frac{1}{2}$

$= \frac{\quad}{8} + \frac{\quad}{2}$

$= \frac{\quad}{8} + \frac{\quad}{8}$

____

4. $4\frac{1}{3} + 1\frac{1}{2}$

$= \frac{\quad}{3} + \frac{\quad}{2}$

$= \frac{\quad}{6} + \frac{\quad}{6}$

____

5. $2\frac{3}{5} + 1\frac{1}{10}$

____

6. $3\frac{1}{6} + 1\frac{1}{12}$

____

7. $2\frac{5}{8} - 1\frac{1}{4}$

____

8. $5\frac{2}{3} - 2\frac{1}{4}$

____

Name ______________________ Date __________ Class __________

LESSON 5C

Homework and Practice

Adding and Subtracting Mixed Numbers

Find each sum or difference. Write each answer in simplest form.

1. $8\frac{6}{11} + 3\frac{1}{3}$

2. $5\frac{4}{5} - 2\frac{9}{13}$

3. $14\frac{1}{14} + 7\frac{6}{7}$

4. $7\frac{3}{4} + 3\frac{3}{8}$

5. $26\frac{1}{2} - 6\frac{10}{13}$

6. $9\frac{1}{4} + 6\frac{3}{14}$

7. $4\frac{10}{11} + 2\frac{1}{2}$

8. $18\frac{1}{2} + 4\frac{1}{11}$

9. $41\frac{4}{5} - 6\frac{6}{15}$

10. $14\frac{1}{5} + 24\frac{2}{3}$

11. $78\frac{3}{7} - 42\frac{4}{11}$

12. $16\frac{4}{5} + 6\frac{1}{7}$

13. $6\frac{3}{8} - 4\frac{2}{3}$

14. $5\frac{3}{10} - 4\frac{1}{5}$

15. $7\frac{3}{4} - 4\frac{1}{3}$

16. $24\frac{2}{3} + 5\frac{2}{9}$

17. $52\frac{5}{8} - 31\frac{7}{12}$

18. $13\frac{1}{12} + 6\frac{5}{6}$

19. John has to walk $3\frac{1}{4}$ blocks to school and Joan has to walk $2\frac{4}{5}$ blocks to school. How far do John and Joan walk combined?

20. If the band plays for 15 minutes, and the first two songs last $7\frac{5}{24}$ minutes total, how much more time do they have to play after the first two songs?

21. A container is holding $26\frac{1}{3}$ gallons of gasoline. The next day, the container only has $19\frac{1}{9}$ gallons. How much of the gasoline was lost?

EXPLORATION

5D Regrouping to Subtract Mixed Numbers

Program of Studies and Core Content

MA-6-NPO-S-NS2, MA-6-NPO-S-NO1, MA-6-NPO-S-NO2, MA-6-AT-S-VE02, MA-06-1.1.2, MA-06-1.3.1, MA-06-5.2.1

A baker starts the day with $2\frac{1}{4}$ lemon cakes and sells $1\frac{1}{2}$ lemon cakes during the day. How much cake is left over?

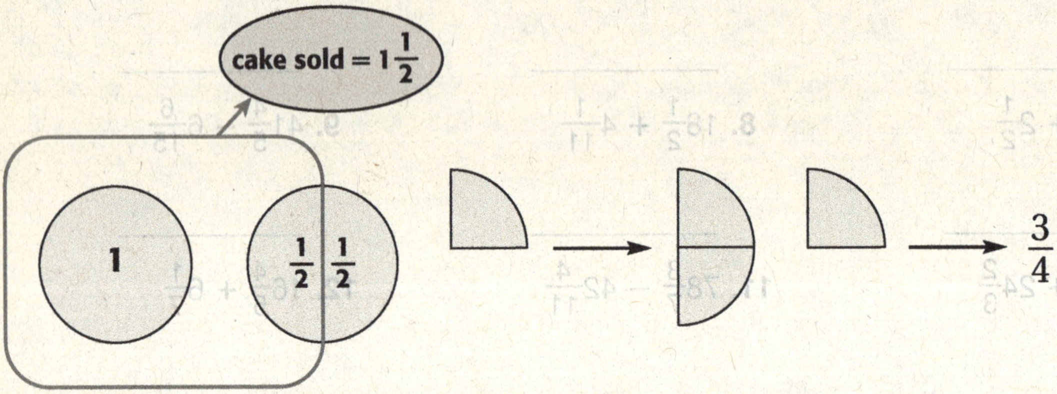

initial amount of cake = $2\frac{1}{4}$ cake left = $\frac{3}{4}$

There is $\frac{3}{4}$ of a cake left over.

Use a model to solve each subtraction problem.

1. $2 - 1\frac{1}{4}$

2. $4 - 1\frac{1}{2}$

3. $5\frac{1}{4} - 3\frac{1}{2}$

4. $1\frac{1}{8} - \frac{3}{4}$

Think and Discuss

5. Discuss your method for subtracting mixed numbers.

6. Explain why the method used to solve the problem about the lemon cakes is called regrouping.

Name ______________________ Date ____________ Class ____________

LESSON 5D

Review for Mastery

Regrouping to Subtract Mixed Numbers

You can use fraction strips to regroup to subtract mixed numbers.

To find $3\frac{1}{4} - 1\frac{3}{4}$, first model the first mixed number in the expression.

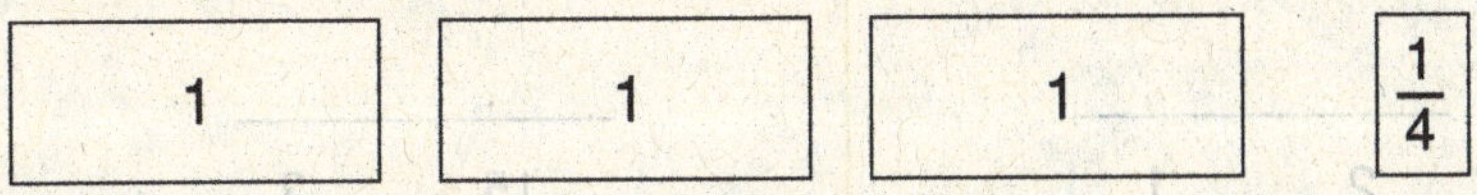

There are not enough $\frac{1}{4}$ pieces to subtract, so you have to regroup.

Trade one one-whole strip for four $\frac{1}{4}$ pieces, because $\frac{4}{4} = 1$.

Now there are enough $\frac{1}{4}$ pieces to subtract. Take away $1\frac{3}{4}$.

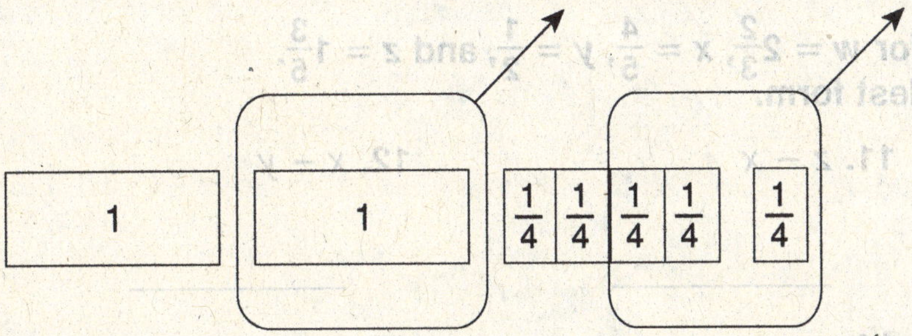

The remaining pieces represent the difference. Write the difference in simplest form.

$3\frac{1}{4} - 1\frac{3}{4} = 1\frac{2}{4} = 1\frac{1}{2}$

Use fraction strips to find each difference. Write your answer in simplest form.

1. $3\frac{1}{4} - 2\frac{3}{4}$ ________ **2.** $3\frac{1}{6} - 1\frac{5}{6}$ ________ **3.** $4\frac{3}{8} - 1\frac{7}{8}$ ________ **4.** $3\frac{1}{3} - 2\frac{2}{3}$ ________

5. $5\frac{5}{12} - 2\frac{7}{12}$ ________ **6.** $3\frac{3}{10} - 1\frac{9}{10}$ ________ **7.** $5\frac{1}{8} - 1\frac{5}{8}$ ________ **8.** $4 - 1\frac{1}{3}$ ________

9. $3\frac{1}{8} - 1\frac{3}{8}$ ________ **10.** $2\frac{1}{8} - 1\frac{7}{8}$ ________ **11.** $3 - 1\frac{1}{4}$ ________ **12.** $6\frac{3}{8} - 2\frac{5}{8}$ ________

Name ______________________ Date __________ Class __________

LESSON 5D

Homework and Practice

Renaming to Subtract Mixed Numbers

Subtract. Write the answer in simplest form.

1. $7 - 5\frac{3}{7}$ ________

2. $3\frac{1}{5} - 1\frac{2}{9}$ ________

3. $22 - 14\frac{3}{16}$ ________

4. $9\frac{1}{5} - 6\frac{7}{9}$ ________

5. $28\frac{2}{5} - 18\frac{1}{4}$ ________

6. $33\frac{15}{16} - 17\frac{3}{8}$ ________

7. $42\frac{4}{5} - 38\frac{6}{13}$ ________

8. $5\frac{2}{5} - \frac{2}{15}$ ________

9. $11\frac{9}{14} - 8\frac{1}{7}$ ________

Evaluate each expression for $w = 2\frac{2}{3}$, $x = \frac{4}{5}$, $y = \frac{1}{2}$, and $z = 1\frac{3}{5}$. Write each answer in simplest form.

10. $w - z$ ________

11. $z - x$ ________

12. $x - y$ ________

13. $w - x$ ________

14. $z - y$ ________

15. $w - y$ ________

16. Jesse has 4 pounds of chocolate. He used $1\frac{8}{11}$ pounds to make a cake. How much chocolate does he have left after he makes the cake?

____

17. Harry grew $3\frac{4}{5}$ inches last year, and $2\frac{1}{2}$ inches this year. How much more has Harry grown last year compared to this year?

____

18. Juan is $6\frac{2}{7}$ feet tall. Glen is $6\frac{4}{9}$ feet tall. Who is taller and by how much?

____

EXPLORATION

Program of Studies and Core Content

MA-6-NPO-S-NO3, MA-6-AT-S-VEO1, MA-6-AT-S-VEO3, MA-6-AT-S-EI1, MA-6-AT-S-EI3, MA-06-1.3.2, MA-06-5.2.2, MA-06-5.3.1

5E Solving Fraction Equations: Addition and Subtraction

You can use mental math to solve addition and subtraction equations that contain fractions. Look at the reasoning used to solve $\frac{7}{9} - x = \frac{2}{9}$.

Since the denominators of the fractions are the same, you can rewrite $\frac{7}{9} - x = \frac{2}{9}$ as a simpler equation: $7 - \blacksquare = 2$.

$7 - \blacksquare = 2$	*Think:* 7 minus what number equals 2?
$7 - 5 = 2$	Use mental math.
$\frac{7}{9} - \frac{5}{9} = \frac{2}{9}$	Write the equation, using the denominator.
$x = \frac{5}{9}$	Write the value of x.

Using the example above as a guide, complete the table below.

	Equation	Simpler Equation	Value of x
1.	$\frac{1}{2} + x = \frac{7}{2}$	$1 + \blacksquare = 7$	
2.	$x - \frac{2}{5} = \frac{1}{5}$	$\blacksquare - 2 = 1$	
3.	$x + \frac{1}{3} = \frac{2}{3}$	$\blacksquare + 1 = 2$	
4.	$x - \frac{1}{3} = \frac{2}{3}$	$\blacksquare - 1 = 2$	

Think and Discuss

5. Explain how you could use mental math to solve the equation $\frac{1}{3} + x = \frac{5}{6}$.

Name ______________________ Date __________ Class __________

LESSON **5E**

Review for Mastery

Solving Fraction Equations: Addition and Subtraction

You can write related facts using addition and subtraction.

$3 + 4 = 7$ $\qquad$ $7 - 4 = 3$

You can use related facts to solve equations.

A. $x + 2\frac{1}{2} = 4$

Think: $4 - 2\frac{1}{2} = x$

$x = 4 - 2\frac{1}{2}$

$x = 3\frac{2}{2} - 2\frac{1}{2}$ $\quad$ Regroup 4 as $3\frac{2}{2}$.

$x = 1\frac{1}{2}$

B. $x - 4\frac{1}{3} = 3\frac{1}{2}$

Think: $3\frac{1}{2} + 4\frac{1}{3} = x$

$x = 3\frac{1}{2} + 4\frac{1}{3}$

$x = \frac{7}{2} + \frac{13}{3}$ $\quad$ Write the mixed numbers as improper fractions.

$x = \frac{21}{6} + \frac{26}{6}$ $\quad$ Write the fractions using a common denominator.

$x = \frac{47}{6}$

$x = 7\frac{5}{6}$ $\quad$ Write the sum as a mixed number.

Use related facts to solve each equation.

1. $x + 3\frac{1}{3} = 7$

$x = 7 - 3\frac{1}{3}$

$x = 6\frac{3}{3} - 3\frac{1}{3}$

$x =$ ______

2. $x - 2\frac{1}{4} = 4\frac{1}{2}$

$x = 4\frac{1}{2} + 2\frac{1}{4}$

$x = \frac{9}{2} + \frac{9}{4}$

$x = \frac{18}{4} + \frac{9}{4}$

$x =$ ______

3. $x + \frac{3}{8} = 5\frac{1}{4}$

$x = 5\frac{1}{4} - \frac{3}{8}$

$x = \frac{21}{4} - \frac{3}{8}$

$x = \frac{42}{8} - \frac{3}{8}$

$x =$ ______

4. $x - \frac{5}{12} = 2\frac{1}{2}$

$x = 2\frac{1}{2} + \frac{5}{12}$

$x = \frac{5}{2} + \frac{5}{12}$

$x = \frac{30}{12} + \frac{5}{12}$

$x =$ ______

5. $x - 1\frac{3}{4} = 7\frac{1}{2}$

______

6. $x - 3\frac{2}{3} = 1\frac{1}{3}$

______

7. $x + 3\frac{1}{2} = 6\frac{1}{4}$

______

8. $x - 2\frac{2}{5} = 1\frac{3}{10}$

______

Name ______________________ Date __________ Class __________

LESSON 5E

Homework and Practice

Solving Fraction Equations: Addition and Subtraction

Solve each equation. Write the answer in simplest form. Check your answers.

1. $g + 1\frac{3}{4} = 5\frac{5}{12} - 3\frac{1}{2}$

2. $2\frac{3}{10} = k - 1\frac{2}{5} + \frac{1}{5}$

3. $b + 8 = 12\frac{1}{4} - 1\frac{1}{2}$

4. $13\frac{2}{5} = h + 6\frac{4}{15}$

5. $9\frac{1}{13} = z + \frac{3}{26}$

6. $x + 1 = 2\frac{3}{7} - 1\frac{1}{7}$

7. $\frac{3}{10} + \frac{1}{5} = d - 2\frac{1}{2}$

8. $7\frac{2}{3} - 7\frac{1}{2} = q + \frac{1}{12}$

9. $6\frac{3}{8} = t - 4\frac{3}{4} - 2\frac{1}{16}$

10. $6\frac{1}{4} + n = 11\frac{1}{3} - 2\frac{1}{5}$

11. $e + 5\frac{4}{7} = 12$

12. $m = 7\frac{3}{5} - 2\frac{3}{10} - 3\frac{3}{20}$

13. In a grapefruit eating contest, Sal ate $5\frac{1}{3}$ grapefruits in 2 minutes. If the world record is $7\frac{1}{6}$ grapefruits in 3 minutes, how many grapefruits would Sal have had to eat in the next minute to tie the record?

14. Explain how you can use the operation addition to help solve the equation $8\frac{1}{2} = z - 13\frac{3}{4}$.

EXPLORATION

Program of Studies and Core Content

MA-6-NPO-S-NS2, MA-6-NPO-S-NO1, MA-6-NPO-S-NO2, MA-6-AT-S-VE02, MA-06-1.1.2, MA-06-1.3.1, MA-06-5.2.1

5F Multiplying Fractions by Whole Numbers

Rosario requires $\frac{3}{4}$ of a 1-pound bag of clay to make one bowl. How many 1-pound bags of clay will she need to make a set of 6 bowls?

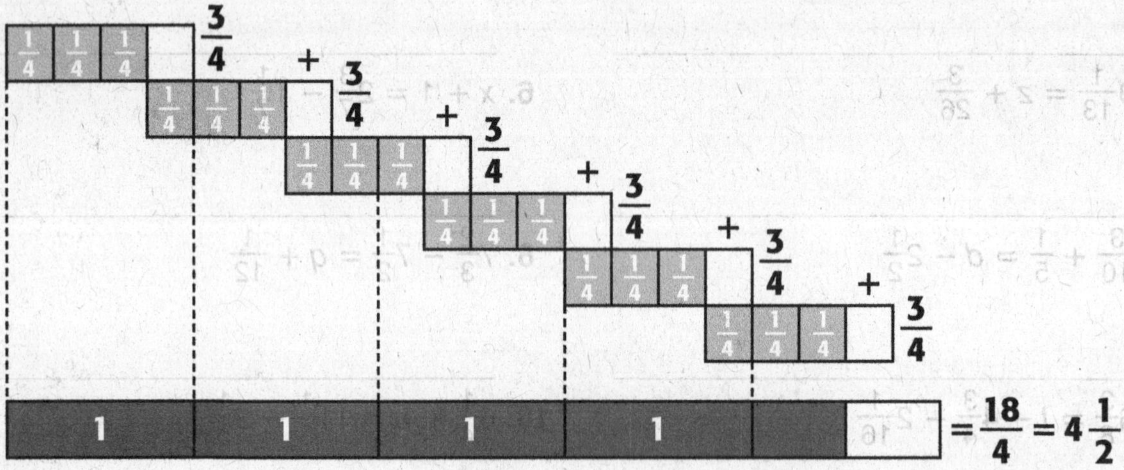

She will need $4\frac{1}{2}$ 1-pound bags of clay.

Draw a model to find each product.

1. $3 \cdot \frac{3}{4}$

2. $4 \cdot \frac{1}{2}$

3. $5 \cdot \frac{2}{3}$

4. $7 \cdot \frac{1}{4}$

Think and Discuss

5. Explain how you know that $3 \cdot \frac{3}{4}$ is less than 3.

Name ______________________ Date ___________ Class ___________

LESSON 5F

Technology Lab

Multiplying Fractions by Whole Numbers

Remember

To multiply fractions by whole numbers:

Write the whole number as a fraction with 1 as the denominator.

Multiply the numerators of the fractions.

Multiply the denominators of the fractions.

Activity

Multiply the following fractions by whole numbers using your calculator.

1. $17 \times \frac{15}{16}$

$= \frac{17}{1} \times \frac{15}{16}$ Rewrite the whole number as a fraction.

Step 1 Multiply the denominators. 1 [×] 16 [ENTER] 16

Step 2 Store the result. [STO>] [ALPHA] A [ENTER]

Step 3 Multiply the numerators. 17 [×] 15 [ENTER] 255

Step 4 Divide the numerator by the denominator by using the stored value. 255 [÷] [ALPHA] A [ENTER] 15.9375

This means that the product of the whole number and the fraction above is 15.9375, when expressed as a decimal. To express the product as a fraction, simply put the new numerator (255) over the denominator (16).

$\frac{17}{1} \times \frac{15}{16} = \frac{255}{16}$

Name ______________________ Date ____________ Class ____________

LESSON 5F

Technology Lab

Multiplying Fractions by Whole Numbers, continued

Think and Discuss

Why is a calculator more useful for multiplying fractions with large numerators and denominators than for multiplying fractions with small numerators and denominators?

____

____

____

____

Try This

Use your calculator to multiply the following whole numbers by $\frac{11}{16}$. Express the product first as a decimal, then as a fraction.

1. 8

____

2. 6

____

3. 244

____

Name ______________________________ Date ____________ Class ____________

LESSON 5F

Review for Mastery

Multiplying Fractions by Whole Numbers

You can use fraction strips to multiply fractions by whole numbers.

To find $3 \cdot \frac{2}{3}$, first think about the expression in words.

$3 \cdot \frac{2}{3}$ means "3 groups of $\frac{2}{3}$."

Then model the expression.

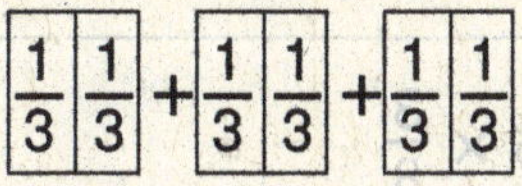

The total number of $\frac{1}{3}$ fraction pieces is 6.

So, $3 \cdot \frac{2}{3} = \frac{2}{3} + \frac{2}{3} + \frac{2}{3} =$

$\frac{6}{3} = 2$ in simplest form.

Use fraction strips to find each product.

1. $4 \cdot \frac{1}{8}$ ____________ **2.** $2 \cdot \frac{2}{5}$ ____________ **3.** $6 \cdot \frac{1}{8}$ ____________ **4.** $8 \cdot \frac{1}{4}$ ____________

You can also use counters to multiply fractions by whole numbers.

To find $\frac{1}{2} \cdot 12$, first think about the expression in words.

$\frac{1}{2} \cdot 12 = \frac{12}{2}$, which means "12 divided into 2 equal groups."

Then model the expression.

The number of counters in each group is the product.

$\frac{1}{2} \cdot 12 = 6.$

Use counters to find each product.

5. $\frac{1}{3} \cdot 15$ ____________ **6.** $\frac{1}{8} \cdot 24$ ____________ **7.** $\frac{1}{4} \cdot 16$ ____________ **8.** $\frac{1}{12} \cdot 24$ ____________

Name ______________________ Date ____________ Class ____________

LESSON **5F**

Homework and Practice

Multiplying Fractions by Whole Numbers

Multiply. Write your answers in simplest form.

1. $5 \times \frac{1}{25}$ ____________

2. $7 \times \frac{2}{21}$ ____________

3. $3 \times \frac{1}{9}$ ____________

4. $3 \times \frac{1}{5}$ ____________

5. $6 \times \frac{3}{8}$ ____________

6. $2 \times \frac{2}{3}$ ____________

7. $7 \times \frac{3}{10}$ ____________

8. $8 \times \frac{4}{7}$ ____________

9. $13 \times \frac{1}{2}$ ____________

10. $4 \times \frac{13}{16}$ ____________

11. $5 \times \frac{4}{9}$ ____________

12. $10 \times \frac{3}{11}$ ____________

13. $6 \times \frac{8}{9}$ ____________

14. $8 \times \frac{7}{8}$ ____________

15. $2 \times \frac{4}{25}$ ____________

Evaluate 4*x* for each value of *x*. Write your answers in simplest form.

16. $x = \frac{1}{4}$ ____________

17. $x = \frac{3}{8}$ ____________

18. $x = \frac{4}{9}$ ____________

19. $x = \frac{6}{7}$ ____________

20. $x = \frac{3}{11}$ ____________

21. $x = \frac{1}{2}$ ____________

22. $x = \frac{4}{5}$ ____________

23. $x = \frac{5}{16}$ ____________

24. You have 20 buckets of sand and you want to make a sandcastle. If the bottom part of the castle takes $\frac{4}{7}$ of the sand, how many buckets do you have left to make the top?

____

Program of Studies and Core Content

MA-6-NPO-S-NS2, MA-6-NPO-S-NO1, MA-6-NPO-S-NO2, MA-6-AT-S-VEO2, MA-06-1.1.2, MA-06-1.3.1, MA-06-5.2.1

5G Multiplying Fractions

You can use paper folding to find the product of two fractions. To find $\frac{3}{4}$ of $\frac{1}{2}$, fold the paper in half vertically to model $\frac{1}{2}$. Then fold it horizontally into four sections to create fourths.

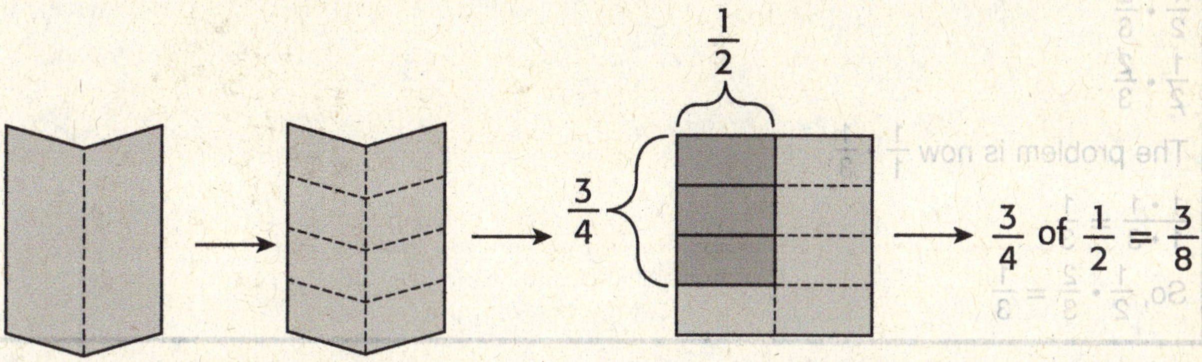

Use paper folding to find each product. Sketch a picture for each product.

1. $\frac{1}{2} \cdot \frac{1}{2}$

2. $\frac{1}{2} \cdot \frac{2}{3}$

3. $\frac{1}{3} \cdot \frac{1}{4}$

4. $\frac{2}{3} \cdot \frac{3}{4}$

Think and Discuss

5. Explain how to multiply two fractions.

Name ______________________ Date __________ Class __________

LESSON 5G

Review for Mastery

Multiplying Fractions

To multiply fractions, multiply the numerators and multiply the denominators.

When multiplying fractions, you can sometimes divide by the GCF to make the problem simpler.

You can divide by the GCF even if the numerator and denominator of the same fraction have a common factor.

$\frac{1}{2} \cdot \frac{2}{3}$

$\frac{1}{\cancel{2}_1} \cdot \frac{\cancel{2}^1}{3}$

The problem is now $\frac{1}{1} \cdot \frac{1}{3}$.

$\frac{1 \cdot 1}{1 \cdot 3} = \frac{1}{3}$

So, $\frac{1}{2} \cdot \frac{2}{3} = \frac{1}{3}$

Is it possible to simplify before you multiply? If so, what is the GCF?

1. $\frac{1}{4} \cdot \frac{1}{2}$ ________

2. $\frac{1}{6} \cdot \frac{3}{4}$ ________

3. $\frac{1}{8} \cdot \frac{2}{3}$ ________

4. $\frac{1}{3} \cdot \frac{2}{5}$ ________

Multiply.

5. $\frac{1}{6} \cdot \frac{3}{5}$ ________

6. $\frac{1}{4} \cdot \frac{1}{3}$ ________

7. $\frac{7}{8} \cdot \frac{4}{5}$ ________

8. $\frac{1}{6} \cdot \frac{2}{3}$ ________

9. $\frac{1}{5} \cdot \frac{1}{2}$ ________

10. $\frac{3}{5} \cdot \frac{1}{4}$ ________

11. $\frac{3}{7} \cdot \frac{1}{9}$ ________

12. $\frac{3}{4} \cdot \frac{1}{2}$ ________

13. $\frac{1}{3} \cdot \frac{6}{7}$ ________

14. $\frac{1}{4} \cdot \frac{2}{3}$ ________

15. $\frac{3}{4} \cdot \frac{1}{3}$ ________

16. $\frac{1}{4} \cdot \frac{1}{8}$ ________

Name ______________________ Date __________ Class __________

LESSON 5G

Homework and Practice

Multiplying Fractions

Multiply. Write each answer in simplest form.

1. $\frac{2}{3} \cdot \frac{3}{8}$ ________

2. $\frac{1}{5} \cdot \frac{1}{16}$ ________

3. $\frac{3}{7} \cdot \frac{1}{4}$ ________

4. $\frac{1}{6} \cdot \frac{2}{9}$ ________

5. $\frac{11}{13} \cdot \frac{1}{8}$ ________

6. $\frac{1}{5} \cdot \frac{5}{6}$ ________

7. $\frac{1}{2} \cdot \frac{4}{9}$ ________

8. $\frac{3}{7} \cdot \frac{3}{4}$ ________

9. $\frac{2}{9} \cdot \frac{3}{11}$ ________

Evaluate the expression $x \cdot \frac{2}{3}$ for each value of x. Write each answer in simplest form.

10. $x = \frac{1}{2}$ ________

11. $x = \frac{9}{11}$ ________

12. $x = \frac{3}{5}$ ________

13. $x = \frac{3}{4}$ ________

14. $x = \frac{8}{13}$ ________

15. $x = \frac{1}{12}$ ________

16. Latoya has $\frac{3}{4}$ of a bag of popcorn. If she eats half of the popcorn at a movie, how much does she have left?

____

17. If Rupert flew $\frac{3}{16}$ of the way around the world, but George only flew $\frac{4}{9}$ as far Rupert, how far around the world did George fly? Simplify your answer.

____

EXPLORATION

Program of Studies and Core Content

MA-6-NPO-S-NS2, MA-6-NPO-S-NO1, MA-6-NPO-S-NO2, MA-06-1.3.1

5H Multiplying Mixed Numbers

You can use paper folding to find products of mixed numbers. To find $\frac{1}{2} \cdot 1\frac{1}{2}$, first fold two sheets of paper in half vertically to represent $1\frac{1}{2}$. To represent $\frac{1}{2}$ of $1\frac{1}{2}$, fold both sheets in half again horizontally.

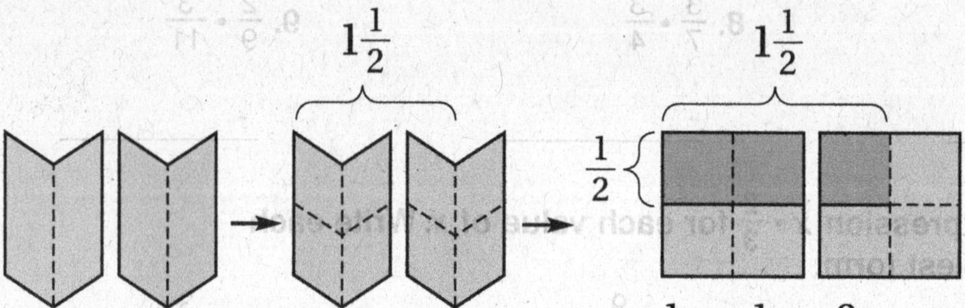

$$\frac{1}{2} \cdot 1\frac{1}{2} = \frac{3}{4}$$

Use paper folding to find each product. Sketch a picture for each product.

1. $\frac{2}{3} \cdot 1\frac{1}{2}$

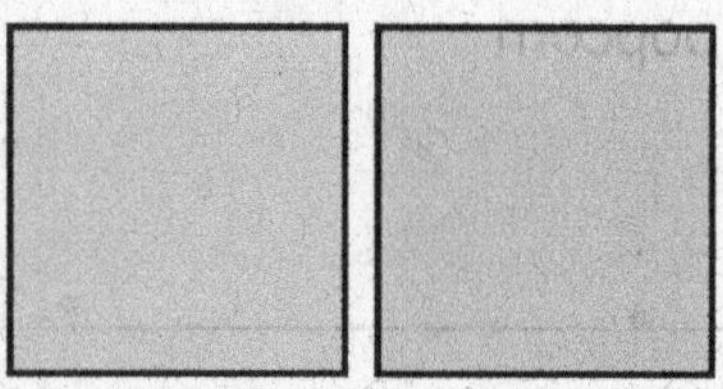

2. $\frac{1}{3} \cdot 1\frac{3}{4}$

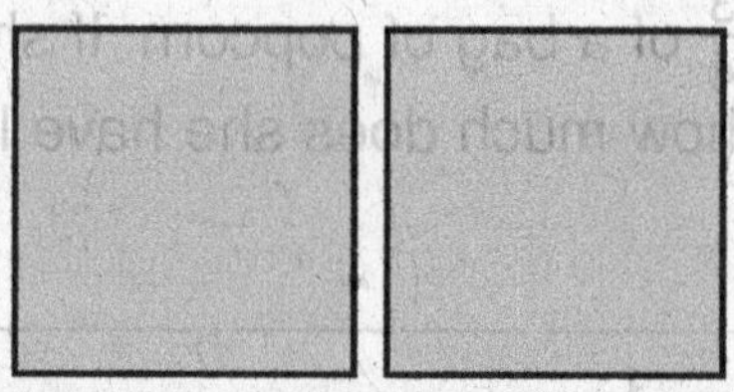

Think and Discuss

3. Explain how to multiply a fraction and a mixed number.

4. Explain why the product of a proper fraction and a mixed number is less than the mixed number.

Name ______________________ Date __________ Class __________

LESSON 5H

Review for Mastery

Multiplying Mixed Numbers

To find $\frac{1}{3}$ of $2\frac{1}{2}$, first change $2\frac{1}{2}$ to an improper fraction.

$2\frac{1}{2} = \frac{5}{2}$

Then multiply as you would with two proper fractions.

Check to see if you can divide by the GCF to make the problem simpler. Then multiply the numerators and multiply the denominators.

The problem is now $\frac{1}{3} \cdot \frac{5}{2}$.

$\frac{1 \cdot 5}{3 \cdot 2} = \frac{5}{6}$

So, $\frac{1}{3} \cdot 2\frac{1}{2}$ is $\frac{5}{6}$.

Rewrite each mixed number as an improper fraction. Is it possible to simplify before you multiply? If so, what is the GCF?

1. $\frac{1}{4} \cdot 1\frac{1}{3}$

   $= \frac{1}{4} \cdot$ ____

   ____

2. $\frac{1}{6} \cdot 2\frac{1}{2}$

   $= \frac{1}{6} \cdot$ ____

   ____

3. $\frac{1}{8} \cdot 1\frac{1}{2}$

   $= \frac{1}{8} \cdot$ ____

   ____

4. $\frac{1}{3} \cdot 1\frac{2}{5}$

   $= \frac{1}{3} \cdot$ ____

   ____

5. $1\frac{1}{3} \cdot 1\frac{2}{3}$

   $\frac{\_}{3} \cdot \frac{\_}{3}$

   ____

6. $1\frac{1}{2} \cdot 1\frac{1}{3}$

   $\frac{\_}{2} \cdot \frac{\_}{3}$

   ____

7. $1\frac{3}{4} \cdot 2\frac{1}{2}$

   $\frac{\_}{4} \cdot \frac{\_}{2}$

   ____

8. $1\frac{1}{6} \cdot 2\frac{2}{3}$

   $\frac{\_}{6} \cdot \frac{\_}{3}$

   ____

9. $3\frac{1}{3} \cdot \frac{2}{5}$

   ____

10. $2\frac{1}{2} \cdot \frac{1}{5}$

    ____

11. $1\frac{3}{4} \cdot 2\frac{1}{2}$

    ____

12. $3\frac{1}{3} \cdot 1\frac{1}{5}$

    ____

Name ______________________ Date __________ Class __________

LESSON **5H**

Homework and Practice

Multiplying Mixed Numbers

Multiply. Write each answer in simplest form.

1. $2\frac{1}{2} \cdot \frac{1}{5}$ ________
2. $1\frac{1}{4} \cdot \frac{4}{9}$ ________
3. $1\frac{4}{7} \cdot \frac{6}{11}$ ________
4. $3\frac{3}{5} \cdot \frac{1}{9}$ ________
5. $1\frac{5}{16} \cdot \frac{4}{7}$ ________
6. $5\frac{2}{3} \cdot \frac{1}{2}$ ________
7. $3\frac{1}{3} \cdot 1\frac{3}{11}$ ________
8. $\frac{1}{8} \cdot 2\frac{4}{5}$ ________
9. $1\frac{10}{11} \cdot \frac{3}{7}$ ________

Find each product. Write each answer in simplest form.

10. $\frac{1}{5} \cdot 2\frac{2}{3}$ ________
11. $\frac{1}{2} \cdot 1\frac{2}{7}$ ________
12. $3\frac{4}{5} \cdot \frac{1}{7}$ ________
13. $1\frac{11}{14} \cdot \frac{1}{5}$ ________
14. $2\frac{3}{4} \cdot 3\frac{1}{2}$ ________
15. $2\frac{6}{11} \cdot 1\frac{1}{3}$ ________

16. Jamie needs $2\frac{4}{5}$ yards of ground to make his flower garden. If you only need $\frac{5}{8}$ as much land to make your garden, how much land do you need?

____

17. Anthony has $2\frac{1}{4}$ pies. If his friends ate $\frac{1}{3}$ of his pies, how many pies does he have left?

____

EXPLORATION

51 Dividing Fractions and Mixed Numbers

Program of Studies and Core Content
MA-6-NPO-S-NS2, MA-6-NPO-S-NO1, MA-6-NPO-S-NO2, MA-06-1.1.2, MA-06-1.3.1

The model shows the quotient $2\frac{1}{2} \div \frac{1}{2}$.

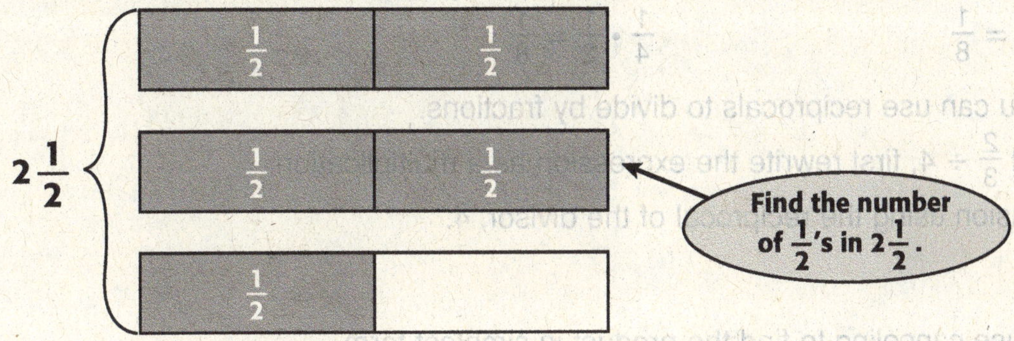

There are 5 halves in $2\frac{1}{2}$, so $2\frac{1}{2} \div \frac{1}{2} = 5$.

Draw a model to solve each division problem.

1. $1\frac{1}{2} \div \frac{3}{4}$

2. $2 \div \frac{2}{3}$

3. $4 \div \frac{2}{3}$

4. $2\frac{1}{4} \div \frac{3}{4}$

Think and Discuss

5. Describe how to model fraction division by using fraction bars.

6. Explain why $3 \div \frac{3}{4} = 4$.

Name ______________________ Date __________ Class __________

LESSON 5I

Review for Mastery

Dividing Fractions and Mixed Numbers

Two numbers are reciprocals if their product is 1. $\frac{2}{3}$ and $\frac{3}{2}$ are reciprocals because $\frac{2}{3} \cdot \frac{3}{2} = \frac{6}{6} = 1$.

Dividing by a fraction is the same as multiplying by its reciprocal.

$\frac{1}{4} \div 2 = \frac{1}{8}$ $\qquad\qquad$ $\frac{1}{4} \cdot \frac{1}{2} = \frac{1}{8}$

So, you can use reciprocals to divide by fractions.

To find $\frac{2}{3} \div 4$, first rewrite the expression as a multiplication expression using the reciprocal of the divisor, 4.

$\frac{2}{3} \cdot \frac{1}{4}$

Then use canceling to find the product in simplest form.

$\frac{2}{3} \div 4 = \frac{2}{3} \cdot \frac{1}{4} = \frac{1}{3} \cdot \frac{1}{2} = \frac{1}{6}$

To find $3\frac{1}{4} \div 1\frac{1}{2}$, first rewrite the expression using improper fractions.

$\frac{13}{4} \div \frac{3}{2}$

Next, write the expression as a multiplication expression.

$\frac{13}{4} \cdot \frac{2}{3}$

$3\frac{1}{4} \div 1\frac{1}{2} = \frac{13}{4} \div \frac{3}{2} = \frac{13}{4} \cdot \frac{2}{3} = \frac{13}{2} \cdot \frac{1}{3} = \frac{13}{6} = 2\frac{1}{6}$

Divide. Write each answer in simplest form.

1. $\frac{1}{4} \div 3$

$\frac{1}{4} \div \frac{\_\_}{1}$

$\_\_ \cdot \_\_$

____

2. $1\frac{1}{2} \div 1\frac{1}{4}$

$\frac{3}{2} \div \frac{\_\_}{4}$

$\_\_ \cdot \_\_$

____

3. $\frac{3}{8} \div 2$

$\frac{3}{8} \div \frac{\_\_}{1}$

$\_\_ \cdot \_\_$

____

4. $2\frac{1}{3} \div 1\frac{3}{4}$

$\frac{\_\_}{3} \div \frac{\_\_}{4}$

$\_\_ \cdot \_\_$

____

5. $\frac{1}{5} \div 2$

____

6. $1\frac{1}{6} \div 2\frac{2}{3}$

____

7. $\frac{1}{8} \div 4$

____

8. $3\frac{1}{8} \div \frac{1}{2}$

____

Name ______________________ Date ____________ Class ____________

LESSON 5I

Homework and Practice

Dividing Fractions and Mixed Numbers

Find the reciprocal.

1. $\frac{10}{11}$ ________

2. $\frac{1}{2}$ ________

3. $\frac{1}{9}$ ________

4. $\frac{2}{5}$ ________

5. $\frac{6}{15}$ ________

6. $\frac{2}{3}$ ________

7. $2\frac{3}{7}$ ________

8. $3\frac{4}{11}$ ________

9. $1\frac{1}{5}$ ________

Divide. Write each answer in simplest form.

10. $\frac{1}{2} \div 2$ ________

11. $2\frac{1}{3} \div 2\frac{4}{5}$ ________

12. $\frac{9}{13} \div \frac{10}{13}$ ________

13. $4\frac{1}{14} \div 2\frac{6}{7}$ ________

14. $\frac{3}{4} \div 4$ ________

15. $\frac{10}{13} \div 9$ ________

16. $\frac{8}{9} \div \frac{2}{15}$ ________

17. $\frac{10}{11} \div 1\frac{1}{2}$ ________

18. $1\frac{1}{4} \div 2\frac{3}{8}$ ________

19. Tina has $7\frac{1}{2}$ cups of bird food. If her bird feeder holds $\frac{1}{2}$ of a cup of bird food, how many times can Tina fill the bird feeder?

____

20. Mrs. Summer's third grade class is putting on a play. If she has $13\frac{1}{8}$ yards of fabric, and each student needs $\frac{7}{8}$ of a yard for a costume, how many students can make costumes?

____

EXPLORATION

Program of Studies and Core Content

MA-6-NPO-S-NO3, MA-6-AT-S-VEO1, MA-6-AT-S-VEO3, MA-6-AT-S-EI1, MA-6-AT-S-EI3, MA-06-1.3.2, MA-06-5.2.2, MA-06-5.3.1

5J Solving Fraction Equations: Multiplication and Division

You can use a number line to solve fraction equations. Look at the reasoning used to solve the equation $\frac{1}{2}n = 50$.

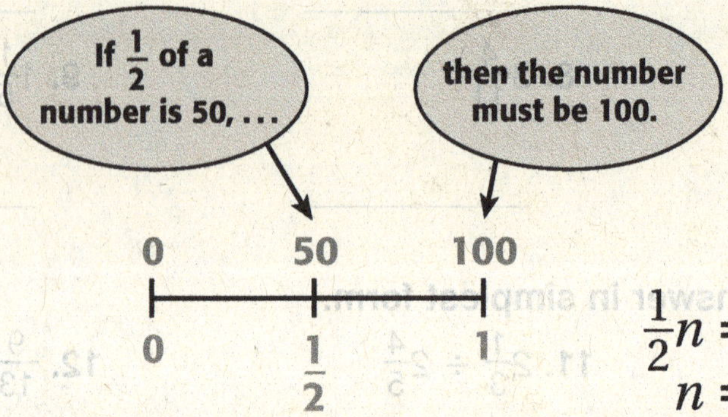

$$\frac{1}{2}n = 50$$
$$n = 100$$

Complete the number line to solve each equation.

1. $\frac{1}{2}n = 125$

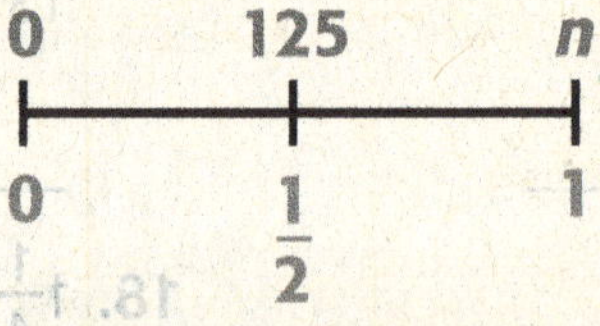

2. $\frac{1}{3}n = 12$

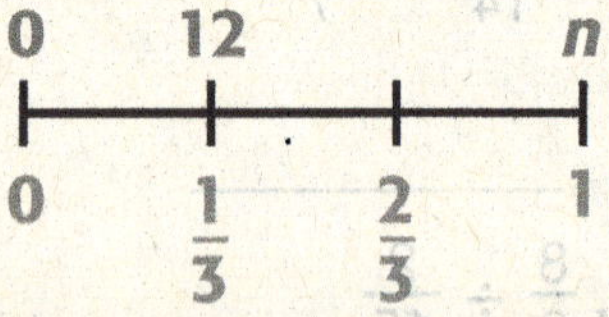

3. $\frac{3}{4}n = 60$

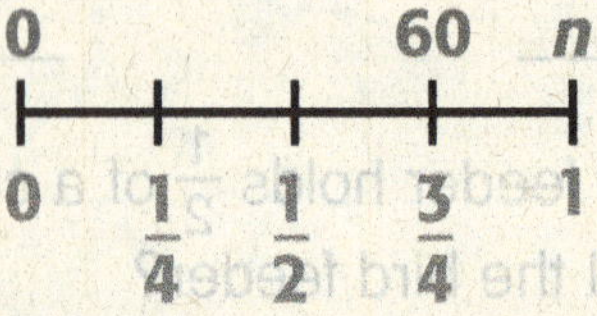

4. $\frac{2}{3}n = 20$

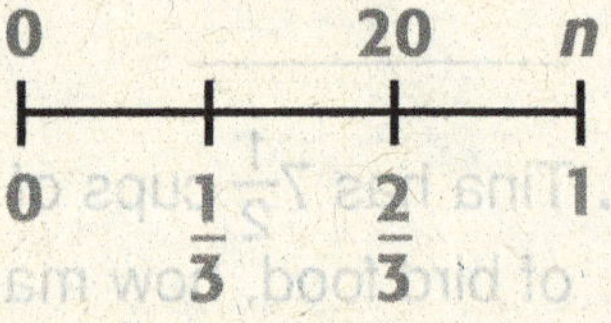

Think and Discuss

5. Describe a real-world situation that could be represented by the equation in number 3.

6. Discuss another way of solving equations that contain fractions and involve multiplication.

Name ____________________ Date __________ Class __________

LESSON 5J

Review for Mastery

Solving Fraction Equations: Multiplication and Division

You can write related facts using multiplication and division.

$3 \cdot 4 = 12 \qquad 4 = 12 \div 3$

You can use related facts to solve equations.

A. $\frac{2}{3} \cdot x = 12$

Think: $12 \div \frac{2}{3} = x$

$x = 12 \cdot \frac{3}{2}$

$x = \frac{12}{1} \cdot \frac{3}{2}$

$x = \frac{36}{2}$

$x = 18$

Check: $\frac{2}{3} \cdot x = 12$

$\frac{2}{3} \cdot 18 \stackrel{?}{=} 12$ Substitute

$\frac{2}{3} \cdot \frac{18}{1} \stackrel{?}{=} 12$

$\frac{36}{3} \stackrel{?}{=} 12$

$12 = 12$ ✔

B. $\frac{2x}{5} = 3$

$\frac{2}{5} \cdot x = 3$

Think: $3 \div \frac{2}{5} = x$

$x = 3 \cdot \frac{5}{2}$

$x = \frac{3}{1} \cdot \frac{5}{2}$

$x = \frac{15}{2}$

$x = 7\frac{1}{2}$

Check: $\frac{2x}{5} = 3$

$\frac{2}{5} \cdot x \stackrel{?}{=} 3$

$\frac{2}{5} \cdot \frac{15}{2} \stackrel{?}{=} 3$ Substitute

$\frac{30}{10} \stackrel{?}{=} 3$

$3 = 3$ ✔

Use related facts to solve each equation. Then check each answer.

1. $\frac{1}{4} \cdot x = 3$ ____________

2. $\frac{3x}{4} = 2$ ____________

3. $\frac{3}{5} \cdot x = \frac{2}{3}$ ____________

4. $\frac{1}{3} \cdot x = 6$ ____________

5. $\frac{2x}{5} = 1$ ____________

6. $\frac{1}{8} \cdot x = 3$ ____________

7. Explain how you can use the operation division to help solve the equation $4x = \frac{2}{3}$.

____

Name ____________________ Date __________ Class __________

LESSON 5J

Homework and Practice

Solving Fraction Equations: Multiplication and Division

Solve each equation. Check your answers.

1. $\frac{1}{6}t = \frac{1}{2}$
2. $\frac{1}{4}w = \frac{1}{2}$
3. $\frac{2}{3}v = 14$
4. $\frac{3}{5}j = \frac{1}{4}$
5. $\frac{3f}{7} = 6$
6. $\frac{1}{6}i = 3$
7. $3x = \frac{9}{14}$
8. $6g = \frac{3}{4}$
9. $\frac{m}{4} = 2$
10. $\frac{2}{3}z = \frac{3}{8}$
11. $2t = \frac{11}{12}$
12. $\frac{1}{13}e = 4$

13. If each child eats $\frac{1}{4}$ of a bag of peanuts and $2\frac{1}{2}$ bags are eaten, how many children are there?

14. Four runners ran an average of $2\frac{1}{3}$ miles. What was the total distance run?

15. A rope braiding machine can make $\frac{10}{13}$ of a foot of rope per second. How many seconds will it take to make 20 feet of rope?

CHAPTER 5

Chapter Summary

Answer these question to summarize the important concepts from Chapter 5 in your own words.

1. How do you find the LCM of a set of numbers?

2. Explain how to add or subtract fractions with unlike denominators using the least common denominator (LCD).

3. Explain how to subtract $4\frac{5}{7}$ from 6.

4. Explain how to divide $4\frac{3}{4}$ by $2\frac{1}{2}$.

For more review of Chapter 5:

- Complete the Chapter 5 Study Guide and Review on pages 282–284 of your textbook.
- Complete the Ready to Go On quizzes on pages 252 and 278 of your textbook.

Name ______________________ Date ________ Class ________

CHAPTER 5

KCCT Prep

Multiple Choice

1. A case contains 48 fruit drinks and $\frac{3}{8}$ of the drinks are lime flavored. How many drinks are NOT lime-flavored?

Ⓐ 24 Ⓒ 32
Ⓑ 30 Ⓓ 36

2. On both Saturday and Sunday, Keiko rode her bike $3\frac{2}{3}$ miles to the library and then rode another $2\frac{2}{5}$ miles to her grandmother's house. How far did Keiko ride her bike altogether on the two days?

Ⓐ $6\frac{1}{15}$ miles Ⓒ $12\frac{2}{15}$ miles
Ⓑ $8\frac{1}{15}$ miles Ⓓ $13\frac{2}{5}$ miles

3. The area of a rectangle is the product of its length and its width. A garden has 3 rectangular plots like the one shown below. What is the total area of the plots?

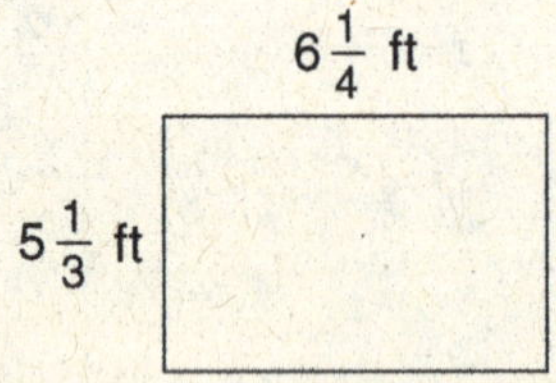

Ⓐ $33\frac{1}{3}$ ft^2 Ⓒ $90\frac{2}{3}$ ft^2
Ⓑ 66 ft^2 Ⓓ 100 ft^2

4. Which number is farthest from 2.3 on a number line?

Ⓐ $2\frac{7}{8}$ Ⓒ $2\frac{1}{3}$
Ⓑ $2\frac{5}{6}$ Ⓓ $2\frac{1}{2}$

5. Which equation has the solution $b = 12$?

Ⓐ $b - 4 = 16$ Ⓒ $24 = \frac{b}{2}$
Ⓑ $26 = b + 14$ Ⓓ $2b = 6$

6. It takes $1\frac{3}{4}$ cups of sauce to make one pizza. Which is the best estimate of the number of cups of sauce that are needed to make $5\frac{1}{2}$ pizzas?

Ⓐ 6 Ⓒ 10
Ⓑ 7 Ⓓ 12

7. Which product is modeled by the figure?

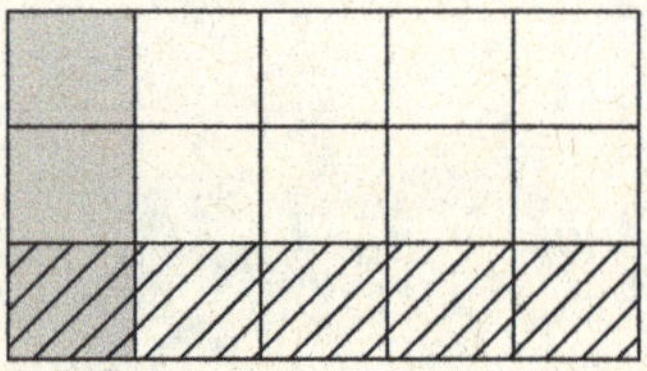

Ⓐ $\frac{1}{3} \times \frac{1}{5} = \frac{1}{15}$ Ⓒ $\frac{1}{3} \times \frac{1}{4} = \frac{1}{12}$
Ⓑ $\frac{1}{2} \times \frac{1}{4} = \frac{1}{8}$ Ⓓ $\frac{1}{3} \times \frac{1}{3} = \frac{1}{9}$

8. During a road trip, Eric drives an average of 51.2 miles every hour. Which is the best estimate of the number of miles Eric drives in 3.8 hours?

Ⓐ 55 miles Ⓒ 150 miles
Ⓑ 100 miles Ⓓ 200 miles

9. The mixed number $1\frac{n}{6}$ is greater than 1.8 but less than 2. What is the value of the numerator n?

Ⓐ 3 Ⓒ 5
Ⓑ 4 Ⓓ 6

10. Carla has $2\frac{1}{2}$ pounds of pecans. She divides the pecans into 3 equal portions. What is the total weight of 2 of the portions?

(A) $\frac{2}{3}$ lb
(C) $1\frac{1}{6}$ lb
(B) $\frac{5}{6}$ lb
(D) $1\frac{2}{3}$ lb

11. Donnell is hiking $6\frac{3}{4}$ miles to a campsite. He hikes $2\frac{2}{3}$ miles, stops for lunch, and then hikes another $1\frac{2}{3}$ miles. How much farther must he hike?

(A) $2\frac{5}{12}$ miles
(C) $3\frac{1}{12}$ miles
(B) $2\frac{11}{12}$ miles
(D) $11\frac{1}{12}$ miles

12. The expression $(5 - 3)^2 + 4 \square 2$ is missing an operation. Which operation gives the expression the greatest value?

(A) +
(C) ×
(B) −
(D) ÷

13. A giraffe is $5\frac{1}{2}$ meters tall. Its neck makes up $\frac{2}{5}$ of its height. What is the length of the giraffe's neck?

(A) $2\frac{1}{5}$ meters
(C) $4\frac{2}{5}$ meters
(B) $2\frac{1}{2}$ meters
(D) $5\frac{1}{10}$ meters

14. The least common multiple of 12 and another number is 48. Which of these could be the other number?

(A) 4
(C) 16
(B) 8
(D) 24

15. A serving of yogurt is $\frac{3}{4}$ cup. A large container holds 6 cups of yogurt. How many servings of yogurt are in 3 containers?

(A) $3\frac{3}{4}$
(C) 24
(B) $9\frac{3}{4}$
(D) 27

16. Which of the following is an example of the Identity Property of Multiplication?

(A) $(4 \times 5) \times 12 = 4 \times (5 \times 12)$
(B) $14 \times 1 = 14$
(C) $17 \times 12 = 12 \times 17$
(D) $2 \times (9 \times 1) = (2 \times 9) \times 1$

17. A number has the prime factorization $2^2 3^{\square}$, but the exponent of the 3 is missing. Which of these could be the number?

(A) 24
(C) 60
(B) 36
(D) 72

Open Response

18. Wendy is putting together a model train. The train consists of 5 boxcars and an engine.

a. Each boxcar is $1\frac{3}{4}$ inches long. What is the total length of the 5 boxcars?

b. The engine is $2\frac{1}{3}$ inches long. What is the total length of the train?

19. The table shows the dimensions of three rectangular flower beds in Jose's garden.

Flower Beds		
Flower	**Length**	**Width**
Dahlias	$2\frac{1}{3}$ m	$1\frac{3}{4}$ m
Tulips	$2\frac{1}{2}$ m	$1\frac{1}{4}$ m
Sunflowers	$2\frac{2}{3}$ m	$1\frac{1}{2}$ m

a. Recall that the area of a rectangle is the length times the width. Which flower bed has the greatest area? Explain how you know.

b. Jose puts a fence around the flower bed for tulips. What is the total length of the fence?

CHAPTER 6

Project

Fraction Operations: Enjoying Nature

Career: Meteorologist

Weather affects our daily activities, and weather information is useful and often necessary. Businesses such as farms, ski resorts, and airlines need to know weather conditions. This information comes from meteorologists, people who study and forecast the weather. They gather data such as temperature, wind speed, and rainfall. They then study this data and make predictions.

Before you start the project, work through the National Park Climates worksheet on the next page.

In this chapter project . . .

You set up your own weather station.

What to do:

- Choose a place near your classroom and set up a thermometer and something to measure precipitation (rain or snowfall), such as a large measuring cup. If available, you can also set up a *hygrometer* to measure relative humidity.
- Collect data from your weather station and record the information in a table.
- Choose an appropriate data display for the information you've collected.

What to present:

- Data displays for the temperature, precipitation, and relative humidity data for the time period for which you collected information.
- A summary of days that stood out from the others, such as the hottest day.
- What a typical day was like for the period you collected data.

Where's the math?

- Calculate the mean, median, mode, or range of your data.
- Use a line graph to show the change in temperature over time.

go.hrw.com
Chapter Project Online
KEYWORD: MR7 Ch6

Name ______________________ Date __________ Class __________

CHAPTER 6

Project Recording Sheet

Collecting and Displaying Data

National Park Climates

Some of the most interesting places in the United States are national parks. This table lists the average monthly temperatures in some U.S. national parks.

Average Monthly Temperatures in National Parks

National Park, State	Month/Temperature (°C)											
	J	**F**	**M**	**A**	**M**	**J**	**J**	**A**	**S**	**O**	**N**	**D**
Badlands, SD	0	2	8	16	22	27	33	32	26	19	9	2
Big Bend, TX	15	18	21	26	29	32	31	31	28	24	19	16
Crater Lake, OR	2	4	6	10	13	19	25	24	20	13	7	2
Everglades, FL	24	24	27	28	30	31	32	32	32	29	27	25
Great Smokey Mountains, TN	9	11	16	21	24	28	29	29	26	21	15	11
Hawaii Volcanoes, HI	19	18	19	19	20	21	21	22	22	22	20	19
Petrified Forest, AZ	8	12	16	21	26	31	34	33	30	23	14	9
Zion, UT	11	14	17	23	28	34	38	36	33	26	17	12

1. What kind of graph would you make to show the monthly temperature for one of the national parks listed in the table? Explain why you choose that type of graph.

____

2. Make a graph of average monthly temperatures for one of the national parks listed in the table.

____

3. What is the difference between the highest January temperature and the lowest January temperature? the highest and lowest July temperatures?

____

Extension: Create your ideal park. Write a description of your park, including its location, geography, climate, wildlife, and vegetation. Create a line graph to show the average monthly temperatures in your park.

EXPLORATION

Program of Studies and Core Content
MA-6-DAP-S-DR4, MA-6-DAP-S-ES1, MA-06-4.1.1, MA-06-5.1.5

6A Make a Table

The table shows the number of medals awarded to the top 13 medal-winning countries during the 2002 Winter Olympics.

1. Compute the total number of medals won by each country.

Country	Gold	Silver	Bronze	Total
Germany	12	16	7	
USA	10	13	11	
Norway	11	7	6	
Canada	6	3	8	
Austria	2	4	10	
Russia	6	6	4	
Italy	4	4	4	
France	4	5	2	
Switzerland	3	2	6	
China	2	2	4	
Netherlands	3	5	0	
Finland	4	2	1	
Sweden	0	2	4	

Think and Discuss

2. Explain why the table is set up the way it is.

3. Describe a different way to organize this data.

Name ______________________________ Date ____________ Class ____________

LESSON 6A

Review for Mastery

Problem Solving Skill: Make a Table

You can make a table to organize data. Then you can use the table to see patterns and draw conclusions.

During the week-long book fair, 324 books were sold. On Monday, 45 books were sold. On Tuesday, students bought 58 books. On Wednesday, 79 books were sold. Sixty-two books were sold on Thursday, and students bought 51 books on Friday.

Day	Books Sold
Monday	45
Tuesday	58
Wednesday	79
Thursday	62
Friday	51

To make a table, arrange the information in order by days so you can see patterns over time. Remember to make headings for each column of the table.

From the table, you can see that the number of books sold increased from Monday to Wednesday, and decreased from Wednesday to Friday.

Use the data to make a table. Then use the table to find a pattern in the data and draw a conclusion.

1. During the championship series, the school basketball team earned 24 points in the first game, 28 points in the second game, 33 points in the third game, 42 points in the fourth game, and 49 points in the last game.

____

____

____

____

2. In the sixth grade, 18 students study Spanish, 35 students study French, 11 students study Latin, and 5 students study no foreign language at all.

____

Name ______________________ Date ____________ Class ____________

LESSON 6A

Homework and Practice

Make a Table

Complete each activity and answer each question.

1. Sandwich Heaven sells different-sized sandwiches. The big eater is 24 inches long. The party is 18 inches long. The foot long is 12 inches long. The half foot is 6 inches long. Use this data to complete the table at right, from the largest to the smallest sandwiches.

2. What pattern do you see in the table's data?

____

____

3. A foot-long olive loaf sandwich at Sandwich Heaven costs $4.75. A foot-long turkey sandwich costs $5.25, and the foot-long roast beef sandwich costs $5.75. The foot-long ham sandwich costs $6.25, and the foot-long chicken sandwich costs $6.75. Use this data to complete the table at right.

4. What pattern do you see in the table's data?

____

____

5. How much does each inch of a foot long ham sandwich from Sandwich Heaven cost? Round your answer to the nearest hundredth of a dollar.

____

6. You and three friends buy 6 foot-long sandwiches from Sandwich Heaven: two turkey, one ham, two chicken, and one olive loaf. If you equally share the total price, how much will you each pay? How many 3-inch pieces will you each get? Round your answer to the nearest hundredth of a dollar.

____

Program of Studies and Core Content
MA-6-DAP-S-CD2, MA-06-4.2.1

6B Mean, Median, Mode, and Range

The table shows an ordered list of all the times in the men's 1,500-meter speed-skating competition in the 2002 Winter Olympics. The fastest time was 1:43.95, or 1 minute 43.95 seconds. Notice that all of the times are only seconds apart.

1:43.95	**1:44.57**	**1:45.26**	1:45.34	1:45.41	1:45.51	1:45.63	1:45.82
1:45.86	1:45.97	1:45.98	1:46.00	1:46.04	1:46.29	1:46.38	1:46.40
1:46.75	1:46.99	1:47.04	1:47.21	1:47.26	1:47.63	1:47.64	1:47.72
1:47.78	1:47.83	1:48.02	1:48.13	1:48.20	1:48.27	1:48.40	1:48.57
1:48.58	1:48.76	1:49.24	1:49.42	1:49.45	1:49.50	1:49.57	1:50.15
1:50.26	1:50.70	1:51.02	1:51.02	1:51.81	1:52.01	1:52.87	

1. How many seconds behind the winner was the second-place skater? the third-place skater?
2. Find the range. (Subtract the fastest time from the slowest time.)
3. Find the median. (The median is the number in the middle of the data set.)
4. Find the mean. (The mean is the average of all the times.)

Think and Discuss

5. **Discuss** how you found the median.
6. **Explain** how the mean compares with the median.

Name ______________________ Date __________ Class __________

LESSON 6B

Hands-On Lab 1

Mean, Median, Mode, and Range

KEY

Mean, median, and mode are all measures of a set of data. You can compute these measures with data from your class.

Activity

Find the mean, median, and mode of your classmates' heights in inches.

1. Survey the class. List all heights below.

____

____

2. Find the mean. Find the sum of the numbers and divide by the number of numbers.

____

____

3. Find the median. List the numbers in order from least to greatest. Then find the middle number of the numbers when listed in order.

____

____

4. Find the mode. Find the number that appears most often in a set of numbers.

____

____

5. Find the range. Find the difference between the least and greatest values in the set.

____

Name ______________________ Date __________ Class __________

LESSON 6B

Hands-On Lab 1

Mean, Median, Mode, and Range, continued

Think and Discuss

1. How do you compute the median for a set of data with an even number of numbers?

____

2. When can there be more than one mode in a set of data?

____

____

3. When will there be no mode in a set of data?

____

____

Try This

1. Find your scores on your last five math tests. List them below.

____

____

2. What is the mean of these scores? _____
3. What is the median of these scores? _____
4. What is the mode of these scores? _____
5. What is the range of these scores? _____
6. What is the total number of pens and pencils you have with you in class? Ask three classmates for their totals as well. List the amounts below.

____

____

7. What is the mean of these numbers? _____
8. What is the median of these numbers? _____
9. What is the mode of these numbers? _______________
10. What is the range of these numbers? _____

Name ______________________ Date __________ Class __________

LESSON 6B

Hands-On Lab 2

Using the Range, Mean, Median, and Mode

The range, mean, median, and mode tell you different things about the data. You can choose which measure is the best, depending on how you want to use the data.

Activity

Ask each student in your class what size their shoes are. Write the shoe sizes below.

____

____

____

____

Find the mean, median, mode, and range of the data.

Mean: ________ Median: ________ Mode: ________ Range: ________

Ken's Traveling Shoe Store is coming to your class to sell shoes. The store will only sell one size. Should the store use the mean, median, mode, or range to decide which size to sell? Explain.

____

____

____

A new store is opening at the mall, and is selling shoes to students your age. The store wants to know how many sizes it should have. Should the store use the mean, median, mode, or range to make this decision? Explain.

____

____

____

If you remove the largest shoe size from your data, how will this affect the mean? Explain.

____

____

____

Name ______________________________ Date ____________ Class ____________

LESSON 6B

Hands-On Lab 2

Using the Range, Mean, Median, and Mode, continued

Think and Discuss

1. If the mean is greater than the median, what does this tell you about the data?

____

____

2. What shoe size would you have to add to the data to increase the mean by 1? Explain.

____

____

Try This

Ms. Jones gave her class a test last week. The grades are listed below.

86 90 73 75 80 82 89 90 85 76 83 86

1. What is the mean? _____

2. What is the median? _____

3. What is the mode? _____

4. What is the range? _____

5. Which measure should Ms. Jones use to convince other teachers that her class did very well on the test? Explain.

____

____

6. Which measure should Ms. Jones use to convince the class that they need to study harder? Explain.

____

____

7. Why is the mean lower than the median?

____

____

8. One student has not taken the test. What grade must this student receive to raise the mean to 84? Explain.

____

____

Name ______________________ Date ___________ Class ___________

LESSON 6B

Review for Mastery

Mean, Median, Mode, and Range

You can find the mean, median, mode, and range to describe a set of data.

Terry's Test Scores	76	81	94	81	78

The **mean** or average is the sum of the items divided by the number of items.

76 + 81 + 94 + 81 + 78 = 410 — First, find the sum of the values.

410 ÷ 5 = 82 — Then divide the sum by the number of values in the set of data.

The mean is 82 points.

The **median** is the middle value of an ordered set of data. If there are two middle values, the median is the mean of those two values.

76, 78, **81**, 81, 94 — Put the values in order first.

The median is 81 points.

The **mode** is the value that occurs most often in a set of data.
The mode is 81 points.

The **range** is the difference between the greatest and least values in the set of data.

94 − 76 = 18 — Use subtraction to find the range.

The range is 18 points.

Find the mean, median, mode, and range of each set of values.

1. 23, 78, 45, 22

mean: ___________

median: ___________

mode: ___________

range: ___________

2. 102, 79, 82, 103, 79

mean: ___________

median: ___________

mode: ___________

range: ___________

3. 56, 99, 112, 112, 56

mean: ___________

median: ___________

mode: ___________

range: ___________

Name ______________________ Date ___________ Class ___________

LESSON 6B

Homework and Practice

Mean, Median, Mode, and Range

Find the range, mean, median, and mode of each data set.

1.

School Jump Rope Records (total jumps)	138	142	144	138	143

____

____

2.

Jessica's History Test Scores	81	97	99	89	91	76

____

____

3.

Weights of Wrestlers (lbs.)	145	210	178	195	153

____

____

4.

Daily Summer Temperatures (°F)	87	93	79	93	89	87	95

____

____

5.

Daily Theater Ticket Sales	121	89	98	101	109	153	162

____

____

6. Kristy has 4 dogs and 2 cats. The mean age of her pets is 3 years. What will their mean age be 10 years from now? Twenty years from now?

____

7. In a class of 19 eighth graders, all but three of the students are 14 years old. There is one student who is 15 and two who are 13. Which two data measurements are the same for the students' ages? What are those measurements?

____

EXPLORATION

6C Bar Graphs

Program of Studies and Core Content
MA-6-DAP-S-DR2, MA-6-DAP-S-DR4, MA-06-4.1.1, MA-06-4.1.4

The bar graph shows the medal totals of the four countries that won the most medals at the 2002 Winter Olympic Games.

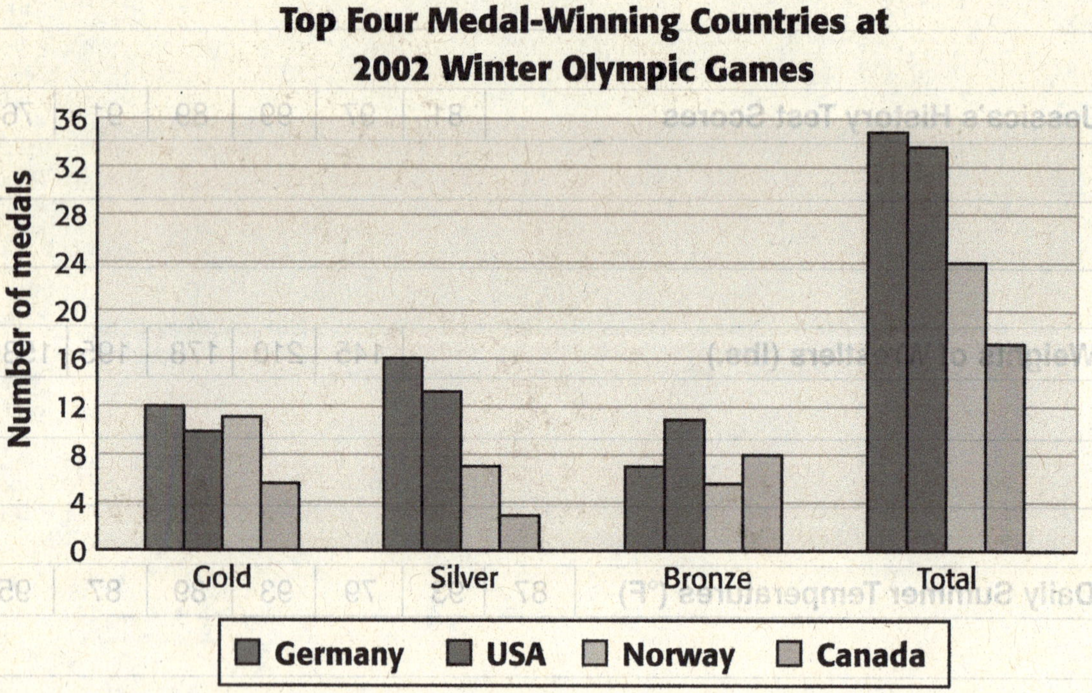

1. Name the country that won
   a. the most gold medals.
   b. the most silver medals.
   c. the most bronze medals.
   d. the most medals overall.
2. What is the total number of medals won by each country?

Think and Discuss

3. **Explain** a different way to display the data.
4. **Discuss** information that is not shown by the bar graph.

Name ______________________________ Date ____________ Class __________

LESSON **6C**

Hands-On Lab 1

Representing Data in Graphs

A **bar graph** displays data with vertical or horizontal bars. Bar graphs can be used to compare data.

Materials: ruler

Activity

To make a bar graph, you first need to decide what intervals to use. When you choose an interval, make sure it is large enough to fit all of the data on the graph. Be careful not to use intervals that are so large the graph looks small.

Annual Number of Sunny Days For Major Western US Cities

City	Number of Sunny Days
Phoenix, AZ	211
Seattle, WA	58
Los Angeles, CA	186
San Francisco, CA	160
Portland, OR	68
Las Vegas, NV	210
San Diego, CA	146
Denver, CO	115
Salt Lake City, UT	125

1. What interval should you use? ________
   Fill in the intervals on the graph below.
2. Label the vertical axis on the graph.
3. Write the names of the cities along the horizontal axis.
4. Use your ruler to draw a bar for each city. The height of the bar should represent the number of sunny days.

Annual Number of Sunny Days
For Major Western US Cities

Name ______________________ Date __________ Class __________

LESSON **6C**

Hands-On Lab 1

Representing Data in Graphs, continued

Think and Discuss

1. How do you pick the intervals for a bar graph?

____

____

____

2. If you want to know which city had the greatest number of sunny days, would it be easier to use the table or the bar graph? Explain.

____

____

____

Try This

Andy works at the grocery store. The table below shows how many hours he and his friends work each week.

Weekly Work Hours

Employee	Hours
Andy	20
Heather	35
Maria	16
Joe	38
Daryl	12

1. Make a bar graph to display the data. Make sure you label the horizontal and vertical axes, and give your graph a title.

Number of Hours Worked By Each Employee Each Week

Name ______________________ Date __________ Class __________

LESSON **6C**

Hands-On Lab 2

Exploring Double Bar Graphs

A **double bar graph** shows two sets of related data.

Materials: ruler, 2 colored pencils

Activity

1. Survey 30 people. Ask them in which month they were born. Then ask in what month each of their brothers and sisters were born. Record your results in the table at the right.

   Since you have two sets of data, one set for girls and one set for boys, you can make a double bar graph.

Season	Girls	Boys
January–March		
April–June		
July–September		
October – December		

2. Choose a color for the bar to represent girls and a color for the bar to represent boys. Color the key in the graph.
3. What interval should you use on the graph? _____
4. Label the intervals along the horizontal axis. Label the vertical axis.
5. For each season, make a bar for the number of girls and a bar for the number of boys. Use your ruler to draw the bars.

Birthdays by Season

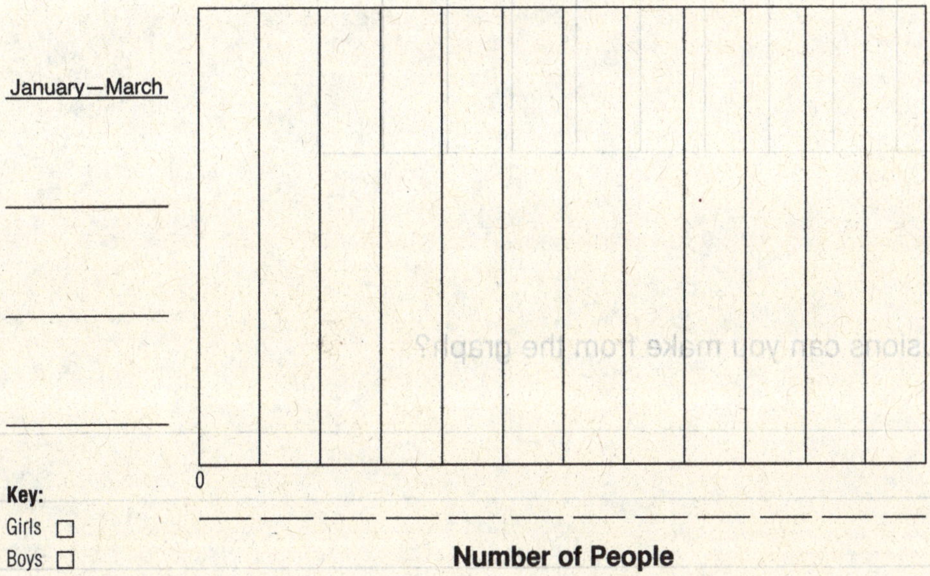

Name ______________________ Date ____________ Class ____________

LESSON 6C

Hands-On Lab 2

Exploring Double Bar Graphs, continued

Think and Discuss

1. When should you use a double bar graph instead of a single bar graph?

____

____

Try This

A scientist surveyed 100 adults and 100 children to find their blood types. The results are shown in the table.

Blood type	Children	Adults
O	44	48
A	40	39
B	12	10
AB	4	3

1. Make a double bar graph to display the data. Make sure you label the horizontal and vertical axes, and give your graph a title.

2. What conclusions can you make from the graph?

____

____

____

Name ______________________ Date __________ Class __________

LESSON 6C

Review for Mastery

Bar Graphs

You can make a bar graph to compare amounts.

Annual Read-a-thon Totals			
Grade	6	7	8
Books Read	86	42	98

To make a bar graph using the data in the table, first choose a scale that includes all of the data values. Next, separate the scale into equal parts, called intervals.

Then draw bars to match the data. The bars should be of equal width and should not touch. Give your graph a title and label its axes.

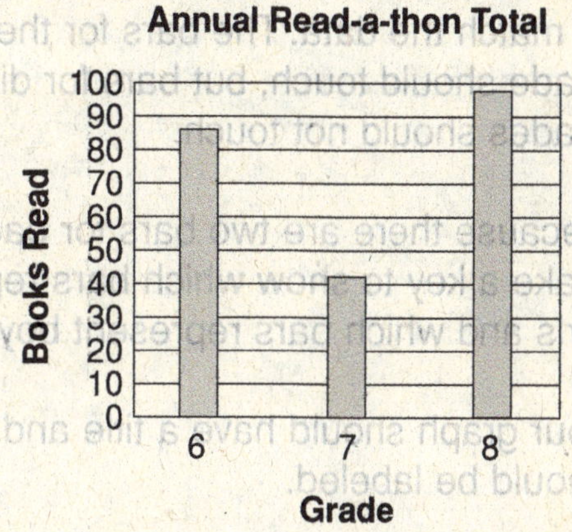

Use the data to make a bar graph.

1.

Canned Food Drive Totals			
Grade	6	7	8
Cans Collected	96	74	62

Look at the bar graph for the Read-a-thon above. Which grade read almost twice as many books as the seventh grade?

The bar for the sixth grade is about twice a long as the bar for the seventh grade. So the sixth grade read almost twice as many books as the seventh grade.

Use the bar graph you made in Exercise 1.

2. How many more items did the sixth grade collect than the eighth grade?

____

Name ______________________ Date ____________ Class ____________

LESSON **6C**

Review for Mastery

Bar Graphs, continued

A double-bar graph shows two sets of related data.

To make a double-bar graph, choose a scale and an interval for the scale. Then draw bars to match the data. The bars for the same grade should touch, but bars for different grades should not touch.

Middle School Students			
Grade	6	7	8
Boys	34	29	25
Girls	28	31	22

Because there are two bars for each grade, make a key to show which bars represent girls and which bars represent boys.

Your graph should have a title and its axes should be labeled.

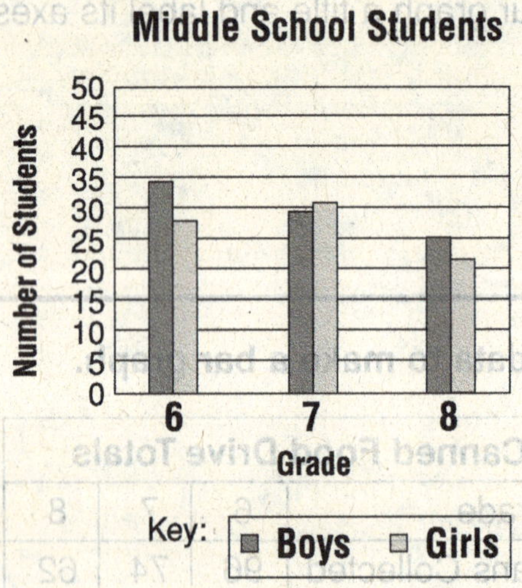

Make a double-bar graph to match the data below. Then answer the question.

3.

Movie Tickets Sold			
	Fri	**Sat**	**Sun**
Adult	136	118	98
Student	84	102	154

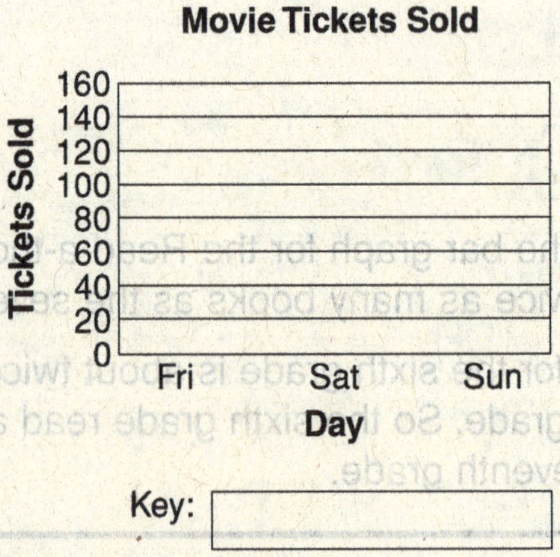

4. On which day was the number of adult tickets sold about the same as the number of student tickets sold?

____

____

Name ______________________ Date __________ Class __________

LESSON 6C

Homework and Practice

Bar Graphs

Use the bar graph to answer each question.

1. In which country did people spend $5,000,000 on toys in 2000?

   ____

2. How much more money was spent on toys in the United States than in Japan?

   ____

3. What was the mean amount of money spent by the five countries on toys in 2000?

   ____

Top Toy-Buying Countries, 2000

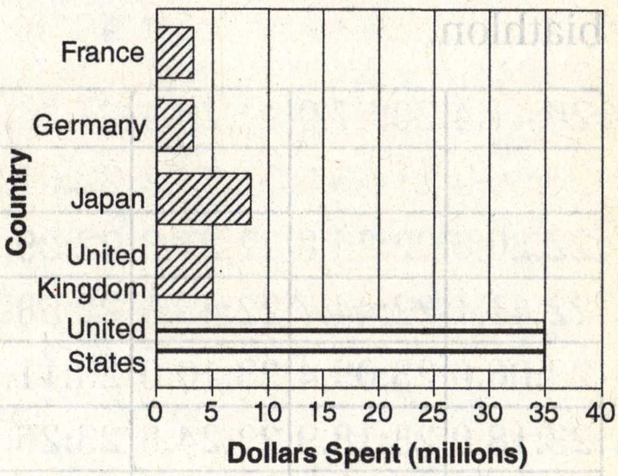

Make a bar graph to compare the data in the table.

Items Sold at Hamburger Shops			
Jerry's		**Mike's**	
Hamburgers	30	Hamburgers	35
Cheeseburgers	40	Cheeseburgers	42
Chili Burgers	50	Chili Burgers	47
Chicken Sandwiches	62	Chicken Sandwiches	57
Hot Dogs	20	Hot Dogs	15

Items Sold at Hamburger Shops

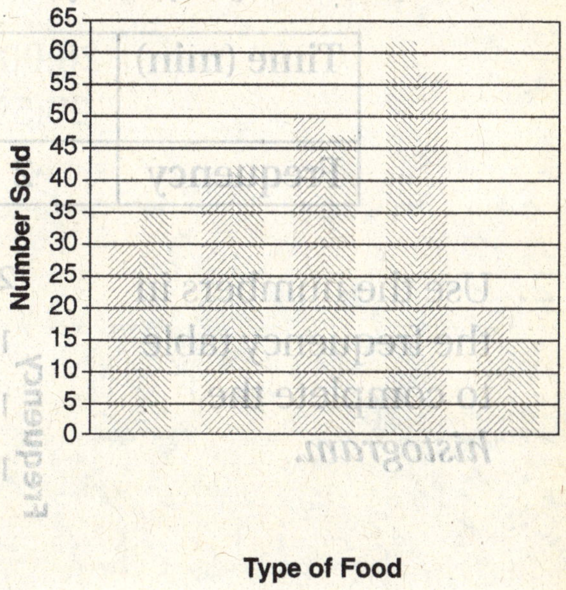

EXPLORATION

Program of Studies and Core Content
MA-6-DAP-S-DR2, MA-6-DAP-S-DR4, MA-06-4.1.1, MA-06-4.1.4

6D Line Plots, Frequency Tables, and Histograms

Below are the top 50 women's times at the 2002 Olympic biathlon.

20:41.4	20:57.0	21:20.4	21:24.1	21:27.9	21:32.1	21:35.7	21:44.2
21:50.3	21:55.6	21:57.0	22:01.7	22:11.9	22:14.9	22:17.7	22:19.7
22:20.6	22:25.8	22:27.3	22:29.9	22:32.1	22:33.5	22:37.7	22:39.9
22:41.1	22:44.7	22:45.5	22:58.3	23:00.0	23:03.5	23:03.8	23:05.0
23:06.6	23:09.4	23:10.0	23:11.2	23:11.3	23:14.2	23:14.6	23:14.7
23:18.0	23:18.9	23:24.6	23:26.5	23:36.8	23:36.9	23:37.4	23:40.9
23:44.1	23:48.7						

1. Complete the *frequency table.*

Time (min)	20:00.0–20:59.9	21:00.0–21:59.9	22:00.0–22:59.9	23:00.0–23:59.9
Frequency	2			

2. Use the numbers in the frequency table to complete the *histogram.*

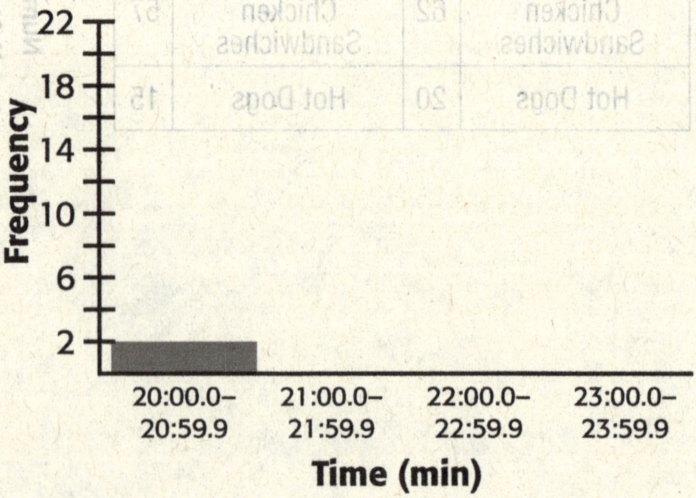

Think and Discuss

3. **Explain** how you completed the histogram in number 2.

Name ______________________________ Date ____________ Class ____________

LESSON 6D

Hands-On Lab

Exploring Line Plots

A **frequency table** tells you how many times each value appears in the data. A **line plot** is a way to display frequency data.

Activity

Students recorded what time they usually wake up in the morning before school. The results are shown below.

5:30	6:15	7:00	6:30	6:00	7:15	7:00
7:15	6:30	6:00	6:45	7:30	6:45	6:30
7:15	7:30	6:45	7:15	6:45	6:30	7:00
7:00	7:15	6:15	7:00	6:45	7:00	6:45

1. Use the tally table below to count the number of students who wake up at each time. For example, if 3 students wake up at one time, the tally will be |||. If 7 students wake up at another time, the tally will be ~~||||~~ ||.

Time	5:30	5:45	6:00	6:15	6:30	6:45	7:00	7:15	7:30
Number of Students									

2. Draw a line plot to represent the data in the tally table above. Use an X for each student. The data for 5:30 is filled in for you.

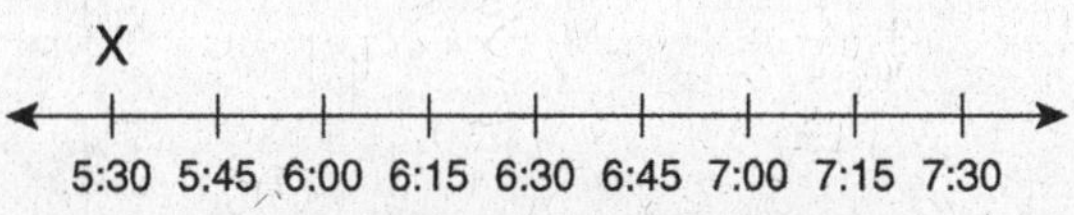

School Morning Wake Up Times

3. What is the mode? ____________________

4. Does the data have any outliers? ____________________

5. Data is symmetrical if one side of a graph looks like the other side. Is this data symmetrical? Explain.

____

____

Name ______________________ Date __________ Class __________

LESSON **6D**

Hands-On Lab

Exploring Line Plots, continued

Think and Discuss

1. Would you expect symmetrical data to be centered at the mean? Explain.

____

____

____

2. Can outliers affect the symmetry of the data? Explain.

____

____

____

Try This

A student surveyed the people on his baseball team to find out their ages. The ages are listed below.

12	12	11	13	12
11	12	13	12	11
14	11	12	12	13
10	11	13	14	13

1. Draw a line plot for the data.

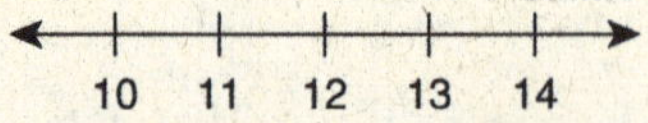

Ages of Baseball Team Members

2. What is the mode? ______________

3. Are there any outliers? Explain. ______________

4. Is the data symmetrical? Explain.

____

____

Name ______________________ Date __________ Class __________

LESSON **6D**

Review for Mastery

Line Plots, Frequency Tables, and Histograms

Julie picked the following cards from a deck.

You can make a tally table to organize the data. Make a row for the numbers. Then for each card, make a tally mark in the appropriate column.

Julie's Cards

2	3	4	5	6
||	||	|||	|	||

1. Make a tally table to organize the data.

Rolls of a Number Cube

2	3	6	5	1	4	1
3	3	5	1	6	1	4

A line plot gives a visual picture of data. To make a line plot of Julie's data, draw a number line. Then use an X to represent each tally mark in the tally table.

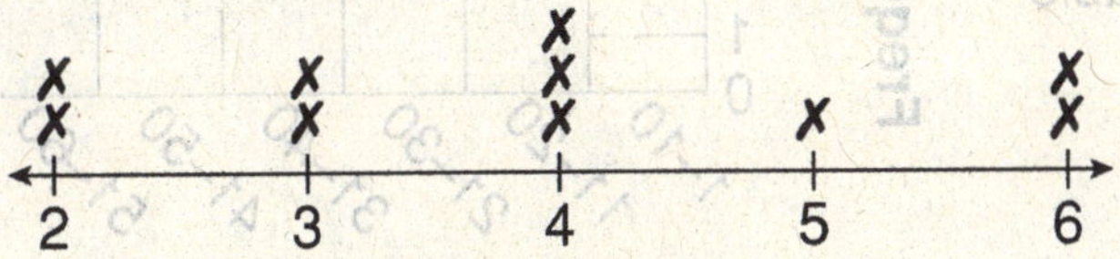

Julie's Cards

2. Make a line plot for the tally table you made Exercise 1.

Name ______________________________ Date ______________ Class ________

LESSON 6D

Review for Mastery

Line Plots, Frequency Tables, and Histograms, continued

Sometimes, you can make a frequency table with intervals or a histogram.

Number of Jumping Jacks Completed in 30 Seconds				
12	28	24	32	35
31	38	55	43	52
42	49	18	22	15
47	37	19	31	37

A frequency table can organize the data with intervals.

Jumping Jacks

Interval	Frequency
1–10	0
11–20	4
21–30	3
31–40	7
41–50	4
51–60	2

A histogram is a bar graph that shows the number of values that occur within each interval.

You make a histogram the same way you make any other bar graph, except that the bars touch. They do not overlap.

Here is a histogram for the frequency table above.

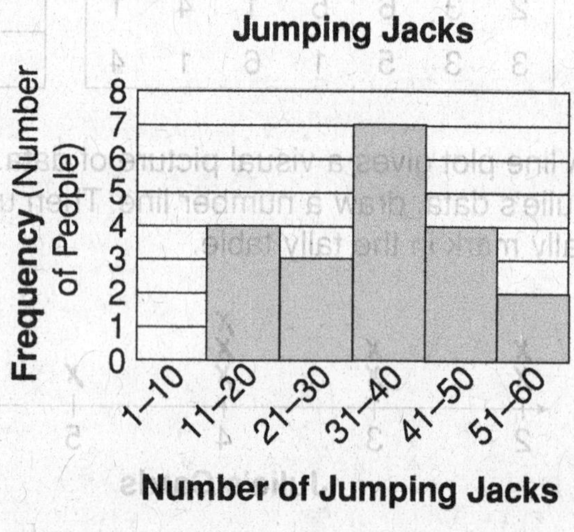

3. Use the data to make a histogram.

Total Books Read by Participants in Summer Reading Program				
5	3	8	7	6
2	9	10	1	2
4	5	7	3	5
3	1	0	10	4
3	5	8	2	1
1	7	0	4	11

Name ______________________ Date ____________ Class ____________

LESSON **6D**

Homework and Practice

Frequency Tables and Histograms

Mrs. Martin said that her employees could choose one day a week to wear jeans in the office. She recorded their votes in the box below.

Tuesday	Wednesday	Friday	Friday	Monday
Monday	Monday	Friday	Tuesday	Thursday
Tuesday	Monday	Monday	Monday	Thursday
Wednesday	Friday	Thursday	Tuesday	Monday

1. Make a tally table to organize the data. Which day got the most votes?

____

2. Use your tally table from Exercise 1 to make a cumulative frequency table. How many employees voted in all?

____

Tally Table for Number of Votes	

Number of Votes		
Day	Frequency	Cumulative Frequency

3. Use the data in the box below to make a frequency table with intervals.

Class Mathematics Test Scores									
79	69	89	100	97	63	78	74	92	94
88	83	85	74	80	58	72	93	90	59

Class Mathematics Test Scores					
Scores					
Frequency					

4. In which range of scores did the most test scores fall?

____

EXPLORATION

6E Line Graphs

Program of Studies and Core Content
MA-6-DAP-S-DR2, MA-6-DAP-S-DR4, MA-06-4.1.1, MA-06-4.1.4

Luis works at a record store. His manager asked him to graph the number of CDs returned each day during one week.

Day	Sun	Mon	Tue	Wed	Thu	Fri	Sat
CDs Returned	10	3	2	4	7	11	14

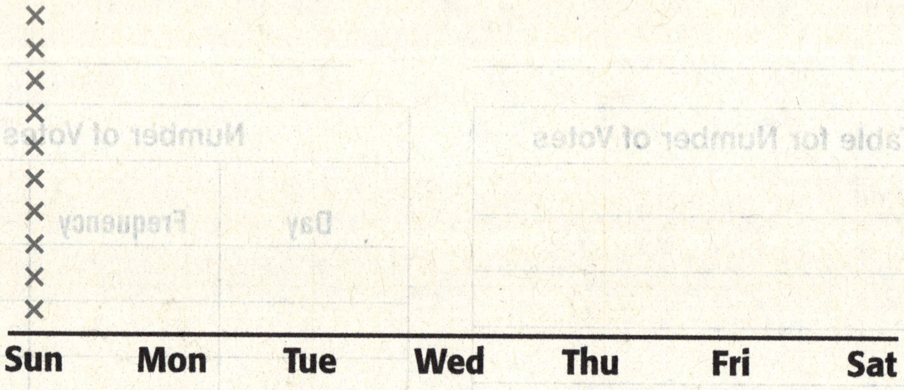

1. Complete the line plot.
2. Draw a vertical line to the left of Sun (Sunday) to form the y-axis. Number this line from 1 to 14.
3. Delete all the ×'s in each stack except the one at the top.
4. Connect the ×'s with line segments. You have constructed a ***line graph.***

Think and Discuss

5. **Explain** how to construct a line graph.
6. **Discuss** some advantages of displaying data on a line graph rather than in a table.

Name ______________________________ Date __________ Class __________

LESSON **6E**

Technology Lab

Create Line Graphs

Activity

The table shows the tide heights at various times. Use the data to make a line graph.

Time (A.M.)	9:00	9:06	9:12	9:18	9:24
Height (ft)	2.41	2.31	2.21	2.10	2.00

1 Enter the data. Put times in the **L1** list and height in the **L2** list. You cannot enter times, so use 0 for 9:00, 1 for 9:06, and so on.

Times: STAT ENTER 0 ENTER 1 ENTER

2 ENTER 3 ENTER 4 ENTER

Heights: ▶ 2.41 ENTER 2.31 ENTER

2.21 ENTER 2.10 ENTER 2.00 ENTER

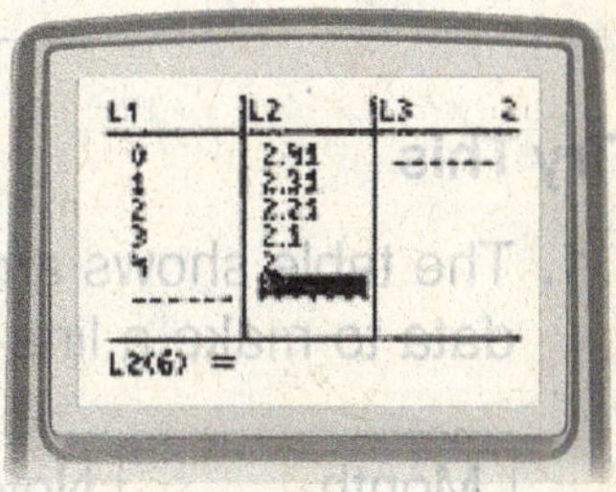

2 Choose a scale. Press WINDOW. The calculator uses **X** for the horizontal axis (time) and **Y** for the vertical axis (height). Set **Xmin** to 0, **Xmax** to 5, **Ymin** to 1, and **Ymax** to 3.

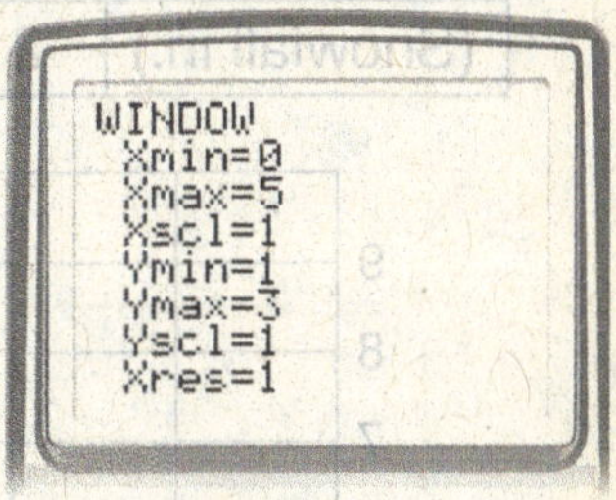

3 Make the graph.

Select Plot 1: 2nd Y= (PLOT) ENTER ENTER.

Use the arrow keys to highlight the line graph icon and press ENTER. Then press GRAPH

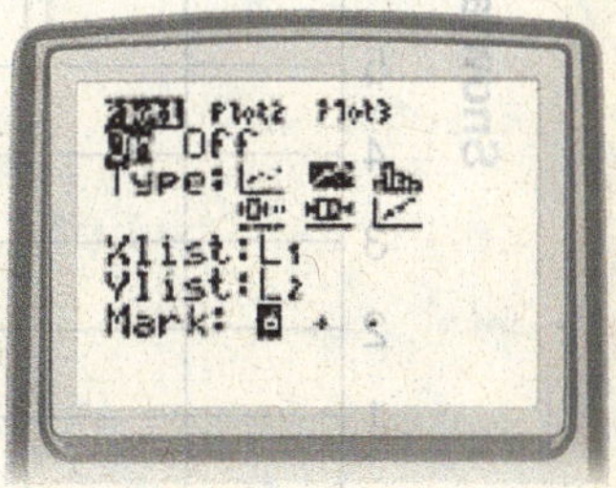

Name ______________________ Date ____________ Class ____________

LESSON 6E

Technology Lab

Create Line Graphs, continued

Think and Discuss

1. What does the line graph tell you about the data? What do you think was happening when the measurements were taken?

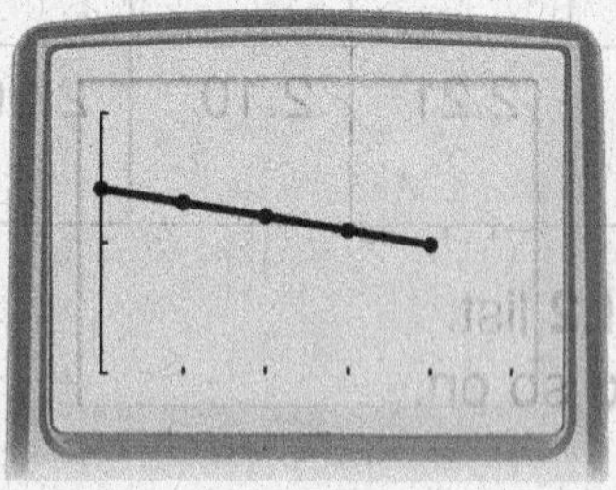

____

____

Try This

1. The table shows amounts of snowfall during one winter. Use the data to make a line graph.

Month	Nov	Dec	Jan	Feb	Mar
(Snowfall in.)	2	6	9	6	2

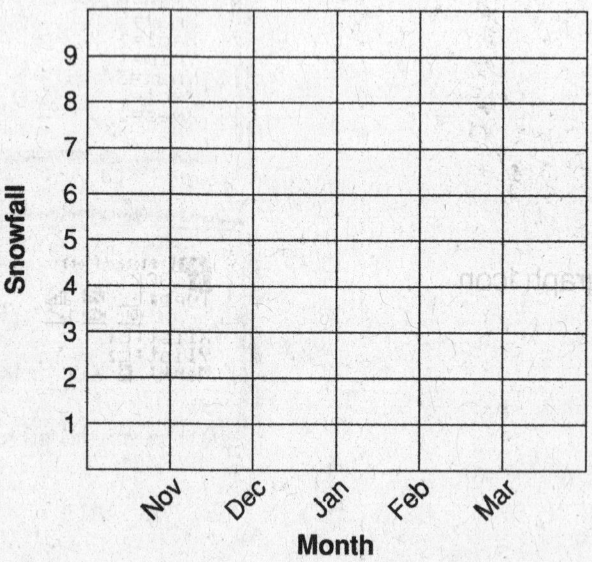

Name ______________________________ Date __________ Class __________

LESSON **6E**

Review for Mastery

Line Graphs

A line graph shows change over time.

You can represent data by making a line graph.

Stock Sales (millions)

Mon	Tue	Wed	Thurs	Fri
1.5	2.0	2.25	1.75	0.5

To make a line graph, make "days" the horizontal axis and "sales" the vertical axis. Label the axes.

Then determine an appropriate scale and interval for each axis.

Think of the data in the table as ordered pairs. Mark a point for each ordered pair. Then connect the points with straight segments.

Make sure your line graph has a title.

Stock Sales

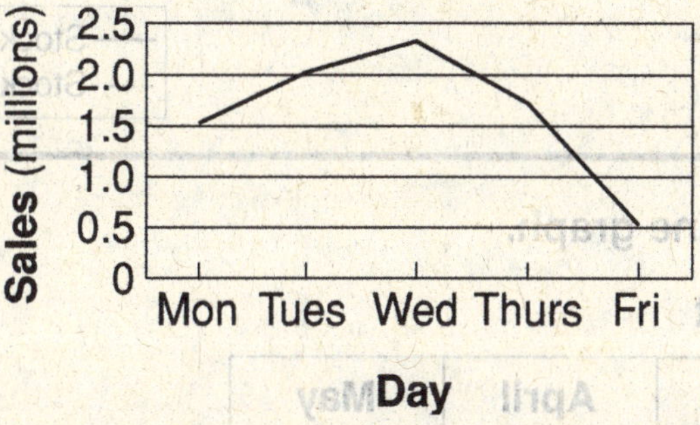

Use the data in the table to make a line graph.

1. **Millie's Savings Account**

Jan	Feb	March	April	May
30	40	35	45	25

Name ____________________ Date ____________ Class ____________

LESSON 6E

Review for Mastery

Line Graphs, continued

Sometimes, you need to make a double-line graph to represent data.

Stock Sales (millions)

Stock	Jan	Feb	March	April	May
A	1.5	2	2.25	1.75	0.5
B	1	2.5	2	1.5	0.75

To make a double-line graph, follow the same steps for making a line graph. Mark and connect points for each of the two sets of data you are displaying. Because there are two sets of data, make a key. Be sure to title the graph and label the axes.

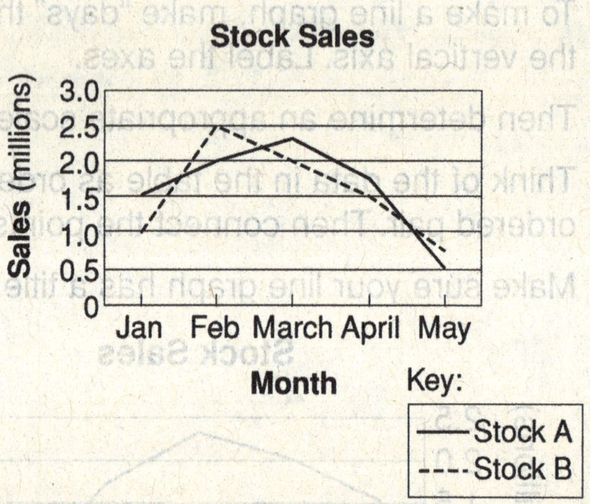

Use the data in the table to make a double-line graph.

2. **Savings Account**

Student	Jan	Feb	March	April	May
Michael	20	10	30	25	35
Janet	30	20	35	25	45

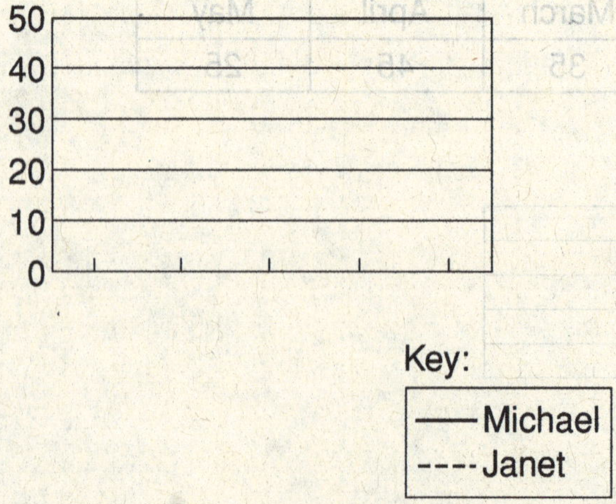

Name ______________________ Date ____________ Class ____________

LESSON 6E

Homework and Practice

Line Graphs

Use the line graph to answer each question.

1. In which year were Joe's average weekly earnings the lowest? the highest?

____

2. In general, how did Joe's average weekly earnings change between 1996 and 2000?

____

____

____

3. In which year did Joe earn about $250 a week?

____

Joe's Average Weekly Earnings

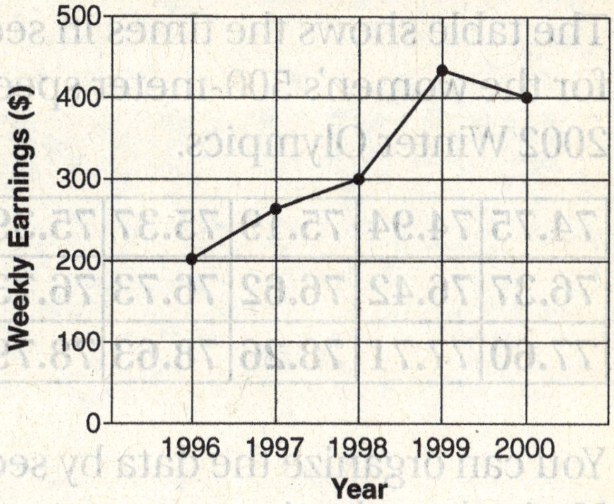

Use the given data to make a line graph and answer each question.

Juan's Hourly Wages	
Year	**Hourly Rate**
1998	$5.15
1999	$5.69
2000	$7.39
2001	$9.23

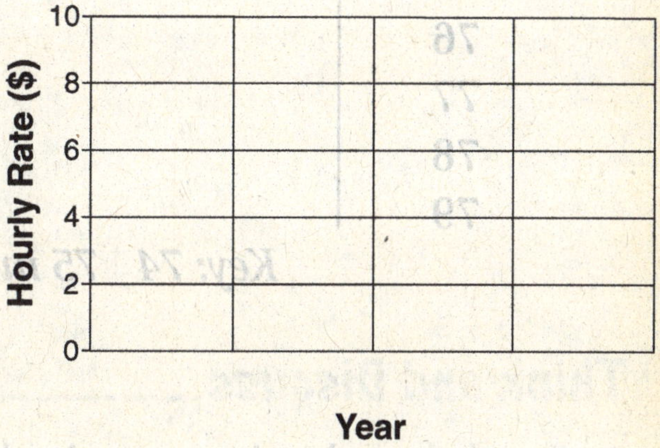

4. Between which two years shown on the graph did Juan's hourly wage change the least?

____

5. How has Juan's hourly wage changed since1998?

____

EXPLORATION

Program of Studies and Core Content

MA-6-DAP-S-DR2, MA-06-4.1.1

6F Stem-and-Leaf Plots

The table shows the times in seconds and hundredths of seconds for the women's 500-meter speed-skating competition in the 2002 Winter Olympics.

74.75	74.94	75.19	75.37	75.39	75.64	75.64	76.17	76.20	76.31
76.37	76.42	76.62	76.73	76.73	76.86	76.92	77.10	77.37	77.60
77.60	77.71	78.26	78.63	78.79	78.89	79.28	79.45	79.45	

You can organize the data by seconds and hundredths of seconds. Notice how the times are grouped using different colors.

1. Complete the ***stem-and-leaf plot.*** The stems represent seconds and the leaves represent hundredths of seconds.

Stems	Leaves
74	75 94
75	
76	
77	
78	
79	

Key: 74 | 75 means 74.75

Think and Discuss

2. **Explain** what it means in this case for a stem to have the most number of leaves.
3. **Explain** what it means in this case for a stem to have the least number of leaves.

Name ______________________ Date __________ Class __________

LESSON 6F

Review for Mastery

Stem-and-Leaf Plots

You can use place value to make a stem-and-leaf plot.

Points Earned in Games During Basketball Season						
27	16	34	29	48	12	33
20	18	42	51	27	32	41

Write the numbers in order from least to greatest.

12 16 18 20 27 27 29 32 33 34 41 42 48 51

List the tens digits in order from least to greatest in the first, or stem, column. Then, for each tens digit, record the ones digit for each data value in order from least to greatest in the second, or leaves, column.

Points Earned

Stem	Leaves
1	2 6 8
2	0 7 7 9
3	2 3 4
4	1 2 8
5	1

Key: 1 | 2 = 12

Make sure your graph has a title and a key.

Use the data to make a stem-and-leaf plot.

1.

Valerie's Test Scores				
62	84	93	88	89
76	68	81	91	88

Valerie's Test Scores

Stem	Leaves

Key: 6 | 2 = ________

2. What is the range? ________

3. What is the median? ________

4. What is the mode? ________

Name ______________________ Date ____________ Class ____________

LESSON 6F

Homework and Practice

Stem-and-Leaf Plots

Complete each activity and answer the questions.

1. Use the data in the table to complete the stem-and-leaf plot below.

Pineville's Normal Monthly Temperatures (°F)											
Jan	Feb	Mar	April	May	June	July	Aug	Sep	Oct	Nov	Dec
49	53	75	82	86	91	93	85	79	62	52	48

Stem	Leaves
4	8 9

Key: 4|8 =

Find each value of the data set.

2. smallest value ____________

3. largest value ____________

4. mean ____________

5. median ____________

6. mode ____________

7. range ____________

Stem	Leaves
1	4 6 8
2	7 9
3	4 7 7
4	0 1 2 9

Key: 1|4 = 14

A back-to-back (or double) stem-and-leaf plot compares two sets of data. The plot at the right compares the heights of the Girls' and Boys' basketball teams.

Girls		Boys
7 5 4	5	6 6 6 9
9 9 7 4 0	6	2 5 7 8 8 9
3 3 2 0	7	0 1

Key: |5|6 means 56 in.
4|5| means 54 in.

8. Which team appears to be taller, on average? Calculate the average height of each team to see if you were correct.

____

9. How many girls are 6 feet or taller? How many boys?

____

10. Are there any players that are the same height, but from different teams? Identify these players by their height.

____

EXPLORATION

Program of Studies and Core Content
MA-06-4.1.1,
MA-06-4.1.4

6G Venn Diagrams

A Venn diagram shows relationships between sets.

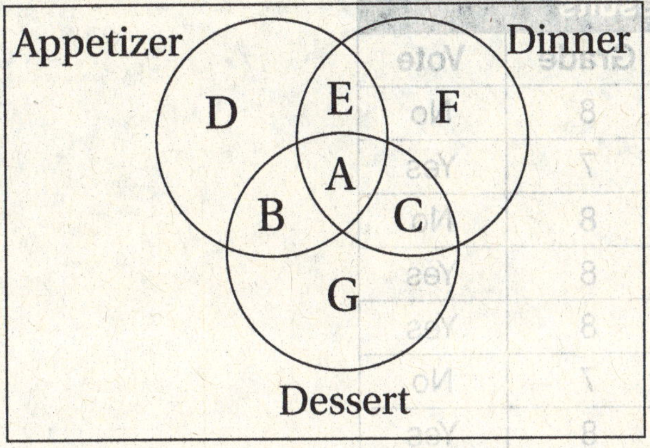

In the Venn diagram, each circle represents customers that ordered an appetizer, a dinner, or a dessert.

Use the intersections to interpret what each section represents. Section C represents customers that ordered dinner and dessert, but not an appetizer.

Determine what each of the following sections represents.

1. Section A ______________________________
2. Section E ______________________________
3. Section G ______________________________
4. Which section represents customers that only ordered dinner?
5. Which section or sections represent customers that ordered dinner, and either a dessert or appetizer (but not both)?
6. The rectangle represents all customers. Describe customers that lie inside the rectangle, but outside the circles.

Think and Discuss

7. **Create** a Venn diagram for two sets: classmates with an "e" in their name, and classmates with an "i" in their name. Place the names in the appropriate sections of the diagram.

Name ______________________________ Date __________ Class __________

LESSON **6G**

Review for Mastery

Venn Diagrams

You can make a Venn diagram to show certain relationships among data. The table below shows how the Grades 7 and 8 tennis teams voted on combining grades into one team.

Voting Results

Grade	Vote	Grade	Vote
7	Yes	8	No
7	No	7	Yes
8	No	8	No
7	Yes	8	Yes
8	No	8	Yes
8	Yes	7	No
7	Yes	8	Yes
8	Yes	7	No
7	Yes	7	Yes
8	No	8	No
7	No	8	Yes
7	Yes	7	Yes

Create a Venn diagram to show the number of 7th-graders that voted yes.

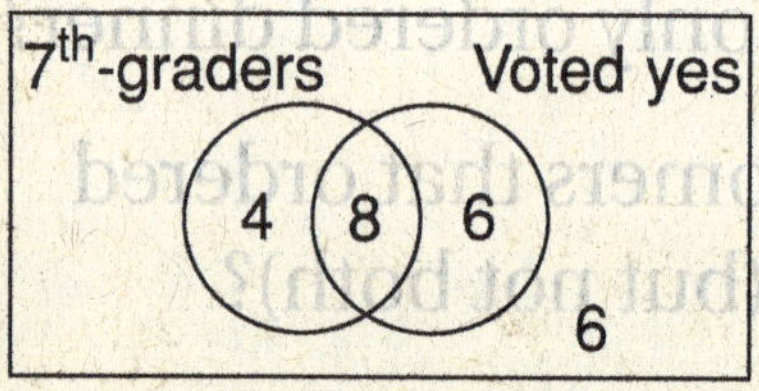

Draw intersecting circles to represent 7th-graders and those that voted yes. The rectangle contains all those that voted.

The intersection shows that 8 7th-graders voted yes.

Use the example above to answer exercises 1–2.

1. Does the Venn diagram provide all the information that the table provides? Explain.

____

____

2. How can you use the Venn diagram to determine the total number of 8th-graders that voted?

____

3. Create a Venn diagram to show the number of 8th-graders that voted no.

Name ______________________ Date __________ Class __________

LESSON 6G

Homework and Practice

Venn Diagrams

Samantha conducted a random survey of 100 shoppers at a local mall. She asked each participant if they watched sports on television. The results of her survey are shown in the Venn diagram. Use Samantha's results to answer exercises 1–4.

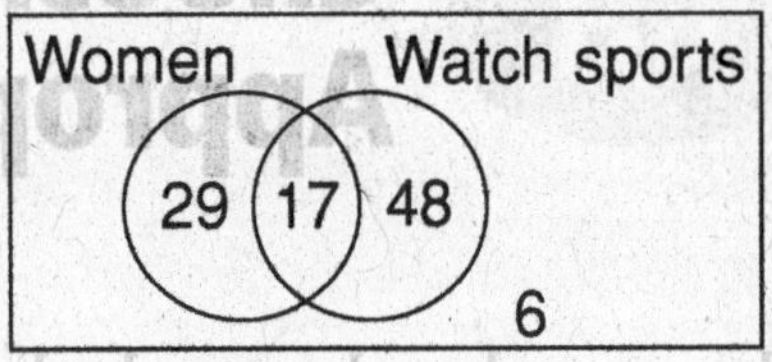

1. How many women were surveyed? ______________

2. How many women indicated they watch sports? ______________

3. How many men indicated they do not watch sports? ______________

4. Suppose Samantha conducted another survey, this time of 500 random men. How many men might she expect to say they watch sports? Explain.

____

____

The Venn diagram shows the number of countries that won gold, silver, or bronze medals at the 2006 Winter Olympics in Turin, Italy. The number of a particular medal that a country received is not indicated. Use the Venn diagram to answer exercises 5–8.

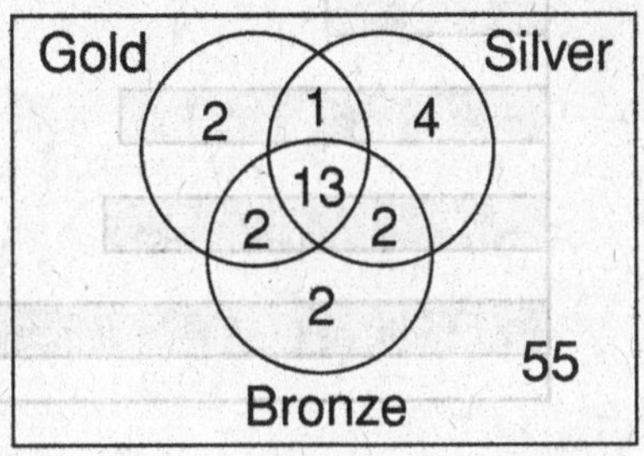

5. How many countries did not medal? ______________

6. How many countries won gold, silver, and bronze medals? ______________

7. How many countries won only one type of medal? ______________

8. Shizuka Arakawa of Japan won a gold medal in figure skating. It was the only medal won by Japan. Is it possible to determine from the Venn diagram how many other countries won 1 medal? Explain.

____

____

EXPLORATION

Program of Studies and Core Content
MA-6-DAP-S-DR1, MA-6-DAP-S-DR3, MA-DAP-S-ES1, MA-06-4.1.2, MA-06-4.1.4

6H Choosing an Appropriate Display

The graphs shown below are not labeled. Match each of the descriptions with the graph that most likely represents the data.

1. The population of a town over the course of several years

2. The areas of the five Great Lakes

3. The test scores of the students in a Spanish class

Graph A

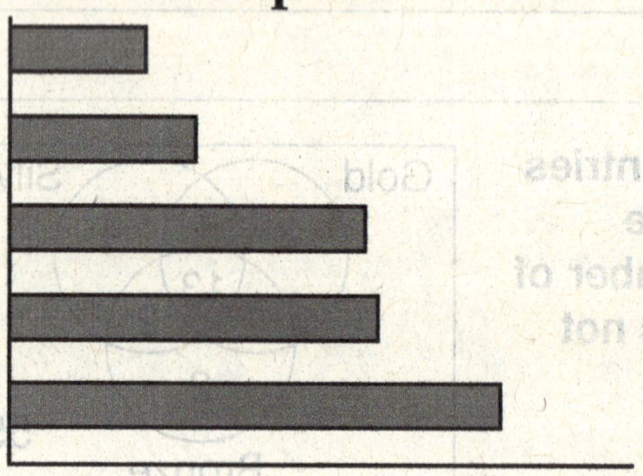

Graph B

Stems	Leaves
6	3 4 7
7	5 5 5 8
8	2 3 5 7 9 9
9	1 4 8

Key: 6 | 3 means 63

Graph C

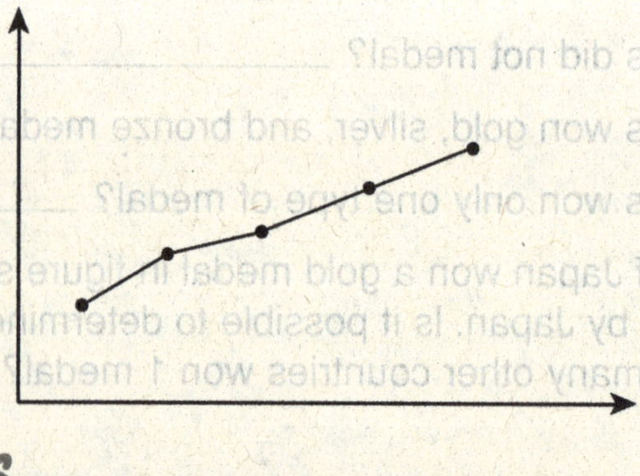

Think and Discuss

4. Explain how you decided which graph matches each description.

5. Explain what you would need to do to complete Graph A.

Name ______________________ Date ____________ Class ____________

LESSON 6H

Hands-On Lab

Multiple Representations of Data

You have learned how to make line plots, bar graphs, line graphs, and stem-and-leaf plots. Depending on the data to be displayed, some types of graphs are more useful than others.

Activity

The data below shows the daily high temperatures in degrees Fahrenheit for 28 days in February.

40	29	28	28	29	28	32
29	33	40	38	38	34	29
35	34	40	40	31	30	29
32	30	33	29	32	32	29

1. If you want to see how the temperature changes during the month, what type of graph would you make? Explain.

____

____

2. Make a stem and leaf plot and a line plot for the data.

Stems	Leaves
2	
3	
4	

2|8 means 28

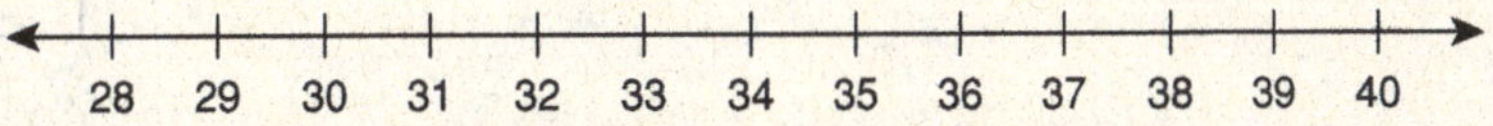

3. What is the mode? ____________________

4. Which display did you use to find the mode? Explain.

____

____

5. What is the median? ____________________

6. Which display did you use to find the median? Explain.

____

____

7. How many days had a high temperature of 33°? Explain how you got your answer.

____

____

Name ______________________________ Date ____________ Class ____________

LESSON 6H

Hands-On Lab

Multiple Representations of Data, continued

Think and Discuss

1. When is a stem-and-leaf plot a good display for data?

____

____

2. When is a line plot a good display for data?

____

____

Try This

A student surveyed pet owners to see what type of pet they have.

1. Decide which type graph is best to display the data, and then make the graph.

Pet	Number of people
Cat	36
Dog	48
Bird	15
Reptile	15
Other	29

2. What conclusions can you make based on your graph? Explain.

____

____

Name ______________________________ Date ____________ Class ____________

LESSON 6H

Review for Mastery

Choosing an Appropriate Display

You can choose the best way to display data by thinking about the data you want to show.

Use a line plot to show the frequency of data on a number line.

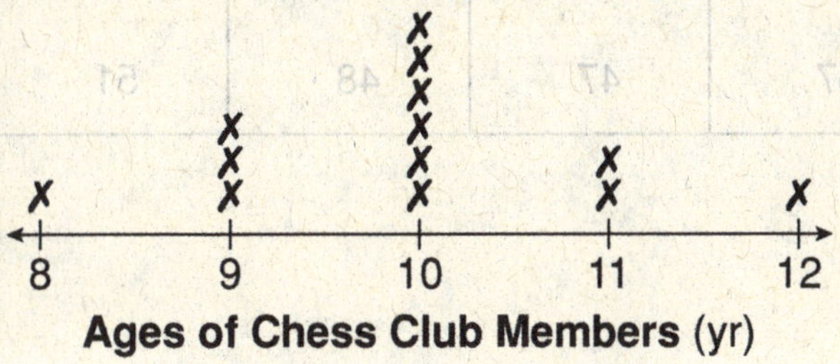

Use a line graph to show changes in data over time.

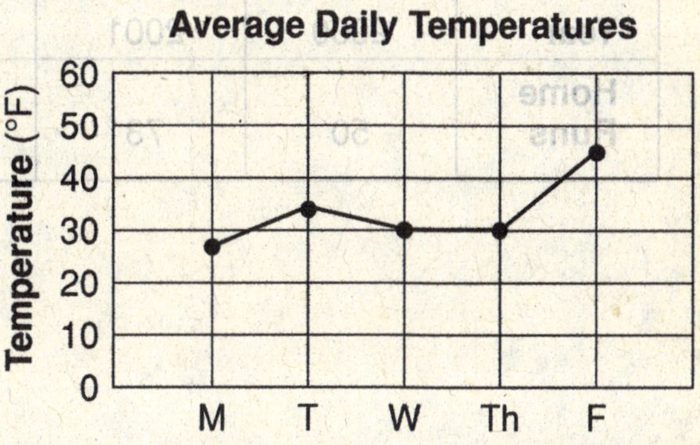

Use a bar graph to display data in separate categories.

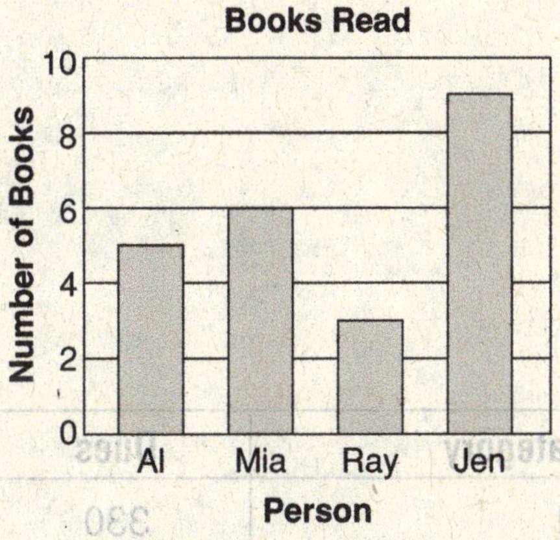

Use a stem-and-leaf plot to show how often data values occur and how they are distributed.

Test Scores

Stem	Leaves
7	0 0 1 3 8
8	1 2 5 8
9	0 0 4 5 5 7

Key: 7|0 = 70

Display the data. Use the most appropriate type of graph.

1.

Month	Mar	Apr	May	June	July
Average Temperature (°C)	12	18	20	23	25

Name ______________________ Date ____________ Class ____________

LESSON 6H

Homework and Practice

Choosing an Appropriate Display

1. The table below shows the number of home runs hit by the home run leader in major league baseball from 2000 to 2005. Which graph would be more appropriate to show the data—a line graph or a bar graph? Draw the more appropriate graph.

Year	2000	2001	2002	2003	2004	2005
Home Runs	50	73	57	47	48	51

2. The table at the right shows how the Hiawassee fishing club spends its annual dues. Which graph would be more appropriate to show the data—a circle graph or a histogram? Draw the more appropriate graph.

Category	Dues
Travel	330
Lodging	750
Tournament Fees	200
End of Season Party	220

Name ______________________ Date __________ Class __________

LESSON 6H

Homework and Practice

Choosing an Appropriate Display, continued

3. Silvia writes down the number of minutes she worked out at the gym for her last 15 visits. The data are given below.

25, 30, 30, 35, 40, 40, 35, 40, 45, 40, 45, 50, 50, 45, 50

a. Silvia would like to graph the data in a way that easily identifies the median and mode workout times. What type of graph should she make? Explain.

____

____

b. Make a graph of the data.

4. The table shows the number of siblings that Lim and his friends have. Which graph would be more appropriate to show the data—a line plot or a bar graph? Draw the more appropriate graph.

Name	Number of Siblings
Lim	1
Biba	0
Dan	2
Carl	1
Melissa	3

CHAPTER 6

Chapter Summary

Answer these question to summarize the important concepts from Chapter 6 in your own words.

1. Explain how to find the median of a data set with an even number of data values.

2. Explain how to make a bar graph from a given data set.

3. Explain the similarities and differences between a bar graph and a histogram.

4. Explain when you should use a line plot, line graph, bar graph, histogram, or a stem-and-leaf plot to display data.

For more review of Chapter 6:

- Complete the Chapter 6 Study Guide and Review on pages 340–342 of your textbook.
- Complete the Ready to Go On quizzes on pages 306 and 336 of your textbook.

Name ______________________ Date ____________ Class ____________

CHAPTER 6

KCCT Prep

Multiple Choice

1. Which location has an ordered pair of the form (a, a), for some number a?

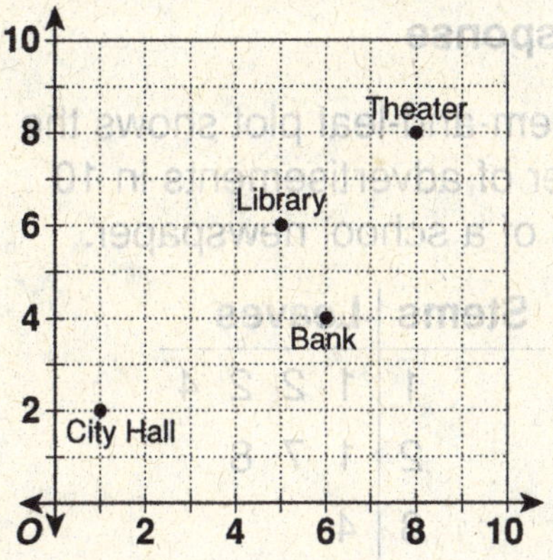

Ⓐ Bank Ⓒ Library
Ⓑ City Hall Ⓓ Theater

2. Tricia took two science quizzes. The mean of her scores is 7.5. One of her scores is 6. What is the other score?

Ⓐ 6 Ⓒ 8.5
Ⓑ 7.5 Ⓓ 9

3. The data set shows the number of wins by the Cleveland Indians baseball team from 1996 to 2006.

99 86 89 97 90 91 74 68 80 93 78

Ben wants to make a stem-and-leaf plot of the data where 7|8 means 78. How many stems will the plot have?

Ⓐ 3 Ⓒ 7
Ⓑ 4 Ⓓ 11

4. The fraction $\frac{n}{15}$ is greater than $\frac{1}{5}$ and less than $\frac{1}{3}$. Which of these could be the value of the numerator n?

Ⓐ 2 Ⓒ 4
Ⓑ 3 Ⓓ 5

5. Kenisha plots the point (3, 7) on a coordinate grid. From there, she moves one unit down and one unit right. At what point does she end up?

Ⓐ (2, 6) Ⓒ (4, 6)
Ⓑ (2, 8) Ⓓ (4, 8)

6. The equation $\square = \frac{m}{3}$ has the solution $m = 45$. What number is missing from the equation?What is the solution of the equation?

Ⓐ 5 Ⓒ 45
Ⓑ 15 Ⓓ 135

7. Trevor makes punch by mixing $2\frac{2}{3}$ cups of cranberry juice with $4\frac{1}{4}$ cups of apple juice. Which is the best estimate of the amount of punch he makes?

Ⓐ 6 cups Ⓒ 8 cups
Ⓑ 7 cups Ⓓ 10 cups

8. Mai collects data about the price of a quart of milk since 1960. Which of the following would be the best way to display this data?

Ⓐ Bar graph
Ⓑ Line graph
Ⓒ Line plot
Ⓓ Venn diagram

9. The prime factorization of a number has the form $2 \cdot 3^n$. Which of these could be the number?

Ⓐ 8 Ⓒ 36
Ⓑ 12 Ⓓ 54

10. The median price of six CD players is \$66. Five of the prices are \$49, \$60, \$75, \$49, and \$72. Which of these CANNOT be the price of the sixth CD player?

Ⓐ \$56 Ⓒ \$73
Ⓑ \$72 Ⓓ \$94

11. Jose plots the points *A, B, C,* and *D* as shown and connects them to form a square. Then he moves the square one unit to the right. Which of these is NOT a point on the square after it is moved?

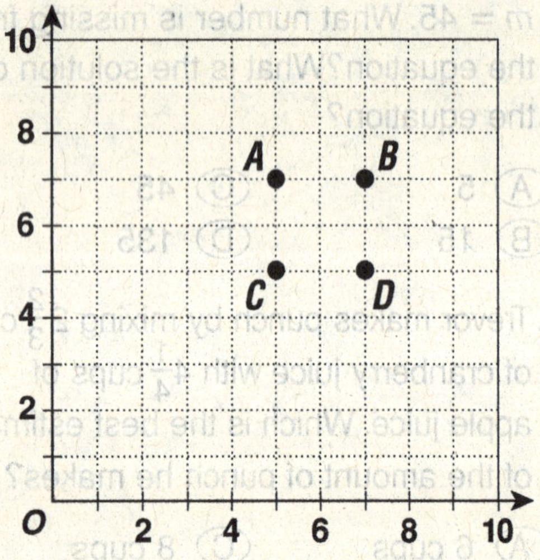

Ⓐ (6, 5) Ⓒ (7, 6)
Ⓑ (6, 7) Ⓓ (8, 5)

12. A frozen pizza weighs 12.4 ounces. Which is the best estimate of the total weight of 3.5 frozen pizzas?

Ⓐ 16 ounces Ⓒ 40 ounces
Ⓑ 38 ounces Ⓓ 43 ounces

13. Which pair of numbers are closest together on a number line?

Ⓐ $4\frac{3}{8}$ and 4.3 Ⓒ $6\frac{2}{3}$ and 6.8
Ⓑ $5\frac{1}{2}$ and 5.6 Ⓓ $7\frac{3}{4}$ and 7.9

14. Leonard has a piece of ribbon that is $3\frac{1}{2}$ ft long. He cuts off a piece that is $1\frac{1}{4}$ ft long. How many $\frac{1}{4}$-inch pieces can he make from the remaining ribbon?

Ⓐ 6 Ⓒ 9
Ⓑ 8 Ⓓ 12

Open Response

15. The stem-and-leaf plot shows the number of advertisements in 10 issues of a school newspaper.

Stems	Leaves
1	1 2 2 4
2	1 7 8
3	4
4	2 3

Key: 1|2 means 12

a. Find the mean, median, and mode.

b. Amy wants to convince her boss that she is doing a good job getting advertisers. Should she describe the data set using the mode or the median? Why?

16. The table shows the ordered pairs for buildings on a map of Roseville.

Buildings in Roseville	
Building	**Ordered Pair**
Department Store	(8, 3)
Concert Hall	(2, 6)
Hospital	(8, 6)

a. Graph and label the three buildings on a coordinate plane.

b. The three buildings and the post office form the four corners of a rectangle. What are the coordinates of the post office? Explain how you know.

CHAPTER 7

Project

Proportional Relationships: Counting Animals

Career: Fisheries Biologist

A fisheries biologist interacts with nature and with people. Fisheries biologists complete surveys, improve habitats, monitor water conditions, and work with land developers. Fisheries biologists often must determine the number of fish in a lake or pond. They use the tag, release, and recapture method to estimate this number.

Before you start the project, work through the Counting the Unknown worksheet on the next page.

In this chapter project . . .

You design a tag, release, and recapture program for an animal that lives in or near your community.

What to do:

- Research what animals live near your community and find estimates for how large the population is.
- Simulate a tag, release, and recapture program by following the same steps outlined on the Counting the Unknown project worksheet. Create counters to represent the animals.
- Divide the class into students who are collectors, and students who influence the population. The students who influence the population can simulate migration, disease, or population booms by adding and removing simulated "animals" from the bag. The students who collect the data should try to guess what is happening to the population based on tag, release, and recapture simulations.

What to present:

- Tables that show the results of the data collection.
- A summary of the predicted population values versus the actual values.

Where's the math?

- Use proportions to predict the animal population.

go.hrw.com
Chapter Project Online
KEYWORD: MR7 Ch7

Name ______________________ Date ____________ Class ____________

CHAPTER 7

Project Recording Sheet

Proportional Relationships

Counting the Unknown

Will the capture-recapture technique used by fisheries biologists help us predict the number of unknown objects? Try this activity and find out.

Begin with a bag filled with centimeter cubes, cardboard squares, or other similar objects with multiple colors. DO NOT LOOK INTO THE BAG.

Step 1 Reach into the bag (without looking) and remove some, but not all, of the objects. Mark the objects and record the number of objects that you marked below.

____

Step 2 Return all objects to the bag and shake the bag to distribute the objects. (The bag now contains both marked and unmarked objects.)

Step 3 Reach into the bag again (still without looking) and remove some of the objects. Record the number of marked objects that you drew in column 2 of the table and the number of objects (marked and unmarked) in column 3. Return the objects to the bag and repeat for at least 5 trials.

Step 4 Total the results. Use the the formula in the table to calculate the predicted total number of objects.

$$\frac{\text{Total Marked Number Recaptured}}{\text{Total Number Recaptured}} = \frac{\text{Original Number Marked}}{\text{Predicted Total Number}}$$

Trial	2 Marked Number Recaptured	3 Number Recaptured
1		
2		
3		
4		
5		
Total		

Open the bag. Compare the actual number of objects with the prediction that you made. Try the activity again. How do your results compare with the first time? What do these results tell you about sampling an unknown population?

EXPLORATION

Program of Studies and Core Content
MA-6-NPO-S-RP2, MA-6-NPO-S-RP3, MA-06-1.4.1

7A Ratios and Unit Rates

A TV network offers the numbers of shows each week shown in the table.

You can compare the numbers of TV shows by using ratios. A *ratio* is a comparison of two quantities that uses division. For example, the ratio of science fiction shows to drama shows is $\frac{3}{14}$, which can also be written 3:14 or 3 to 14.

Type of TV Show	Number of Shows
Comedy	14
Drama	14
Science fiction	3
Game show	7
Talk show	15
News	14
Morning show	10
Late-night show	5
Sports	6

Find each ratio.

1. comedy shows to game shows
2. game shows to news shows
3. morning shows to late-night shows
4. talk shows to sports shows

Think and Discuss

5. **Discuss** whether the ratios in numbers 1–4 compare part to part, part to whole, or whole to part.
6. **Discuss** whether order is important when calculating ratios. (*Hint:* Is $\frac{\text{news}}{\text{sports}}$ equivalent to $\frac{\text{sports}}{\text{news}}$?)

Name ______________________ Date ____________ Class ____________

LESSON **7A**

Review for Mastery

Ratios and Unit Rates

A ratio is a comparison of two quantities by division.

To compare the number of times vowels are used to the number of time consonants are used in the word "mathematics," first find each quantity.

Number of times vowels are used: 4
Number of times consonants are used: 7

Then write the comparison as a ratio, using the quantities in the same order as they appear in the word expression. There are three ways to write a ratio.

$\frac{4}{7}$	4 to 7	4:7

Write each ratio.

1. days in May to days in a year

____

2. sides of triangle to sides of a square

____

Equivalent ratios are ratios that name the same comparison.

The ratio of inches in a foot to inches in a yard is $\frac{12}{36}$. To find equivalent ratios, divide or multiply the numerator and denominator by the same number.

$\frac{12}{36} = \frac{12 \div 3}{36 \div 3} = \frac{4}{12}$ $\qquad$ $\frac{12}{36} = \frac{12 \cdot 2}{36 \cdot 2} = \frac{24}{72}$

So, $\frac{12}{36}$, $\frac{4}{12}$, and $\frac{24}{72}$ are equivalent ratios.

Write three equivalent ratios to compare each of the following.

3. 8 triangles to 12 circles

____

4. 20 pencils to 25 erasers

____

5. 5 girls to 6 boys

____

6. 10 pants to 14 shirts

Name ______________________________ Date ____________ Class ____________

LESSON **7A**

Review for Mastery

Ratios and Unit Rates, continued

A rate is a comparison of two quantities that have different units of measure.

Suppose a bus travels 150 miles in 3 hours. The rate could be written as $\frac{150 \text{ miles}}{3 \text{ hours}}$.

When the second term of a rate is 1 unit, the rate is a unit rate.

To write $\frac{150 \text{ miles}}{3 \text{ hours}}$ hours as a unit rate, divide each term by 3.

$$\frac{150 \text{ miles}}{3 \text{ hours}}$$

$$= \frac{150 \text{ miles} \div 3}{3 \text{ hours} \div 3}$$

$$= \frac{50 \text{ miles}}{1 \text{ hour}}$$

The unit rate is $\frac{50 \text{ miles}}{\text{hour}}$.

Find each unit rate.

7. $\frac{40 \text{ books}}{2 \text{ shelves}}$ ____________

8. $\frac{36 \text{ students}}{6 \text{ groups}}$ ____________

9. $\frac{300 \text{ seconds}}{5 \text{ minutes}}$ ____________

10. $\frac{54 \text{ miles}}{2 \text{ gallons}}$ ____________

11. $\frac{4 \text{ miles}}{20 \text{ minutes}}$ ____________

12. $\frac{\$1.29}{3 \text{ pounds}}$ ____________

13. $\frac{72 \text{ hours}}{3 \text{ days}}$ ____________

14. $\frac{42 \text{ trading cards}}{6 \text{ packs}}$ ____________

Name ______________________ Date ____________ Class ____________

LESSON **7A**

Homework and Practice

Ratios and Unit Rates

Use the table to write each ratio.

Animals at the Aquarium	
Sharks	17
Eels	7
Sea Turtles	11
Penguins	6
Alligators	3

1. alligators to penguins ____________
2. eels to sea turtles ____________
3. sharks to alligators ____________
4. penguins to eels ____________
5. sea turtles to sharks ____________
6. Write three equivalent ratios to compare the number of diamonds with the number of spades in the box.

____

♠♠♠♠♠♠♠♠♠♠♠♠♦♦♦♦

Use the table to write each ratio as a fraction.

Soccer Team Stats	Kings	Hawks
Wins	18	13
Losses	10	15

7. Hawks' wins to Hawks' losses

____

8. Kings' losses to Kings' wins

____

9. Kings' losses to Hawks' losses

____

10. Hawks' wins to Kings' losses ____________
11. A 5-lb sack of potatoes costs $4.32. An 8-lb sack of potatoes costs $6.78. Use unit rates to determine which is the better deal.

____

12. Jessie charges Jim $45.00 for 3 hours of tutoring. Mike charges Jim $90.00 for 5 hours of tutoring. Who offers Jim the lower hourly rate of pay?

____

EXPLORATION

Program of Studies and Core Content
MA-6-NPO-S-RP2, MA-6-NPO-S-RP3, MA-06-1.4.1, MA-06-4.1.1

7B Using Tables to Explore Equivalent Ratios and Rates

McMillans Restaurant is celebrating its 50th anniversary by offering 3 hamburgers for $2.

1. Sue makes this table for quick reference at the drive-up window. Complete the table.

Number of Hamburgers	3	6	9		15		21
Total Cost ($)	2	4					

2. The weekend goal is to sell 1,200 hamburgers. How much money will the restaurant receive if it reaches its goal of 1,200 hamburgers? Complete the table to help you answer this question.

Number of Hamburgers Sold	300	600	900	1,200
Amount Received ($)	200			

Think and Discuss

3. Describe any patterns you notice in the table for Exercise 1.

4. Explain how you could find the amount of money the restaurant would receive if it sold 1,500 hamburgers.

Name ____________________ Date __________ Class __________

LESSON 7B

Review for Mastery

Using Tables to Explore Equivalent Ratios and Rates

You can use a table to find ratios equivalent to $\frac{1}{4}$.
Write the numerator in the top box for the original ratio.
Write the denominator in the bottom box for the original ratio.
Then multiply the numerator and the denominator by 2, 3, and 4.

Original ratio	**1 · 2**	**1 · 3**	**1 · 4**
1	2	3	4
4	8	12	16
	4 · 2	**4 · 3**	**4 · 4**

Use each new numerator and denominator to write an equivalent ratio.

So, the ratios $\frac{2}{8}$, $\frac{3}{12}$, and $\frac{4}{16}$ are equivalent to $\frac{1}{4}$.

1. Use the table to find three ratios equivalent to $\frac{2}{5}$.

2			
5			

Equivalent ratios: $\frac{2}{5}$, ________, ________, and ________

You can use a table to find ratios equivalent to 3 to 8.
Write the first number in the top box for the original ratio.
Write the second number in the bottom box for the original ratio.
Then multiply both numbers by 2, 3, and 4.

Original ratio	**3 · 2**	**3 · 3**	**3 · 4**
3	6	9	12
8	16	24	32
	8 · 2	**8 · 3**	**8 · 4**

Use each new top number and bottom number to write an equivalent ratio.

So, the ratios 6 to 16, 9 to 24, and 12 to 32 are equivalent to 3 to 8.

2. Use the table to find three ratios equivalent to 4 to 10.

4			
10			

Equivalent ratios: 4 to 10, ________, ________, and ________

Name ____________________ Date __________ Class __________

LESSON 7B

Homework and Practice

Using Tables to Explore Equivalent Ratios and Rates

Use a table to find three equivalent ratios.

1. $\frac{3}{5}$ ____________________

2. 4 to 3 ____________________

3. $\frac{5}{6}$ ____________________

4. 8 : 10 ____________________

5. 7 to 2 ____________________

6. 8/5 ____________________

7. 27 to 81 ____________________

8. 4 : 9 ____________________

9. Reel-Fun movie theater shows double-features on Saturday afternoons. The table shows the total cost for several groups to go see a double-feature.

Number in Group	8	6	4	2
Total Cost ($)	54	40.50	27	13.50

How much do you predict it would cost for a group of 5 people to see a double-feature?

____

10. The table shows how much flour is needed to make some pancakes.

Flour (cups)	3	9	15	18	24
Number of Pancakes	13	39	65	78	104

How many pancakes do you predict you could make with 12 cups of flour?

____

Name ______________________ Date __________ Class __________

LESSON 7B

Homework and Practice

Using Tables to Explore Equivalent Ratios and Rates, cont.

Use a table to find the unit price for each item.

11.

Bananas (lb)	2.4	3.1	3.5	3.8
Cost ($)	1.92	2.48	2.80	3.04

Unit Price: ______________ per ______________

12.

Number of Games Bowled	3	6	7	10
Cost ($)	6.75	13.50	15.75	22.50

Unit Price: ______________ per ______________

13.

Number of Songs Downloaded	8	12	20	30
Cost ($)	7.92	11.88	19.80	29.70

Unit Price: ______________ per ______________

Use the table for exercises 14–16.

distance (ft)	0.15	0.6	1.95	2.7	3
time (s)	3.5	14	45.5	63	

14. Use a calculator to verify the ratios for each pair of values in the table are equal. Write it to 4 decimal places.

____

15. Rewrite the ratio 0.6 to 14 by multiplying both numbers by 10. Write the ratio as a fraction.

____

16. Write the fraction from Exercise 15 in lowest terms. Use your answer to fill in the missing value in the table.

____

Program of Studies and Core Content
MA-6-NPO-S-RP3,
MA-06-1.4.1

7C Proportions

An automobile assembly line finishes 3 cars every 2 hours.

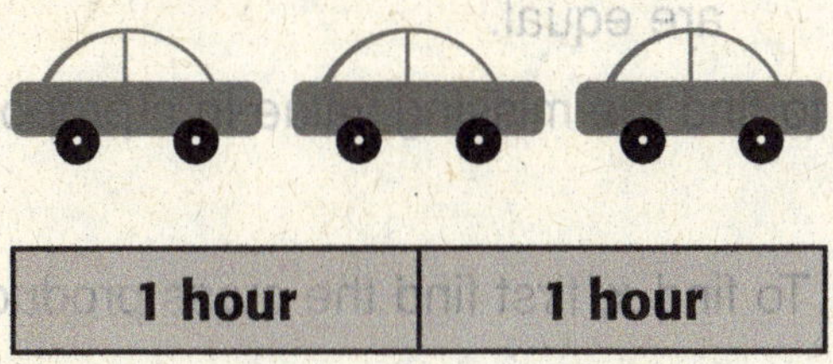

1. Use the diagram to determine how many cars are finished each hour.

2. Use the diagram to determine approximately how long it takes to finish 1 car.

3. If it takes 2 hours to finish 3 cars, how many hours does it take to finish

   a. 6 cars? **b.** 9 cars? **c.** 12 cars?
   d. 4 cars? **e.** 8 cars? **f.** 16 cars?

Think and Discuss

4. **Discuss** how you used the diagram to solve numbers 1 and 2.
5. **Explain** how you solved numbers 3a–3f.

Name ______________________ Date __________ Class __________

LESSON 7C

Review for Mastery

Proportions

A proportion is an equation that shows two equivalent ratios.

$\frac{3}{4} = \frac{9}{12}$ is an example of a proportion.

$3 \cdot 12 = 36$ and $4 \cdot 9 = 36$. The cross products of proportions are equal.

You can use cross products to find the missing value in a proportion.

$$\frac{3}{x} = \frac{12}{48}$$

$12 \cdot x = 3 \cdot 48$ To find x, first find the cross products.

$12x = 144$

Think: $144 \div 12 = x$ Then use a related math sentence to solve the equation.

$x = 12$

So, $\frac{3}{12} = \frac{12}{48}$.

Find the cross products to solve each proportion.

1. $\frac{x}{8} = \frac{3}{4}$

   $x \cdot 4 =$ ______

   ______

2. $\frac{2}{3} = \frac{x}{6}$

   $2 \cdot 6 =$ ______

   ______

3. $\frac{2}{5} = \frac{4}{x}$

   $2 \cdot x =$ ______

   ______

4. $\frac{6}{x} = \frac{1}{3}$

   $6 \cdot 3 =$ ______

   ______

5. $\frac{3}{8} = \frac{12}{x}$ ______

6. $\frac{3}{5} = \frac{6}{x}$ ______

7. $\frac{x}{8} = \frac{2}{16}$ ______

8. $\frac{2}{9} = \frac{4}{x}$ ______

9. $\frac{3}{4} = \frac{15}{x}$ ______

10. $\frac{1}{2} = \frac{x}{30}$ ______

11. $\frac{x}{5} = \frac{24}{30}$ ______

12. $\frac{25}{35} = \frac{5}{x}$ ______

Name ______________________________ Date __________ Class __________

LESSON **7C**

Homework and Practice

Proportions

Find the missing value in each proportion.

1. $\frac{18}{33} = \frac{n}{11}$ ____________

2. $\frac{2}{26} = \frac{3}{n}$ ____________

3. $\frac{n}{5} = \frac{8}{10}$ ____________

4. $\frac{16}{5} = \frac{n}{20}$ ____________

5. $\frac{3}{2} = \frac{9}{n}$ ____________

6. $\frac{4}{n} = \frac{24}{18}$ ____________

7. $\frac{16}{9} = \frac{n}{18}$ ____________

8. $\frac{15}{3} = \frac{45}{n}$ ____________

9. $\frac{4}{84} = \frac{1}{n}$ ____________

Write a proportion for each model.

10. 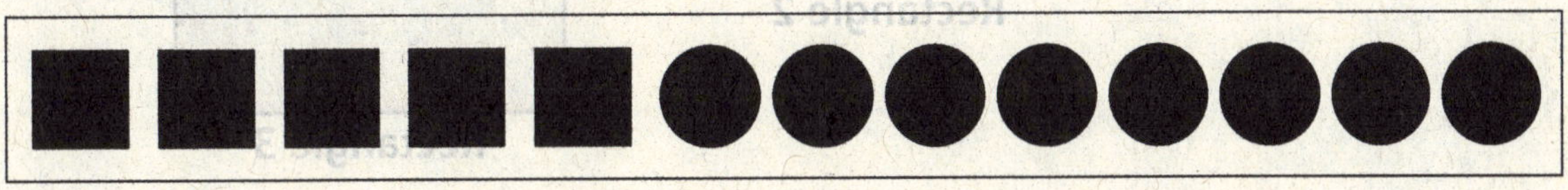

____

11. 

____

12. Jonathan decided to pledge \$1.45 for every dollar donated in the charity auction. If the charity auction takes in a total of 75 dollars, how much money will Jonathan donate?

____

13. Gerald scored 15 points for his basketball team for every 25 minutes he played in the season. If he played 200 minutes in the season, how many points did he score?

____

EXPLORATION

Program of Studies and Core Content

MA-6-G-S-SR6, MA-6-M-S-MPA2, MA-06-3.1.5

7D Similar Figures

Similar rectangles have the same shape but may be different sizes.

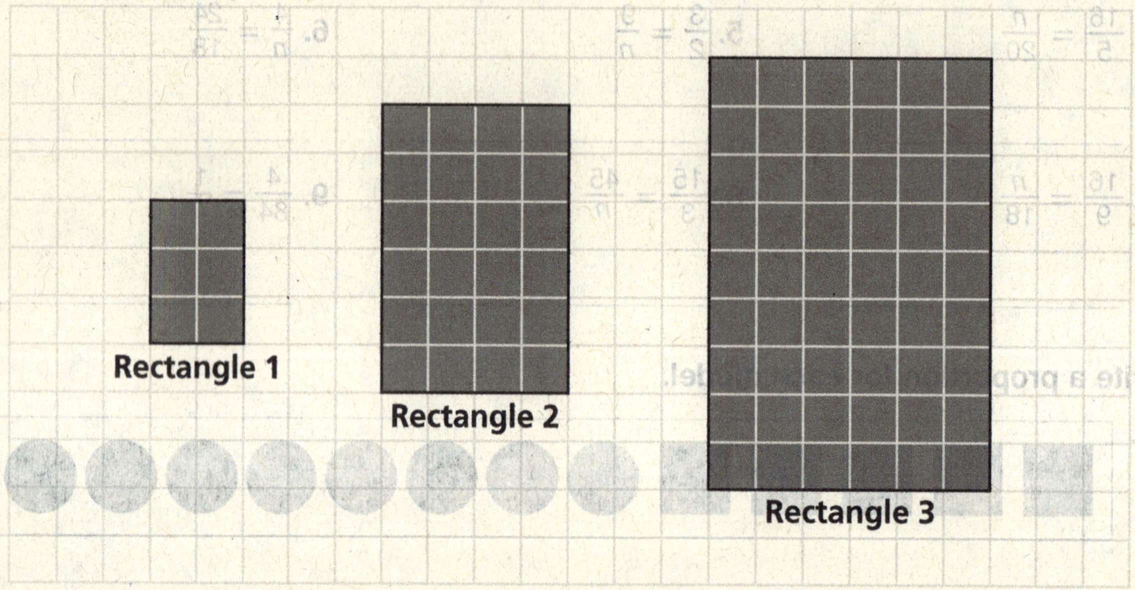

Determine the length and width of each rectangle and find each ratio.

	Ratio 1	Ratio 2
1.	$\frac{\text{length of rectangle 1}}{\text{length of rectangle 2}} =$	$\frac{\text{width of rectangle 1}}{\text{width of rectangle 2}} =$
2.	$\frac{\text{length of rectangle 2}}{\text{length of rectangle 3}} =$	$\frac{\text{width of rectangle 2}}{\text{width of rectangle 3}} =$
3.	$\frac{\text{length of rectangle 1}}{\text{length of rectangle 3}} =$	$\frac{\text{width of rectangle 1}}{\text{width of rectangle 3}} =$

Think and Discuss

4. Describe the pattern between ratio 1 and ratio 2.

5. Explain why the three rectangles are similar.

Name ______________________________ Date __________ Class __________

LESSON 7D

Hands-On Lab 1

Constructing Similar Figures

Two triangles are similar if their corresponding angles are congruent and the lengths of their corresponding sides are in proportion.

Activity 1

Use a metric ruler to measure each side of the right triangle below. Write the length of each side in millimeters to the nearest millimeter.

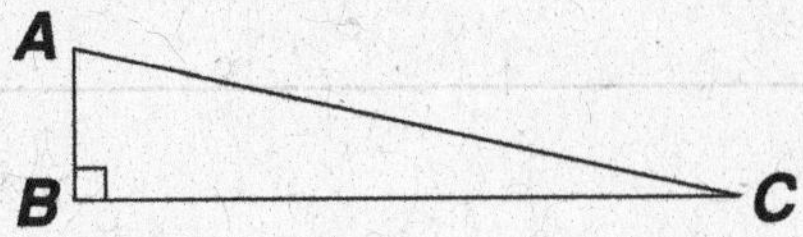

1. *AB*: __________ **2.** *BC*: __________ **3.** *AC*: __________

In the space below, you will draw a new triangle, ΔDEF, with sides that are two times the length of the corresponding sides of triangle ΔABC.

4. First, draw a vertical segment that is two times the length of $\overline{AB}$. Label the endpoints *D* and *E*. What is the length of $\overline{DE}$? __________

5. Second, draw a segment, $\overline{EF}$, perpendicular to $\overline{DE}$ that is two times the length of $\overline{BC}$. What is the length of $\overline{EF}$? __________

6. Next, draw a segment that connects *D and F*. To find the length of $\overline{DF}$, measure with the ruler.

Write the length of $\overline{DF}$________________________.

7. How does the length of $\overline{DF}$ compare to the length of $\overline{AC}$?

________________________.

Name ______________________ Date ____________ Class ____________

LESSON 7D

Hands-On Lab 1

Constructing Similar Figures, continued

8. Explain why triangles ΔABC and ΔDEF are similar triangles.

____

____

9. List the pairs of corresponding angles that the congruent.

____

Think and Discuss

1. What are some other geometric shapes, besides triangles and rectangles, which could be similar?

____

____

2. How could you use a protractor and ruler to find out if any two shapes are similar?

____

____

Name ______________________ Date __________ Class __________

LESSON 7D

Hands-On Lab 1

Constructing Similar Figures, continued

Try This

You will use a protractor and a metric ruler to draw a triangle, ΔTUV, similar to triangle ΔXYZ. Make the length of each side of triangle ΔTUV 1.5 times the length of the corresponding side of triangle ΔXYZ.

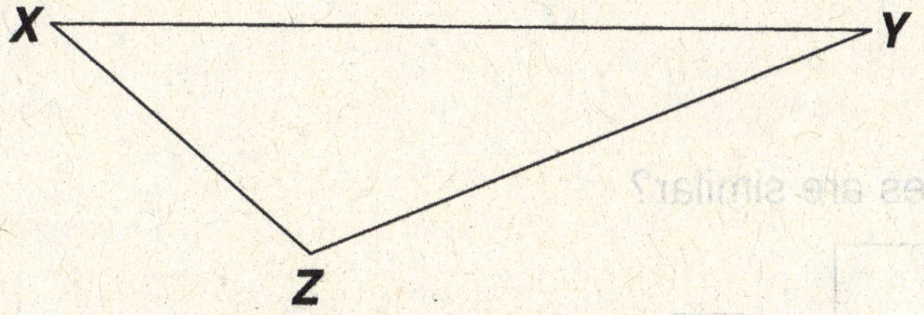

First, find the lengths of each side of triangle ΔXYZ.

1. $\overline{XY}$ __________ **2.** $\overline{YZ}$ __________ **3.** $\overline{XZ}$ __________

Next, find the measure of each angle.

4. $m\angle X$ __________ **5.** $m\angle Y$ __________ **6.** $m\angle z$ __________

7. Draw triangle ΔTUV in the space below. Label the side lengths.

Name ______________________ Date ____________ Class ____________

LESSON 7D

Hands-On Lab 2

Measuring Similar Figures

Similar figures have the same shape, but can be different sizes. Two similar figures have corresponding angles congruent and corresponding sides in proportion.

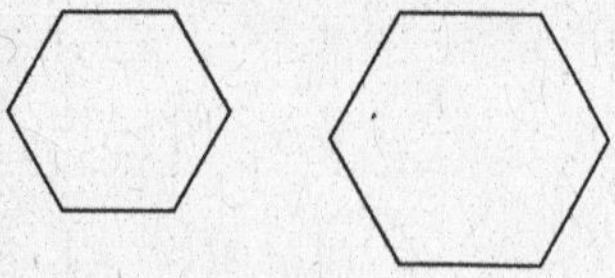

Activity 1

Which of the following rectangles are similar?

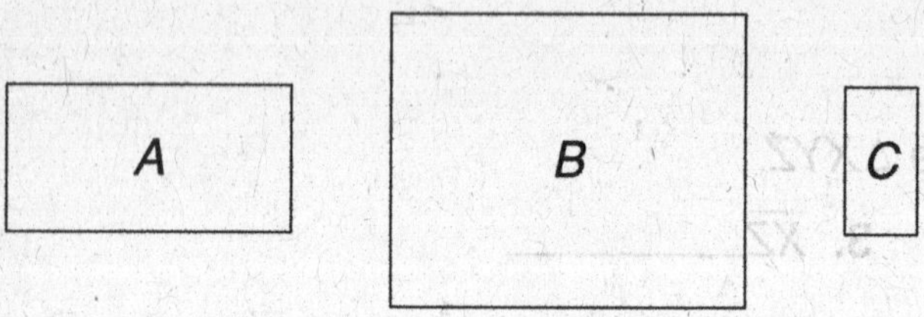

Compare the angles of each rectangle.

Observe that all angles are right angles, therefore all angles are congruent.

Measure the lengths of each rectangle with an inch ruler.

Rectangle *A* is 1 inch by 0.5 inch.
Rectangle *B* is 1.25 inches by 1 inch.
Rectangle *C* is 0.25 inch by 0.5 inch.

Create ratios of the sides. Match the longer side of the rectangle with the longer side of another rectangle.

Compare Rectangle *A* and Rectangle *B*.

$\frac{1}{1.25} \stackrel{?}{=} \frac{0.5}{1}$; 1 ≠ 0.625, the sides are not in proportion.

Compare Rectangle *B* and Rectangle *C*.

$\frac{1.25}{0.5} \stackrel{?}{=} \frac{1}{0.25}$; since 0.3125 ≠ 0.5, the sides are not in proportion.

Compare Rectangle *A* and Rectangle *C*.

$\frac{1}{0.5} \stackrel{?}{=} \frac{0.5}{0.25}$; since 0.25 = 0.25, the sides are proportional.

Determine which rectangles are similar.

Since Rectangles *A* and *C* have congruent angles and sides in proportion, they are similar.

Name ______________________ Date __________ Class __________

LESSON 7D

Hands-On Lab 2

Measuring Similar Figures, continued

Think and Discuss

1. What can you do to Rectangle *C* to make it look more similar to Rectangle *A*?

____

2. How can you decide which sides are corresponding with similar figures?

____

____

Activity 2

Below are three pairs of similar triangles. Match each triangle to its similar partner. Use a protractor to measure angles and a centimeter ruler to measure sides.

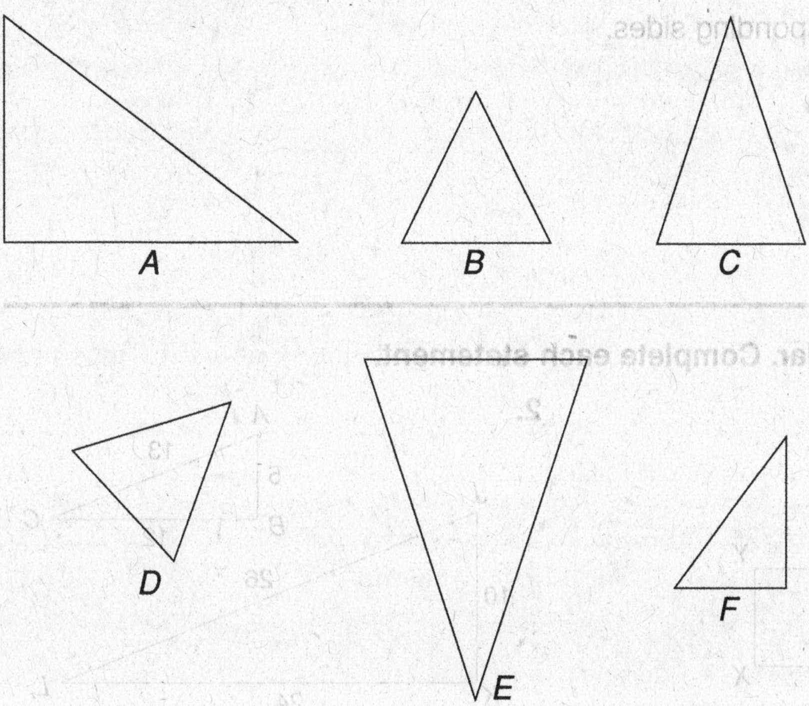

____

____

____

Name ______________________ Date ____________ Class ____________

LESSON 7D

Review for Mastery

Similar Figures

Similar figures have the same shape but are different sizes.

Similar figures have corresponding sides and corresponding angles. Corresponding sides are proportional. Corresponding angles are congruent.

Look at the similar triangles below.

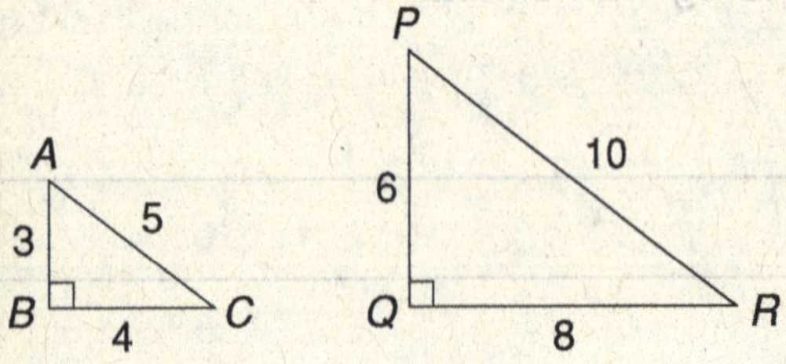

$\overline{AB}$ corresponds to $\overline{PQ}$. $\angle A$ corresponds to $\angle P$.
$\overline{BC}$ corresponds to $\overline{QR}$. $\angle B$ corresponds to $\angle Q$.
$\overline{AC}$ corresponds to $\overline{PR}$. $\angle C$ corresponds to $\angle R$.

Compare the ratios of corresponding sides.

$$\frac{AB}{PQ} = \frac{3}{6} = \frac{1}{2}$$

$$\frac{BC}{QR} = \frac{4}{8} = \frac{1}{2}$$

$$\frac{AC}{PR} = \frac{5}{10} = \frac{1}{2}$$

The figures shown are similar. Complete each statement.

1.

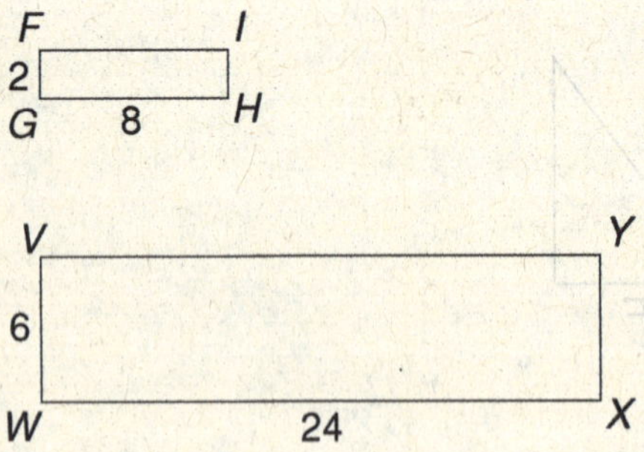

$\overline{GH}$ corresponds to ________.

$$\frac{FG}{\square} = \frac{GH}{WX}$$

2.

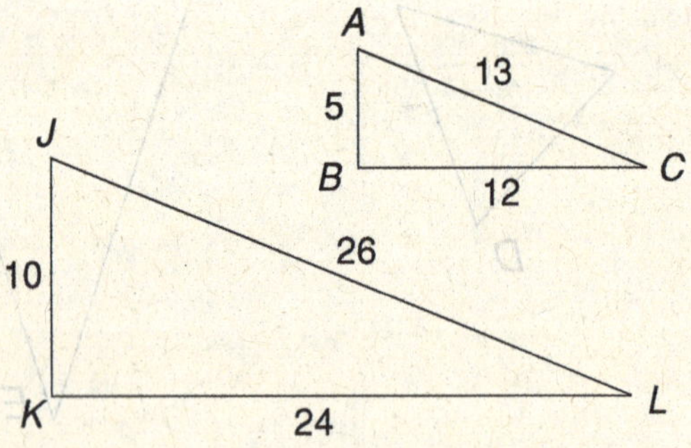

$\angle K$ corresponds to ________.

The ratio of corresponding sides of ΔABC to ΔJKL is ________.

Name ______________________ Date __________ Class __________

LESSON **7D**

Homework and Practice

Similar Figures

The two triangles are similar. Find the missing information.

1.

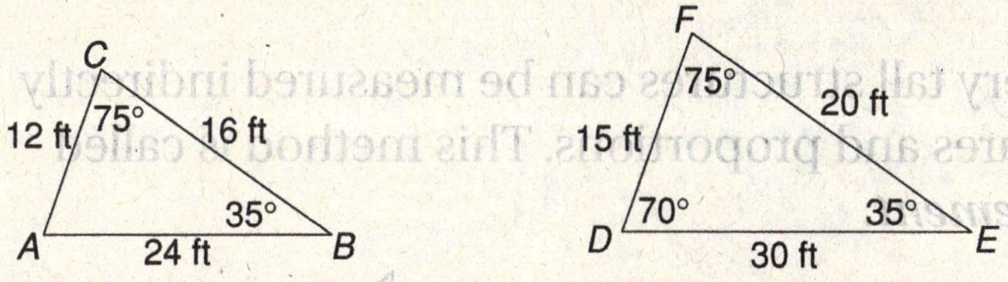

m∠*A* __________ perimeter of Δ*ABC* __________

$\frac{AC}{DF}$ __________ perimeter of Δ*DEF* __________

2. Find the ratio of the perimeters of Δ*ABC* and Δ*DEF*. How does it compare to the ratio of corresponding sides?

____

Are the figures similar? Explain.

3.

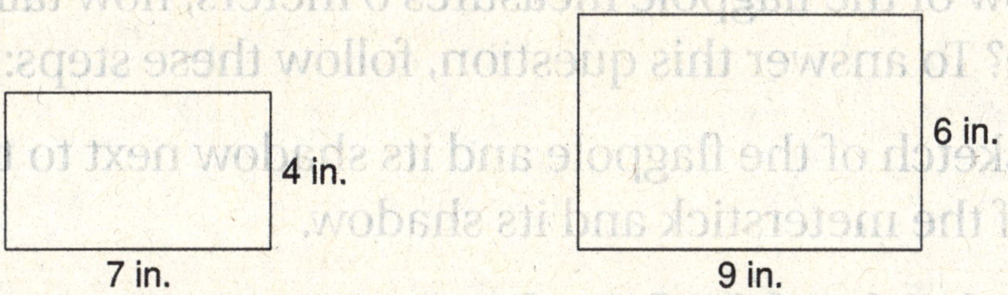

____

4.

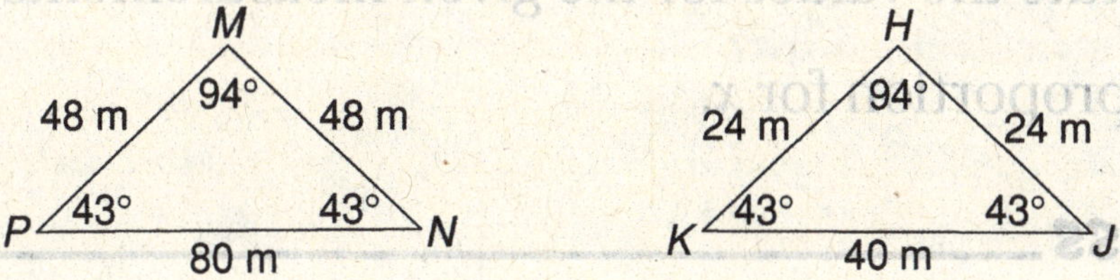

____

5. Lenny's dog had puppies. He saw a pen that is a triangular shape with the sides measuring 6 feet, 12 feet, and 9 feet. He wants his pen to be a similar triangle with the largest side measuring 24 feet. Determine the perimeter of the new pen.

____

EXPLORATION

Program of Studies and Core Content

MA-6-NPO-S-RP3, MA-6-G-S-SR6, MA-06-1.4.1, MA-06-3.1.5

7E Indirect Measurement

The heights of very tall structures can be measured indirectly using similar figures and proportions. This method is called ***indirect measurement.***

Augustine and Carmen want to measure the height of the school's flagpole. To do this, they go outside and hold a meterstick upright. The meterstick casts a shadow that measures 50 cm.

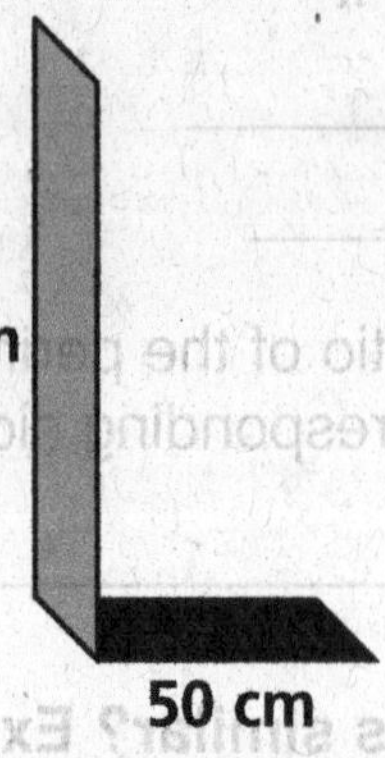

1. If the shadow of the flagpole measures 6 meters, how tall is the flagpole? To answer this question, follow these steps:

a. Draw a sketch of the flagpole and its shadow next to the sketch of the meterstick and its shadow.

b. Label the height of the flagpole x.

c. Write the proportion $\frac{\text{height of flagpole}}{\text{shadow of flagpole}} = \frac{\text{height of meterstick}}{\text{shadow of meterstick}}$, and substitute the values for the given measurements.

d. Solve the proportion for x.

Think and Discuss

2. Explain how you solved the proportion for x in number **1d.**

3. Explain whether you could have solved the problem by writing the proportion as $\frac{\text{shadow of flagpole}}{\text{height of flagpole}} = \frac{\text{shadow of meterstick}}{\text{height of meterstick}}$.

Name ______________________________ Date ___________ Class ___________

LESSON 7E

Technology Lab

Indirect Measurement

Use your computer tools to draw a diagram that helps you model indirect measurement.

Activity

It's spirit week at school and your team wants to decorate the light poles in the front of the building with strings of pennants in your school's colors. The pennants are strung together and sold by the foot. To know how many feet of pennants to buy, you need to know the height of the light poles.

1. The student who will be used for a comparison is 5 feet and 11 inches tall. The shadow cast by the student is 15 feet long. The shadow cast by the light pole is 88 feet long. You will use drawing tools to create and label a model of the student and a light pole.

2. To make the drawing tools easily accessible, click on the View menu and select the Toolbars option. From there, be certain Drawing is selected.

3. Draw the vertical line for the light pole. To draw a straight line, choose the Line command, click in your document and drag the line across the screen. Then, draw the horizontal line for the shadow. This line should start at the bottom of your first line. Then make lines for the student and the student's shadow.

4. Be sure to make your lines the right lengths. For example, the line for the student's shadow should be longer than the line for the pole. Also, the line for the student should be shorter than the line for the pole.

5. To make this drawing a useful problem-solving tool for math, the illustration should be labeled. Choose Text Box from the Insert menu. Type the label information into the box. Click and drag on the box handles to resize and reposition the words within the illustration.

Name ______________________________ Date __________ Class __________

LESSON 7E

Technology Lab

Indirect Measurement, continued

6. Find the ratio of height to shadow to determine the height of the light.

The ratios you constructed should look like this. Remember to change all units to inches.

$$\frac{\text{5 feet 11 inches}}{\text{15 feet}} = \frac{h\text{ feet}}{\text{88 feet}} \qquad \frac{\text{71 inches}}{\text{180 inches}} = \frac{h\text{ inches}}{\text{1056 inches}}$$

____

Think and Discuss

1. There are six light poles in the front of school. Will you be able to buy enough flags for each of the poles if they cost $0.80 per foot and you have $155.00 to spend on the project?

____

____

Try This

1. Complete the following ratios to compute the height of the missing side of these right triangles.

$$\frac{5}{20} = \frac{h}{120}$$

____

$$\frac{h}{9} = \frac{7}{21}$$

____

2. On a separate piece of paper, draw an illustration and show the ratio for a redwood tree that is 200 feet in height that casts a shadow 500 feet long.

Name ______________________ Date __________ Class __________

LESSON **7E**

Review for Mastery

Indirect Measurement

If you cannot measure a length directly, you can use indirect measurement. Indirect measurement uses similar figures and proportions to find lengths.

The small tree is 8 feet high and it casts a 12-foot shadow. The large tree casts a 36-foot shadow.

The triangles formed by the trees and the shadows are similar. So, their heights are proportional.

To find the height of the large tree, first set up a proportion. Use a variable to stand for the height of the large tree.

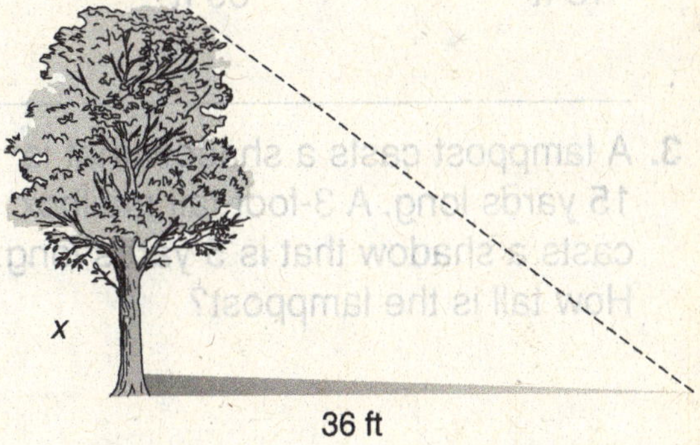

$\frac{8}{12} = \frac{x}{36}$	Write a proportion using corresponding sides.
$8 \cdot 36 = 12 \cdot x$	The cross products are equal.
$12x = 288$	x is multiplied by 12.
$\frac{12x}{12} = \frac{288}{12}$ $x = 24$	Divide both sides by 12.

So, the height of the tall tree is 24 feet.

Use indirect measurement to find the missing heights.

1.

____

2.

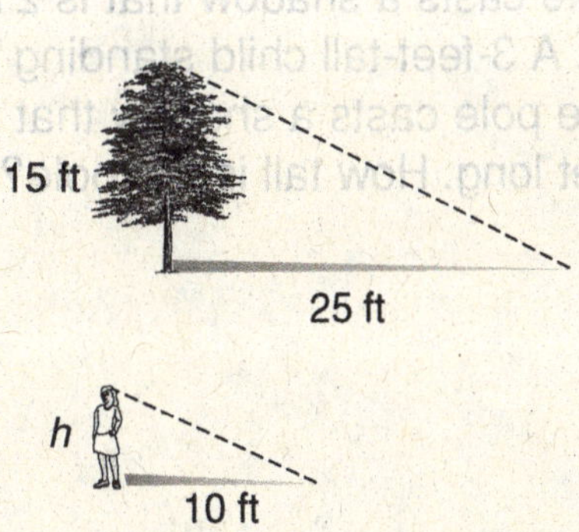

____

Name ______________________ Date ____________ Class ____________

LESSON 7E

Homework and Practice

Indirect Measurement

Write the correct answer.

1. Use similar triangles to find the height of the building.

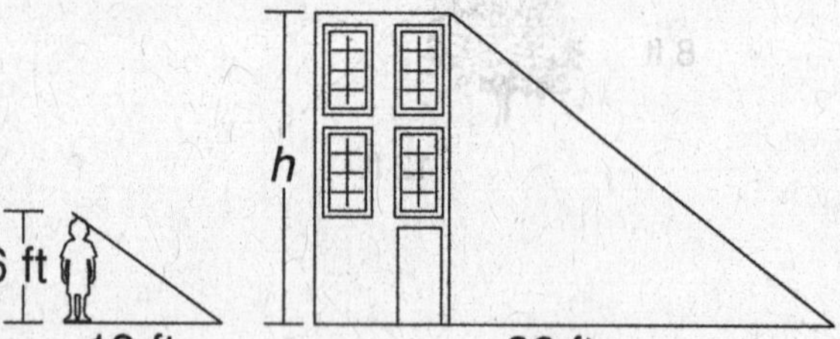

____

2. Use similar triangles to find the height of the tree.

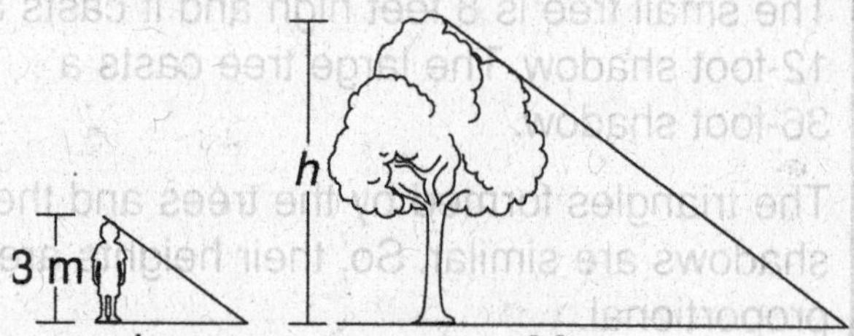

____

3. A lamppost casts a shadow that is 15 yards long. A 3-foot-tall mailbox casts a shadow that is 5 yards long. How tall is the lamppost?

____

4. An 8-foot-tall statue stands in the park and casts a shadow that is 16 feet long. A dog stands next to it and is 3 feet tall. How long is the dog's shadow?

____

5. A building casts a shadow that is 420 meters long. At the same time, a person who is 2 meters tall casts a shadow that is 24 meters long. How tall is the building?

____

6. On a sunny day around noon, a tree casts a shadow that is 12 feet long. At the same time, a person who is 6 feet tall standing beside the tree casts a shadow that is 2 feet long. How tall is the tree?

____

7. A pole casts a shadow that is 21 feet long. A 3-feet-tall child standing next to the pole casts a shadow that is 9 feet long. How tall is the pole?

____

8. Jeremy has two trophies next to each other sitting in the window of his room. His football trophy is 7 inches tall and his basketball trophy is 13 inches tall. As the light shines in, the basketball trophy's shadow measures 26 inches. How long is the football trophy's shadow?

____

EXPLORATION

7F Scale Drawings and Maps

KY Program of Studies and Core Content
MA-6-NPO-S-RP3, MA-06-1.4.1

A *scale* is a ratio between two sets of measurements. For example, the scale 2 in:1 mi means that 2 inches on a scale drawing represents 1 mile.

1. Each letter of the Hollywood sign measures 50 ft tall and 30 ft wide. Use the rectangles below to sketch a scale drawing of the Hollywood sign. The side lengths of each square inside each rectangle represent 10 ft.

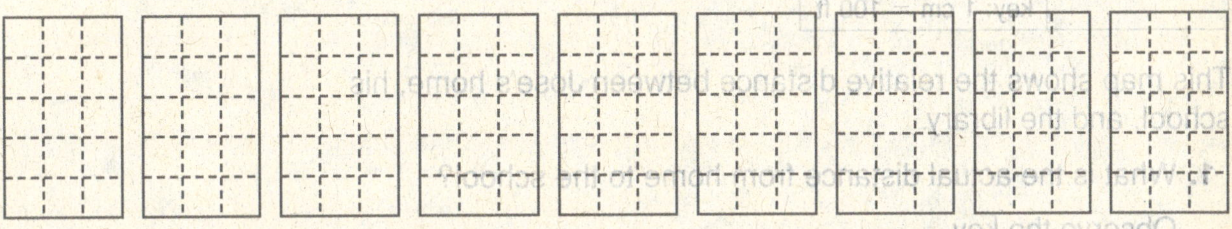

2. If the total width of the Hollywood sign is approximately 450 ft, what is the approximate distance between each pair of neighboring letters?

Think and Discuss

3. Explain how you found the answer in number 2.

4. Discuss other examples of scale drawings.

Name ______________________ Date __________ Class __________

LESSON 7F

Hands-On Lab

Measuring on Maps

KEY

1 in. = 2 miles

A key or scale on a map tells how the distances on a map translate into actual distances. If you see the key above on a map, you would count 2 miles for every inch measured on the map.

Activity

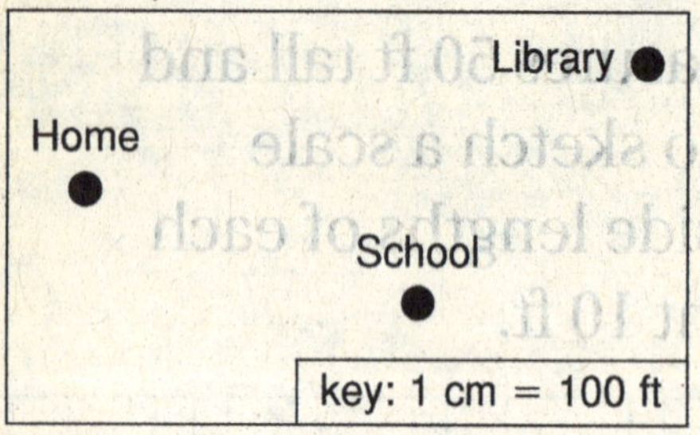

This map shows the relative distance between Jose's home, his school, and the library.

1. What is the actual distance from home to the school?

Observe the key.

The key says 1 centimeter is equivalent to 100 feet.

Measure the distance from home to school.

Since the key is in centimeters, use your centimeter ruler. The distance is 3 cm.

Compute the actual distance with a proportion.

$$\frac{1 \text{ cm}}{100 \text{ ft}} = \frac{3 \text{ cm}}{x \text{ ft}}$$

$x = 100 \times 3$ Cross multiply.

$x = 300$ The actual distance is 300 feet.

2. A park (not shown on the map) is 450 feet from Jose's home. How far away would it be placed from his home on the map?

Solve with a proportion.

$$\frac{1 \text{ cm}}{100 \text{ ft}} = \frac{x \text{ cm}}{450 \text{ ft}}$$

$100x = 450$ Cross multiply.

$x = 4.5$

The distance on the map would be 4.5 centimeters.

Name ______________________________ Date ____________ Class ____________

LESSON **7F**

Hands-On Lab

Measuring on Maps, continued

Think and Discuss

1. Why is it important for the scale to be shown on a map? Give an example.

____

____

2. The distance between New York City and San Francisco is 4,156 kilometers. If you were to draw a map of the 48 contiguous United States on an 8 1/2″ by 11″ piece of paper, what would be a reasonable scale? Explain your reasoning.

____

____

____

Try This

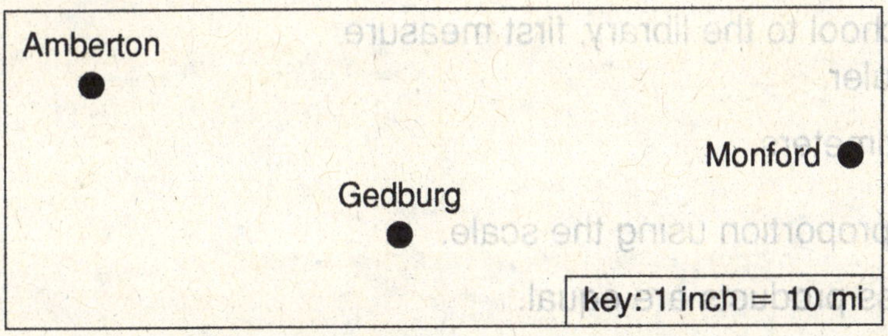

1. What is the actual distance from Amberton to Gedburg?

____

2. What is the actual distance from Gedburg to Monford?

____

3. Ketty is 35 miles from Monford. If Ketty were placed on the map, how far would it be placed from the dot representing Monford?

____

Name ______________________ Date ____________ Class ____________

LESSON 7F

Review for Mastery

Scale Drawings and Maps

A scale drawing is a drawing of a real object that is proportionally smaller or larger than the real object.

A scale is a ratio between two sets of measurements. In the map below, the scale is 2 cm: 0.5 km. This means that each centimeter on the map represents 0.25 kilometer.

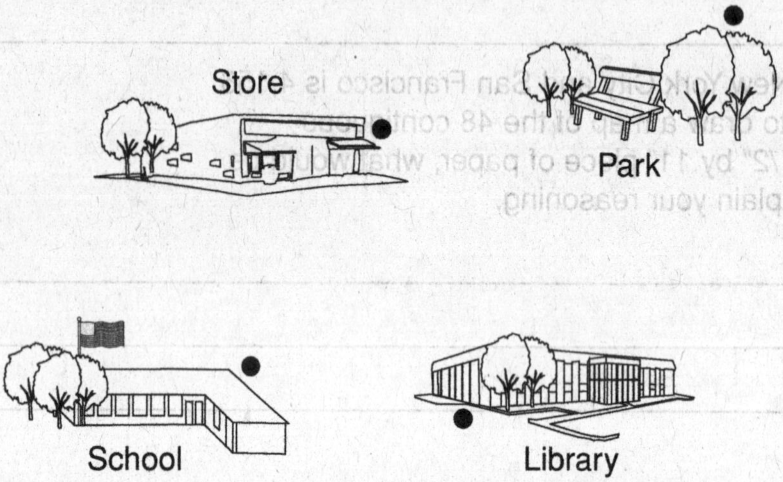

To find the actual distance from school to the library, first measure the distance on the map using a ruler.

The distance on the map is 3 centimeters.

$\frac{2\text{ cm}}{0.5\text{ km}} = \frac{3\text{ cm}}{x\text{ km}}$	Write a proportion using the scale.
$2 \cdot x = 0.5 \cdot 3$	The cross products are equal.
$2x = 1.5$	x is multiplied by 2.
$\frac{2x}{2} = \frac{1.5}{2}$	Divide both sides by 2.
$x = 0.75$	

The distance from school to the library is 0.75 kilometer.

Use the map to find each actual distance.

1. from the store to the library

____

2. from the park to the store

____

3. from the park to the library

____

4. from the park to the school

____

Name ______________________ Date __________ Class __________

LESSON 7F

Homework and Practice

Scale Drawings and Maps

Use the map to answer the questions.

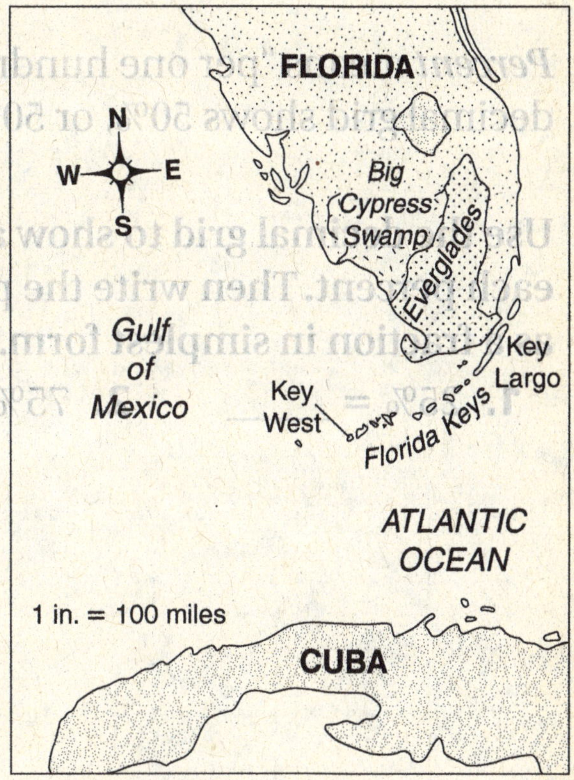

1. On the map, the distance between Key West and Key Largo is 1 inch. What is the actual distance in miles?

   ____

2. On the map, the distance between Everglades and Cuba is $1\frac{3}{10}$ inch. What is the actual distance?

   ____

3. Use a ruler to measure the distance between Big Cypress Swamp and Key Largo on the map. About how far is the actual distance?

   ____

4. Some people want to construct a connection from Cuba to mainland Florida. It would be 250 miles long! If it were shown on this map, how many inches long would it be?

   ____

Use the scale drawing to answer each question.

5. This scale drawing is of the lighthouse on Key West, originally built in 1825. What is the actual height of the smaller building just outside of the lighthouse?

   ____

6. The original lighthouse was 66 feet tall. It was rebuilt at its present height after a hurricane destroyed it in 1846. How tall would half of the original lighthouse be in this scale drawing?

   ____

EXPLORATION

Program of Studies and Core Content

MA-6-NPO-S-NS1, MA-6-NPO-S-NS5, MA-6-NPO-S-RP1, MA-06-1.1.1, MA-06-1.1.2, MA-06-1.4.1, MA-06-1.1.3

7G Percents

Percent means "per one hundred." The decimal grid shows 50%, or 50 out of 100.

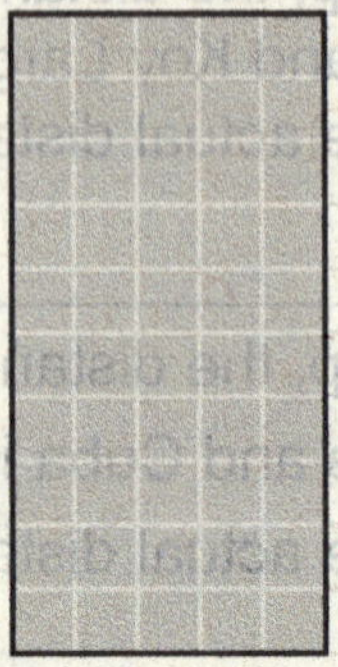

Use the decimal grid to show a model of each percent. Then write the percent as a fraction in simplest form.

1. 25% = ____

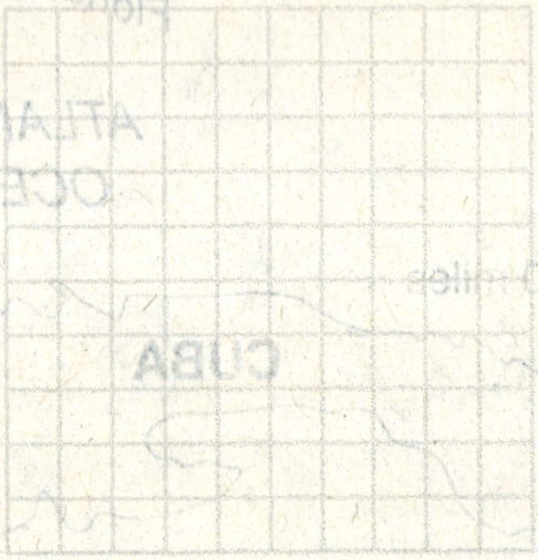

2. 75% = ____

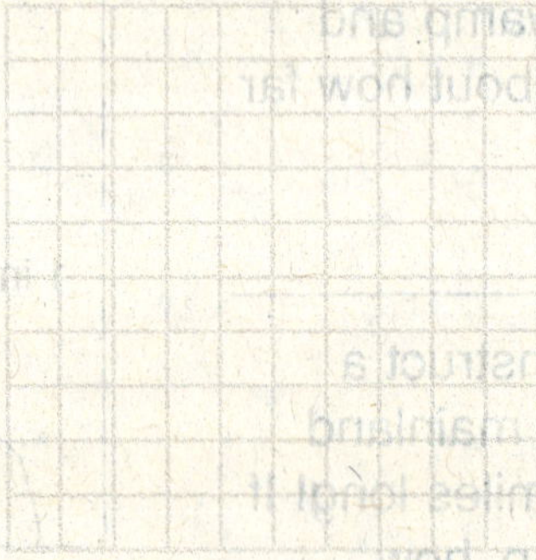

3. 80% = ____

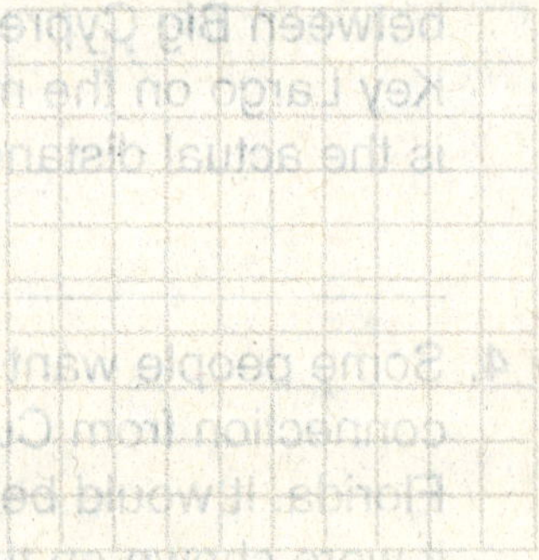

Determine the percent modeled by each decimal grid, and then write it as a fraction in simplest form.

4.

5.

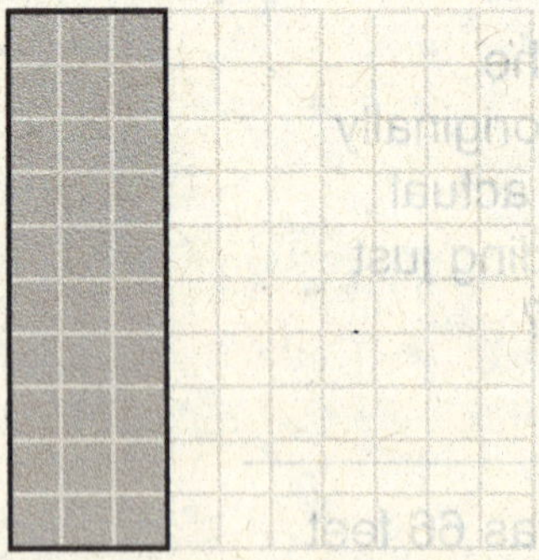

6.

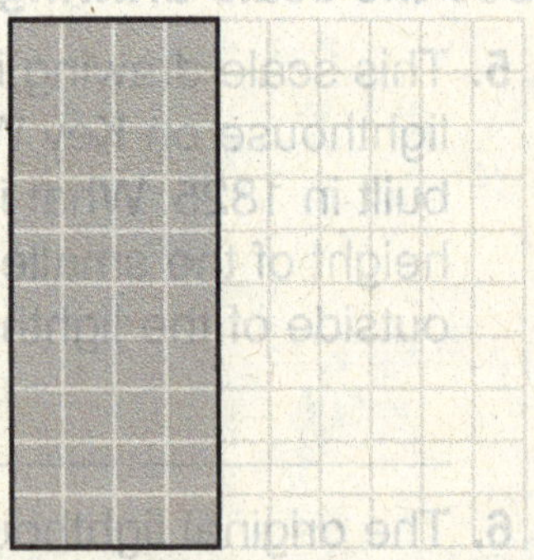

Think and Discuss

7. Explain how to write a percent as a fraction.

Name ______________________________ Date ___________ Class ___________

LESSON **7G**

Hands-On Lab

Modeling Percents with Coins

Activity

A coin has a value that is less than or equal to a dollar, so the value of coin can be represented as a percentage of a dollar.

The word "cent" represents $\frac{1}{100}$ or one percent of a dollar.

So $25¢ = \frac{25}{100} = 25\%$

You can model this percentage using a 10 × 10 square grid.

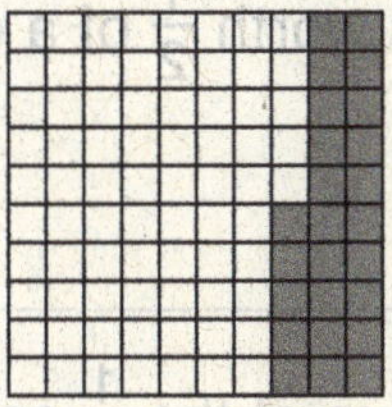

Write the value of each coin as a fraction of a dollar. Then use the 10 × 10 square grid to model the percentage of a dollar.

1. Nickel 5¢ = ______/100

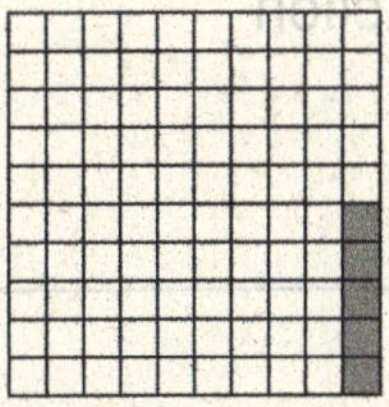

= ______%

2. Dime 10¢ = ______/100

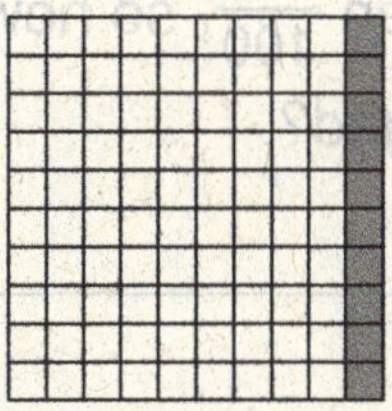

= ______%

3. Penny 1¢ = __________/100

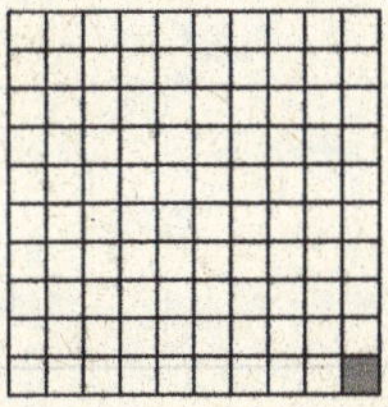

= ______%

4. Dollar 100¢ = __________/100

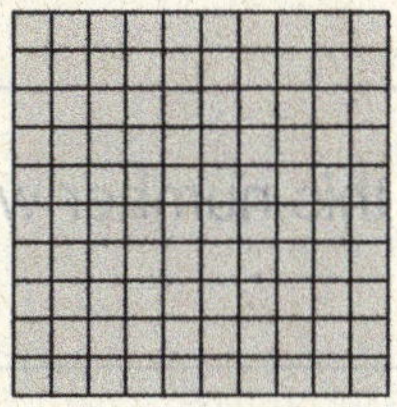

= ______%

Name ______________________ Date __________ Class __________

LESSON 7G

Hands-On Lab

Modeling Percents with Coins, continued

Think and Discuss

1. How could you model 37% of a dollar with coins?

____

____

2. When can you have more than 100% of a quantity?

____

____

Try This

1. Imagine you had to design a coin that was worth $\frac{1}{2}$ of a cent. Draw a picture of your $\frac{1}{2}$ cent coin below.

____

2. Since $\frac{1}{2}$ cent = 0.5 cent, represent the value of the $\frac{1}{2}$ cent coin as a fraction of a dollar.

____

3. This fraction is less than $\frac{1}{100}$, so how might you represent this fraction on a 10 × 10 square grid?

____

____

4. Rewrite the fraction with a denominator of 1000.

____

5. How would you model this number with base 10 blocks?

____

Name ______________________ Date __________ Class __________

LESSON 7G

Review for Mastery

Percents

A percent is a ratio of a number to 100. Percent means "per hundred."

To write 38% as a fraction, write a fraction with a denominator of 100.

$\frac{38}{100}$

Then write the fraction in simplest form.

$\frac{38}{100} = \frac{38 \div 2}{100 \div 2} = \frac{19}{50}$

So, $38\% = \frac{19}{50}$.

Write each percent as a fraction in simplest form.

1. 43% **2.** 72% **3.** 88% **4.** 35%

_______ _______ _______ _______

To write 38% as a decimal, first write it as fraction.

$38\% = \frac{38}{100}$

$\frac{38}{100}$ means "38 divided by 100."

$$\begin{array}{r} 0.38 \\ 100\overline{)38.00} \\ -300 \\ 800 \\ -800 \\ 0 \end{array}$$

So, 38% = 0.38.

Write each percent as a decimal.

5. 64% **6.** 92% **7.** 73% **8.** 33%

_______ _______ _______ _______

Name ______________________ Date ____________ Class ____________

LESSON 7G

Homework and Practice

Percents

Write each percent as a fraction in simplest form.

1. 28% ____________

2. 63% ____________

3. 14% ____________

4. 75% ____________

5. 30% ____________

6. 98% ____________

7. 5% ____________

8. 90% ____________

9. 16% ____________

Write each percent as a decimal.

10. 42% ____________

11. 96% ____________

12. 1% ____________

13. 13% ____________

14. 56% ____________

15. 9% ____________

Order the percents from least to greatest.

16. 92%, 8%, 52%, 62% ____________________

17. 56%, 9%, 97%, 76% ____________________

18. Marlene correctly answered 76% of the questions on her science test. What fraction of the test questions did she answer correctly? Write your answer in simplest form.

____

19. Jeremy ran $\frac{2}{5}$ of his total running goal for the day. What percent of his total running goal did Jeremy complete?

____

EXPLORATION

7H Percents, Decimals, and Fractions

Program of Studies and Core Content

MA-6-NPO-S-NS1, MA-6-NPO-S-NS5, MA-6-NPO-S-RP1, MA-06-1.1.1, MA-06-1.1.3

To report what percent of their fund-raising goal has been reached, a charity uses the number-line model below.

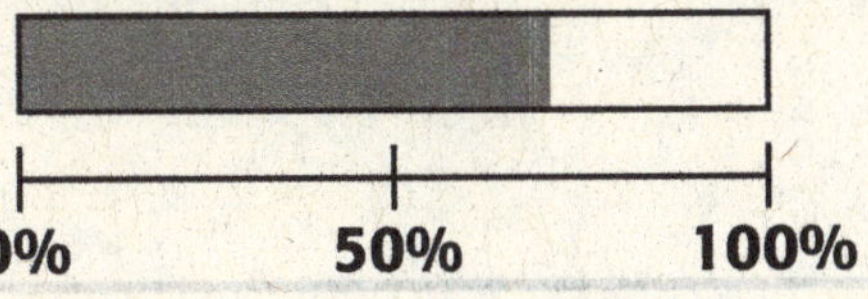

1. Has the charity reached about 50%, about 75%, or about 100% of its goal?

Complete each number-line model by writing percents above the line and the corresponding fractions below the line.

2. 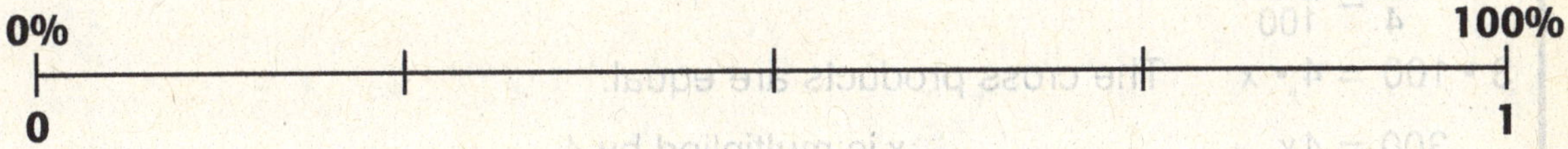

3. 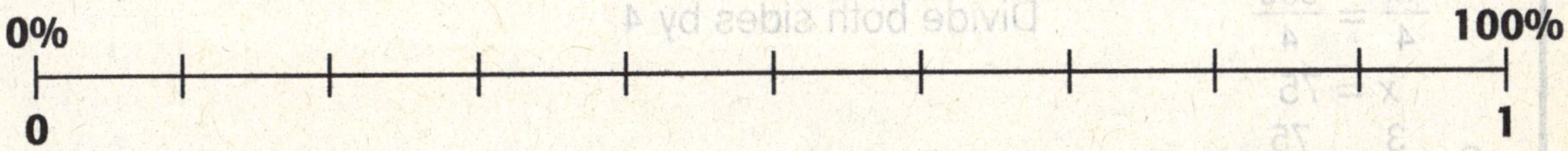

Think and Discuss

4. Explain how you matched percents with fractions in numbers 2 and 3.

5. Explain how you could label the number lines with decimals.

6. Suppose the charity set a goal of $2 million, but actually raised $3 million. Discuss what percent of the goal would describe this situation.

Name ______________________ Date __________ Class __________

LESSON 7H

Review for Mastery

Percents, Decimals, and Fractions

You can write decimals as percents.

To write 0.5 as a percent, multiply the decimal by 100%.

0.5 • 100% = 50%

To multiply a number by 100, move the decimal point two places to the right.

0.50

So, 0.5 = 50%.

Write each decimal as a percent.

1. 0.8 ________ **2.** 0.64 ________ **3.** 0.075 ________ **4.** 0.29 ________

You can solve a proportion to write a fraction as a percent.

To write $\frac{3}{4}$ as a percent, first set up a proportion.

$\frac{3}{4} = \frac{x}{100}$

$3 \cdot 100 = 4 \cdot x$ The cross products are equal.

$300 = 4x$ x is multiplied by 4.

$\frac{4x}{4} = \frac{300}{4}$ Divide both sides by 4.

$x = 75$

So, $\frac{3}{4} = \frac{75}{100}$

$\frac{75}{100} = 75\%$, So, $\frac{3}{4} = 75\%$.

Write each fraction as a percent.

5. $\frac{4}{5}$ ________ **6.** $\frac{9}{10}$ ________ **7.** $\frac{1}{8}$ ________ **8.** $\frac{7}{25}$ ________

9. $\frac{1}{4}$ ________ **10.** $\frac{5}{6}$ ________ **11.** $\frac{3}{4}$ ________ **12.** $\frac{1}{5}$ ________

Name ______________________ Date __________ Class __________

LESSON 7H

Homework and Practice

Percents, Decimals, and Fractions

Write each decimal as a percent.

1. 0.17 ______

2. 0.85 ______

3. 0.64 ______

4. 0.33 ______

5. 0.19 ______

6. 0.8 ______

7. 0.23 ______

8. 1.02 ______

9. 0.25 ______

Write each fraction as a percent.

10. $\frac{10}{20}$ ______

11. $\frac{1}{25}$ ______

12. $\frac{2}{10}$ ______

13. $\frac{3}{5}$ ______

14. $\frac{6}{8}$ ______

15. $\frac{2}{20}$ ______

Compare. Write <, >, or =.

16. 32% ______ $\frac{1}{3}$

17. 0.8 ______ $\frac{6}{7}$

18. 0.05 ______ 5%

19. $\frac{1}{23}$ ______ 0.05

20. $\frac{5}{13}$ ______ 40%

21. $\frac{5}{20}$ ______ 20%

22. Michael completed six of eight questions on his test. What percent of his test does he still need to complete?

______

23. After reading a book for social science class, 100 students were asked whether or not they enjoyed it. One fifth of the students did not like the book. How many students liked the book?

______

EXPLORATION

Program of Studies and Core Content
MA-6-NPO-S-RP1, MA-06-1.1.2

7I Percent Problems

You can use a number line to find the percent of a number.

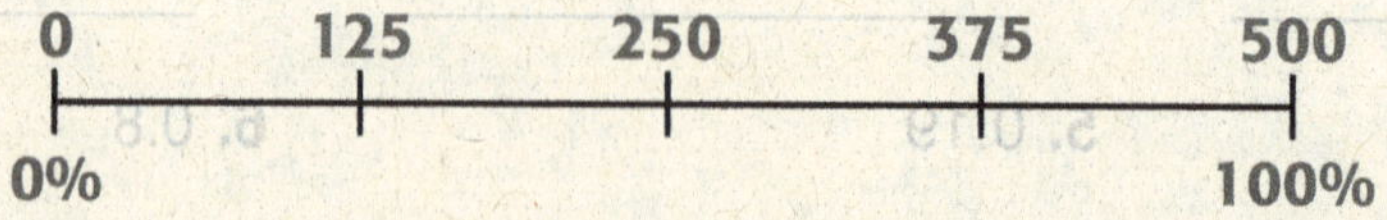

Use the number-line model above to complete each problem.

1. $\frac{125}{500}$ = ______% **2.** $\frac{250}{500}$ = ______% **3.** $\frac{375}{500}$ = ______%

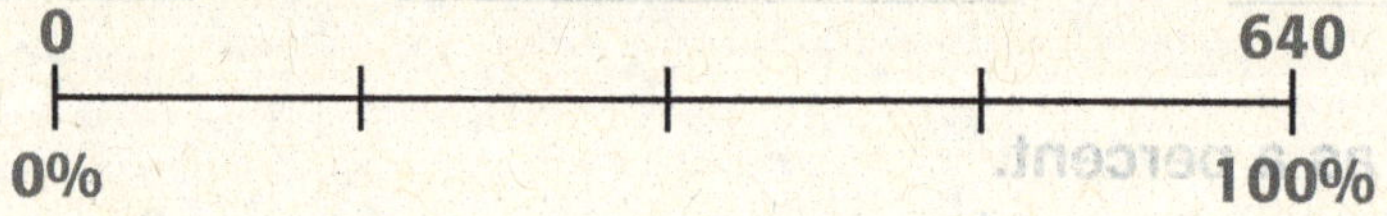

Label the number-line model above to find each percent of 640.

4. 50% of 640 is ______.

5. 25% of 640 is ______.

6. 75% of 640 is ______.

Think and Discuss

7. Discuss what it means to find 100% of a number. (*Hint:* What is 100% of 640?)

8. Explain how you can use number-line models to solve percent problems.

Name ______________________________ Date ____________ Class ____________

LESSON 71

Technology Lab
Percent Problems

Sometimes you know the whole and the percent, but you do not know the part that the percent represents.

Activity 1

There are 50 students in a class. Every student plays a sport: 40% play football, 30% play soccer, 20% play basketball, and 10% play baseball.

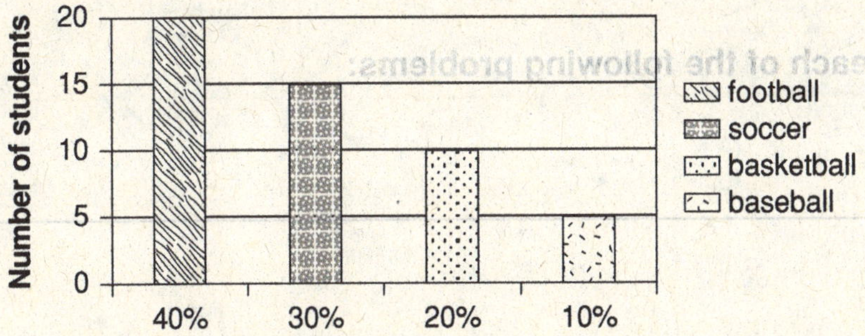

1. Look at the bar graph. From the graph you can see that 20 students play football; 15 play soccer; 10 play basketball; and 5 play baseball.
2. The lengths of the bars change by 5 with every increase of 10%. So, you can say that 10% of 50 is 5.

Activity 2

You can also use proportions to find the number of students (*S*) who play football.

$$S = \frac{50 \cdot 40}{100} = 20$$

On your calculator use the keystrokes: 50 [×] 40 [÷] 100 [ENTER]

1. A class has 60 students; 40% of them play football. How many students play football? 60 [×] 40 [÷] 100 [ENTER]

   24 students play football.

2. A class has 40 students; 25% play soccer. How many students play soccer? 40 [×] 25 [÷] 100 [ENTER]

   10 students play soccer.

Name ______________________ Date ____________ Class ____________

LESSON 7I

Technology Lab

Percent Problems, continued

Think and Discuss

1. A class has 20 students: 50% play football and 60% play basketball. How many students play both football and basketball?

____

____

Try This

Use your calculator to solve each of the following problems:

1. 4% of 50

____

2. 30% of 150

____

3. 75% of 350

____

4. 120% of 200

____

5. 12% of 44

____

Name ______________________________ Date ____________ Class ____________

LESSON 71

Review for Mastery

Percent Problems

You can use proportions to solve percent problems.

To find 25% of 72, first set up a proportion.

$$\frac{25}{100} = \frac{x}{72}$$

$25 \cdot 72 = 100 \cdot x$ Next, find cross products.

$1{,}800 = 100x$

$\frac{100x}{100} = \frac{1{,}800}{100}$ Then solve the equation.

$x = 18$

So, 18 is 25% of 72.

Use a proportion to find each number.

1. Find 3% of 75. **2.** Find 15% of 85. **3.** Find 20% of 50. **4.** Find 6% of 90.

__________ __________ __________ __________

You can use multiplication to solve percent problems.

To find 9% of 70, first write the percent as a decimal.

$9\% = 0.09$

Then multiply using the decimal.

$0.09 \cdot 70 = 6.3$

So, 9% of 70 = 6.3.

Use multiplication to find each number.

5. Find 80% of 48. **6.** Find 6% of 30. **7.** Find 40% of 120. **8.** Find 20% of 98.

__________ __________ __________ __________

9. Find 70% of 70. **10.** Find 35% of 120. **11.** Find 9% of 50. **12.** Find 40% of 150.

__________ __________ __________ __________

Name ______________________ Date ____________ Class ____________

LESSON 71

Homework and Practice

Percent Problems

Find the precent of each number.

1. 7% of 65 ________

2. 56% of 30 ________

3. 95% of 275 ________

4. 66% of 66 ________

5. 14% of 89 ________

6. 51% of 698 ________

7. 65% of 102 ________

8. 103% of 57 ________

9. 3% of 200 ________

10. 48% of 210 ________

11. 87% of 78 ________

12. 5.5% of 600 ________

13. 23% of 326 ________

14. 1% of 250 ________

15. 80% of 5 ________

16. 2% of 15 ________

17. 29% of 410 ________

18. 15% of 400 ________

19. Fred ordered 15 pizzas for a party. Three-fifths of those pizzas were just cheese. How many pizzas were just cheese?

____

20. Jenny sold 380 boxes of cookies for her Girl Scout troop. Fifty-five percent of the cookies she sold were oatmeal cinnamon boxes. How many boxes of oatmeal cinnamon cookies did she sell?

____

21. Keon jumps rope for 20 minutes. If he has completed 80% of his routine, how many minutes will Keon jump rope in all?

____

22. Michelle ran 460 miles this month preparing for a race. Thirty-five percent of the miles she ran were during bad weather. How many miles did Michelle run in bad weather?

____

EXPLORATION

Program of Studies and Core Content
MA-6-NPO-S-E1, MA-6-NPO-S-RP1, MA-06-1.2.1

7J Using Percents

Stores that go out of business often offer big discounts on purchases. In such situations, 50% off sales are common.

Estimate the discount for each item at 50% off. Then calculate the actual discount.

	Item	Price	Estimated Discount	Actual Discount
1.	Shirt	$39.95		
2.	DVD player	$288.99		
3.	Speakers	$239.95		
4.	TV	$1,035.29		
5.	MP3 player	$247.99		

Think and Discuss

6. Discuss the estimation strategies you used.

7. Explain whether a one-time 50% discount is equivalent to two consecutive 25% discounts. (*Hint:* Use $100.00 as the base amount.)

Name ______________________ Date __________ Class __________

LESSON 7J

Technology Lab

Uses of Percents

When you eat in a restaurant, you may want to leave a tip. How will you know how much to leave?

Activity 1

A tip table will suggest how much tip to leave.

Tip Table (based on 15% gratuity)

Amount	Tip
$2.00	$0.30
$3.00	$0.45
$4.00	$0.60
$5.00	$0.75
$6.00	$0.90
$7.00	$1.05
$8.00	$1.20
$9.00	$1.35
$10.00	$1.50

If your meal costs $7.00, the tip should be $1.05.

Your meal costs $3.00. What is your total cost?

Total = $3.00 + $0.45 = $3.45

The total cost is $3.45.

Activity 2

You can also use proportions to find the tip (T) for a $7.00 meal.

$$T = \frac{15}{100} \cdot 7 = 1.05$$

On your calculator use the keystrokes: 15 [÷] 100 [×] 7 [ENTER]

1. Use your calculator to find the tip for $11.60. 15 [÷] 100 [×] 11 [.] 60 [ENTER]

   The tip is $1.74.

2. What is the tip for $8.80? 15 [÷] 100 [×] 8 [.] 80 [ENTER]

   The tip is $1.32.

Name ______________________ Date __________ Class __________

LESSON 7J

Technology Lab

Uses of Percents, continued

Think and Discuss

1. Look at the changes in the cost of the meal and the tips. Based on these changes, can you figure out what the tip is for $11.00? How?

____

____

____

Try This

1. Make your own tip table for a 15% tip. Use a calculator or a spreadsheet.

Tip Table (based on 15% gratuity)

Amount	Tip
$12.00	
$15.00	
$18.00	
$21.00	
$24.00	
$27.00	
$30.00	
$33.00	
$36.00	

2. Your meal total is $33.00. You decide to tip your waiter 15% of the meal. How much tip will you leave?

____

3. Your barber charges you $21.00 for a haircut. If you tip him 15%, how much tip will you leave?

____

Name ______________________________ Date ____________ Class ____________

LESSON **7J**

Review for Mastery

Using Percents

There are many uses for percents.

Common Uses of Percents

Discounts	A **discount** is an amount that is subtracted from the regular price of an item. discount = regular price • discount rate
Tips	A **tip** is an amount added to a bill. tip = total bill • tip rate
Sales Tax	**Sales tax** is an amount added to the price of an item. sales tax = purchase price • sales tax rate

Rachel is buying a sweater that costs $42. The sales tax rate is 5%. About how much will the total cost of the sweater be?

You can use fractions to find the amount of sales tax.

First round $42 to $40.

Think: 5% is equal to $\frac{1}{20}$.

So, the amount of tax is about $\frac{1}{20} \cdot \$40$.

The tax is about $2.00.

Then find the sum of the price of the sweater and the tax.

$42 + $2.00 = $44.00

Rachel will pay about $44.00 for the sweater.

Solve each problem.

1. About how much would you pay for a meal that costs $29.75 if you left a 15% tip?

____

2. About how much do you save if a book whose regular price is $25.00 is on sale for 10% off?

____

3. About how much would you pay for a box of markers whose price is $5.99 with a sales tax rate of 9.5%?

____

Name ______________________ Date __________ Class __________

LESSON 7J

Homework and Practice

Uses of Percents

Write the correct answer.

1. Michelle and Rhonda eat dinner at a fancy restaurant. Their bill came to $105.36. They gave their waiter a tip that was 22% of the bill. About how much money did they give the waiter for his tip?

____

2. Christine buys a necklace that costs $18.62 and a ring that costs $7.98. She uses a 35%-off coupon on the purchase. If Christine pays with a $20 dollar bill, how much change should she get back?

____

3. The church's goal for the fundraiser was $18,000. They exceeded the goal by 8%. How much money did the church raise?

____

4. Kelly decided to get sandwiches for herself and some friends. The cashier said that if she spent over $10 she would get 15% off. The total bill was $13.27. How much was the total bill after the discount?

____

5. Lauren had a 26%-off coupon for the flower store. She bought a bouquet of roses that are regularly priced at $51.23. How much did Lauren spend on the bouquet of roses after using her coupon?

____

6. Joe bought two pizzas for $16.78. He wants to give the delivery person an 18% tip. How much money should Joe give the delivery person?

____

7. A television costs $258.98. The sales tax rate is 8.5%. How much will it cost to buy the television?

____

8. Milton bought a bike that costs $84.21. The sales tax rate is 6%. How much did Milton pay for the bike altogether?

____

9. Tana bought $74.12 worth of CDs at the music store. She gets a 19% discount since she is an employee at the store. What was the actual price of the CDs she bought before the discount?

____

10. Ronald saves $10.00 to buy a new book. The book's original price was $15, but it was on sale for 35% off with a 6% sales tax. Does he have enough money to buy the book? Explain.

____

____

CHAPTER 7

Chapter Summary

Answer these question to summarize the important concepts from Chapter 7 in your own words.

1. Explain how to find an equivalent ratio.

2. Explain how to find the value of n in the proportion $\frac{5}{20} = \frac{n}{35}$.

3. On a map the distance between two cities is 3.5 inches. The scale on the map is 1 inch = 10 miles. Explain how to find the actual distance between the two cities.

4. Explain how to write 45% as a fraction.

5. Explain how to write 0.08 as a percent.

For more review of Chapter 7:

- Complete the Chapter 7 Study Guide and Review on pages 404–406 of your textbook.
- Complete the Ready to Go On quizzes on pages 378 and 398 of your textbook.

Name ______________________ Date __________ Class __________

CHAPTER 7 **KCCT Prep**

Multiple Choice

1. Which decimal is closest to 6%?

Ⓐ 0.005　　Ⓒ 0.5
Ⓑ 0.07　　Ⓓ 7.0

2. The Hawks' mean score in two football games was 26. They scored 18 points in the first game. How many points did they score in the second game?

Ⓐ 8　　Ⓒ 34
Ⓑ 22　　Ⓓ 44

3. The scale of a map is 1 in:12 mi. The distance between Springfield and Mason on the map is 10 inches. If Kevin drives at 60 mi/h, how long does it take him to drive from Springfield to Mason?

Ⓐ 0.5 h　　Ⓒ 1.5 h
Ⓑ 1 h　　Ⓓ 2 h

4. After Karen added 4 new jars of spices to her spice rack, she had no more than 18 jars altogether. Which inequality represents the original number of jars?

Ⓐ $x < 14$　　Ⓒ $x < 22$
Ⓑ $x \leq 14$　　Ⓓ $x \leq 22$

5. Casey downloaded 308 songs for her MP3 player. Of these songs, about 35% were dance songs. Which is the best estimate of the number of dance songs that Casey downloaded?

Ⓐ 10　　Ⓒ 100
Ⓑ 50　　Ⓓ 150

6. The figure shows a pattern of tiles on the top of a table. Jared writes the ratio of black tiles to white tiles as n:6. What is the value of n?

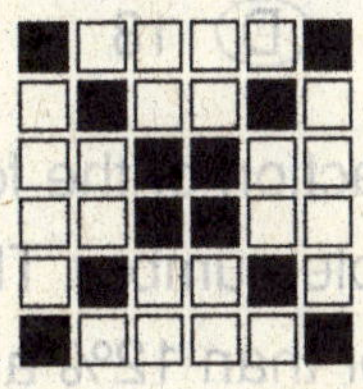

Ⓐ 1　　Ⓒ 3
Ⓑ 2　　Ⓓ 12

7. There are 408 students at Cesar Chavez Middle School. About 5% of them are in the soccer club. Which is the best estimate of the number of students in the soccer club?

Ⓐ 5　　Ⓒ 20
Ⓑ 10　　Ⓓ 40

8. Which of these is NOT equal to the others?

Ⓐ 2.5%　　Ⓒ 0.25
Ⓑ $\frac{1}{4}$　　Ⓓ $\frac{3}{12}$

9. A restaurant uses $5\frac{1}{4}$ ounces of pasta for each serving of linguini. Which is the best estimate of the amount of pasta needed to make 6 servings?

Ⓐ 30 ounces　　Ⓒ 33 ounces
Ⓑ 32 ounces　　Ⓓ 36 ounces

10. A machine can make 10 widgets in 6 minutes. A worker sets the machine to make 15 widgets. A bell rings every 30 seconds while the widgets are being made. How many times does the bell ring while the 15 widgets are being made?

Ⓐ 9 Ⓒ 15
Ⓑ 12 Ⓓ 18

11. Frank wrote a fraction of the form $\frac{n}{20}$, where n is a whole number. The fraction is greater than 12% and less than 20%. Which of these is the value of n?

Ⓐ 1 Ⓒ 3
Ⓑ 2 Ⓓ 4

12. Nicole drives 153 miles in 3.1 hours. What is the best estimate of her unit rate?

Ⓐ 20 mi/hr Ⓒ 50 mi/hr
Ⓑ 30 mi/hr Ⓓ 60 mi/hr

13. Paul uses graph paper to make a 10-by-10 grid of small squares. Then he uses a pencil to shade 40 of the small squares. What percent of the small squares are shaded?

Ⓐ 0.4% Ⓒ 40%
Ⓑ 4% Ⓓ 100%

14. The ratio of girls to boys at a summer camp is $\frac{2}{3}$. There are 20 children at the camp. How many girls are there?

Ⓐ 2 Ⓒ 12
Ⓑ 8 Ⓓ 16

15. Which of the following is greatest for the data set shown below?

3 3 3 5 5 6 7 8 8 10

Ⓐ Mean Ⓒ Mode
Ⓑ Median Ⓓ Range

Open Response

16. Tomas is planning to drive along a straight highway from Mapleton to Greenville. His map has a scale of 1 in:16 mi. The distance between Mapleton and Greenville on the map is 5 inches.

a. Tomas claims that the actual distance from Mapleton to Greenville is at least 100 mi. Do you agree or disagree? Explain.

b. Tomas plans to stop for a break after he drives 70% of the distance. How many miles of driving will he do before he takes his break?

17. Debra is comparing three brands of paper towels. Brand A has 112 sheets and costs $2.24, Brand B has 165 sheets and costs $1.65, and Brand C has 144 sheets and costs $2.88.

a. What is the price per sheet for Brand A? Show your work.

b. Which brand of paper towels is the best deal? Why?

c. Brand C offers a larger roll of paper towels with 188 sheets. It has the same unit rate as the roll with 144 sheets. What is the price of the larger roll?

18. An amusement park has a total of 50 rides. Of the 50 rides, $\frac{3}{10}$ are roller coasters.

a. How many roller coasters are there at the amusement park?

b. Of the roller coasters, $\frac{4}{5}$ require riders to be taller than 48 inches. What percent of the roller coasters require riders to be taller than 48 inches? Explain.

c. How many roller coasters do not require riders to be taller than 48 inches?

CHAPTER 8

Project
Geometric Relationships: Tangram Tangle

Career: Artist

Artists help us to see the world in new ways. They use their creativity in many different kinds of careers. Artists might design graphics for websites, draw cartoons, design textiles and furniture, paint murals, or even illustrate courtroom scenes. Artists work with many materials, such as different kinds of paints, paper, stone, metal, stained glass, and tile.

Before you start the project, work through the Tile Artistry worksheet on the next page.

In this chapter project . . .

You design your own tangram set.

What to do:

- Learn the names and properties of basic geometric shapes.
- Decide what shapes to use in your tangram set.
- Create your set using cardboard or construction paper. Begin with a large square or rectangle, and then carefully cut it into the shapes you chose.

What to present:

- Puzzles for your classmates. Trade sets with a partner and try to solve the puzzles your partner designed with his or her unique tangram set.
- Solutions to your own tangram puzzles.

Where's the math?

- Learn to identify squares, rectangles, triangles, and other geometric shapes.
- Use adjacent and complementary angles.

go.hrw.com
Chapter Project Online
KEYWORD: MR7 Ch8

Name ______________________ Date __________ Class ________

Project Recording Sheet

CHAPTER 8

Geometric Relationships

Tile Artistry

Complete the table.

Triangle	Shape	Description
Equilateral	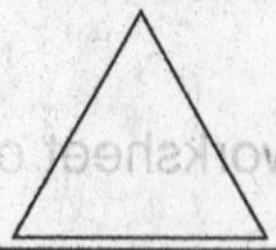	All sides and angles are the same.
Isosceles		
Scalene		
Obtuse		
Acute		
Right		

Create each of the following designs.

1. A design made up of congruent polygons.
2. A design made up of similar, but not congruent polygons.
3. A design made up of two kinds of polygons. (The pieces should fit together without gaps or overlaps.)
4. A design made up of irregular polygons.

Extension: Use a Tangram set to construct the figures below. Both figures can be made by using all seven pieces with no overlaps.

EXPLORATION

Program of Studies and Core Content
MA-6-G-S-SR3,
MA-06-3.1.1

8A Building Blocks of Geometry

Geometry can be used to describe the physical world around us. Check the box of the geometry term that each real-world item represents.

		Point	Line Segment	Plane
1.	A freckle			
2.	A strand of hair			
3.	A poster			
4.	A pixel on your calculator screen			
5.	A period at the end of a sentence			
6.	A guitar string			
7.	The minute hand of a clock			
8.	A computer screen			

Think and Discuss

9. Describe the characteristics of the items that you classified as *points* in the table above.

10. Describe the characteristics of the items that you classified as *line segments* in the table above.

Name ______________________ Date __________ Class __________

LESSON **8A**

Review for Mastery

Building Blocks of Geometry

Here are terms that can help you understand geometry.

A point is an exact location. A point has no size.

•*A* point *A* or *A* Use a capital letter to name a point.

A line is a straight path that extends without end in opposite directions. A line has infinite length, but no width.

A *B* $\overleftrightarrow{AB}$ or $\overleftrightarrow{BA}$ Two points name a line.

A plane is a flat surface that extends without end in all directions. A plane has infinite length and width, but no depth.

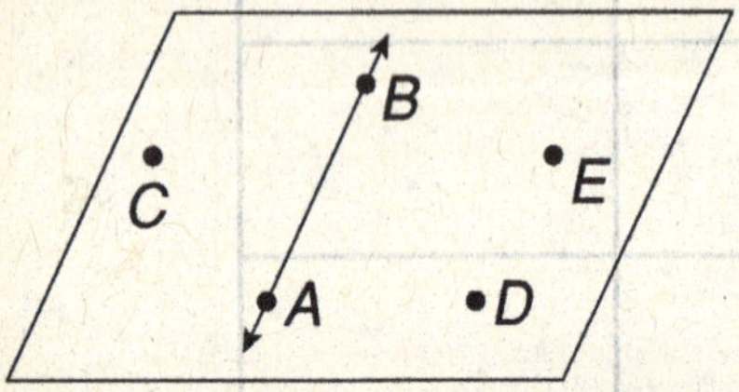

CDE	*EDC*	A plane is named by
DEC	*CED*	3 points on the plane
ECD	*DCE*	that are not on a line.

A line segment has two endpoints. The length of a line segment can be measured.

J *K* $\overline{JK}$ or $\overline{KJ}$ A line segment is named by its endpoints.

A ray has one endpoint and extends without end in one direction.

X *Y* $\overrightarrow{XY}$ A ray is named by its endpoint first and another point on the ray.

Use the diagram to name each geometric figure.

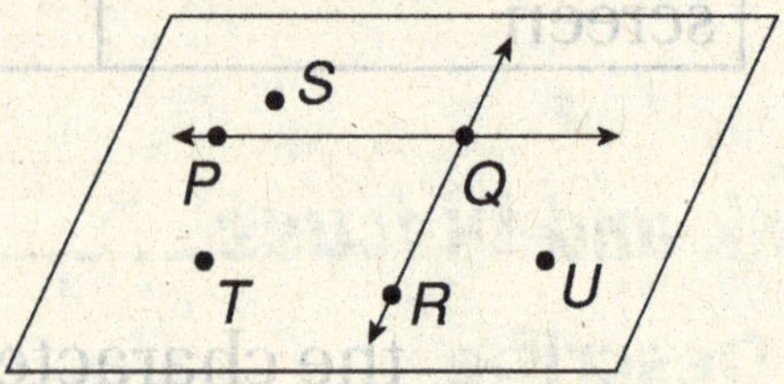

1. two points ______________

2. two lines ______________

3. a plane ______________

4. two line segments ______________

5. two rays ______________

6. a point shared by two lines ______________

Name ______________________ Date __________ Class __________

LESSON 8A

Homework and Practice

Building Blocks of Geometry

Use the diagram to name each geometric figure.

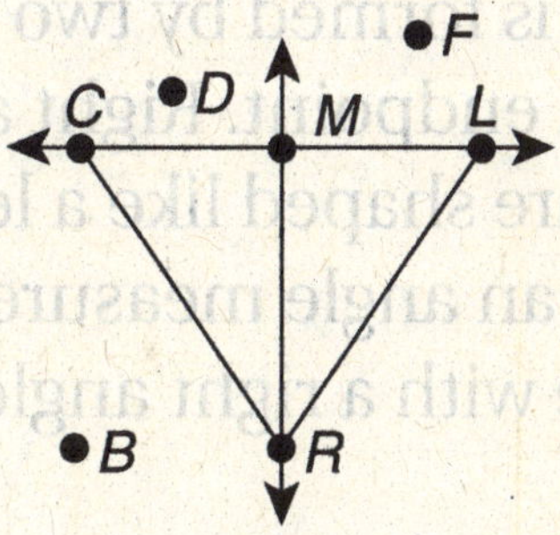

1. two points

____

2. a plane

____

3. a line segment

____

4. a point shared by two lines

____

5. a line

____

Use the diagram to give a possible name for each figure.

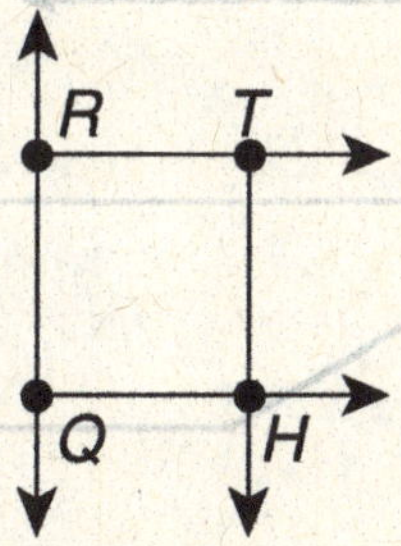

6. two different ways to name a line

____

7. four different names for rays

____

8. another name for $\overline{RT}$

____

9. Is the following statement always true, sometimes true, or never true? Explain your reasoning. A line segment can be longer than a line.

____

____

10. Is the following always true, sometimes true, or never true? A line always contains a line segment and a ray.

____

____

EXPLORATION

Program of Studies and Core Content

MA-6-M-S-MPA2, MA-6-M-S-MPA4, MA-6-G-S-SR3, MA-06-3.1.1

8B Measuring and Classifying Angles

An angle is formed by two rays that have a common endpoint. Right angles measure 90° and are shaped like a letter L. You can estimate an angle measure by comparing the angle with a right angle.

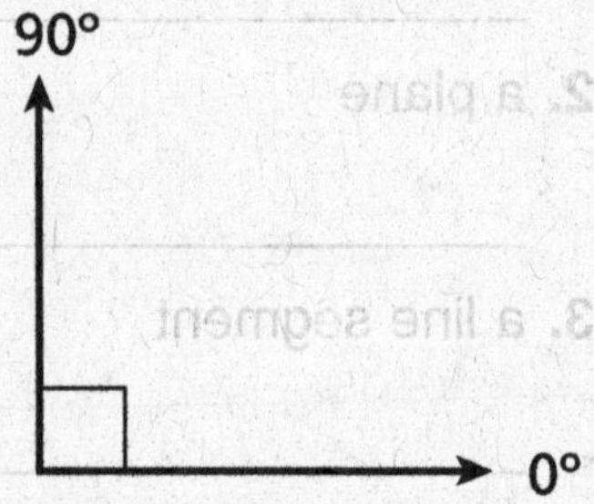

Estimate the measure of each angle. Then measure the angle with a protractor.

		Estimate	Actual
1.	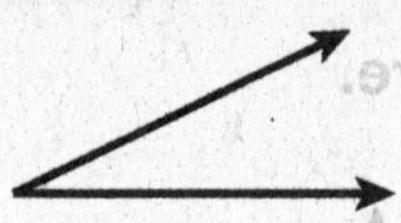		
2.			
3.	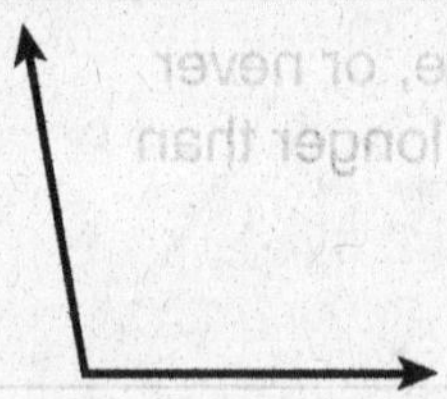		

Think and Discuss

4. Discuss how you estimated the angle measures.

Name ______________________ Date ____________ Class ____________

LESSON **8B**

Review for Mastery

Measuring and Classifying Angles

An angle is formed by two rays with a common endpoint, called the vertex.

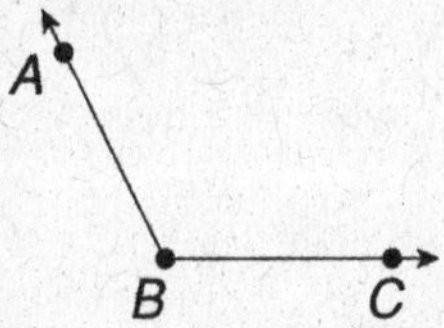

You can name an angle in different ways.

By its vertex OR By its vertex and one point from each ray.

$\angle B$ — $\angle ABC$, $\angle CBA$ Vertex letter in the middle.

Give three ways to name each angle.

1.

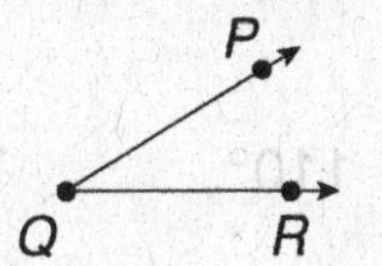

____

2.

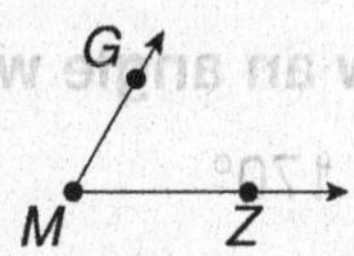

____

You can use a protractor to measure an angle.

To measure the angle, place the center point of the protractor on the vertex, K. Make sure that the protractor is placed so that $\overrightarrow{KL}$ passes through 0°. The measure of the angle is where $\overrightarrow{KJ}$ crosses the protractor.

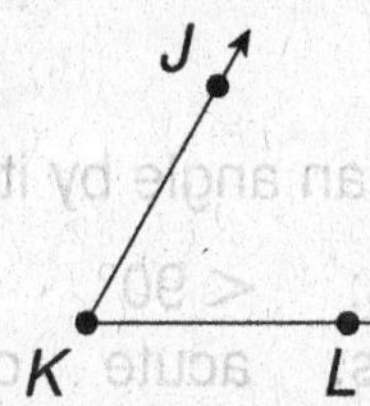

The measure of the angle is 60 degrees.

Write $m\angle K = 60°$.

Use a protractor to measure each angle.

3.

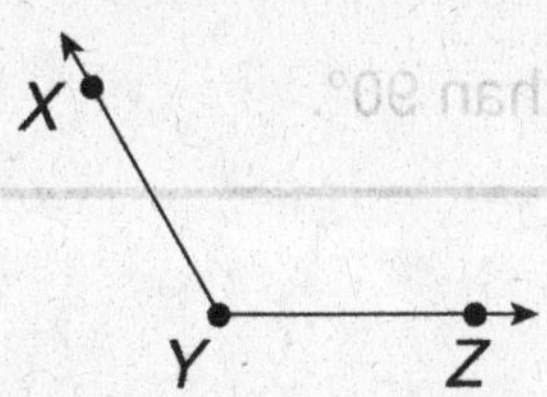

____

4.

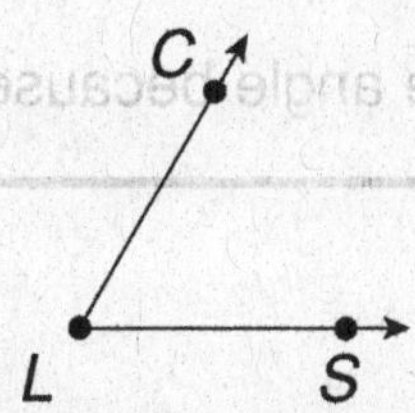

____

Name ____________________ Date ____________ Class ____________

LESSON 8B

Review for Mastery

Measuring and Classifying Angles, continued

You can use a protractor to draw an angle.

To draw $\angle PQR$ that measures 145°, first draw a base ray. Label the endpoint *Q*. Next, arrange the protractor so that its center is on *Q* and the base ray passes through 0°. Then, make a mark at 145° above the scale on the protractor. Use a straightedge to draw a ray from *Q* through the mark you made.

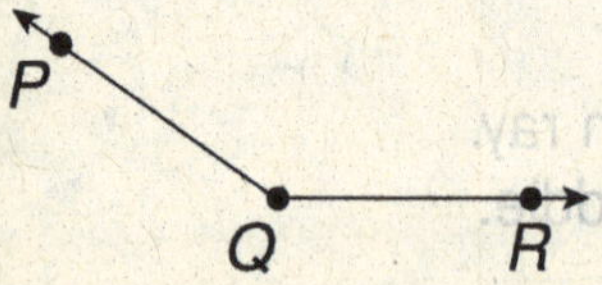

Label one point on a ray *P* and a point on the other ray *R*.

Use a protractor to draw an angle with each given measure.

5. 50° **6.** 170° **7.** 85° **8.** 110°

You can classify an angle by its measure.

If the measure is:	< 90°	> 90°	= 90°	= 180°
Then the angle is:	acute	obtuse	right	straight

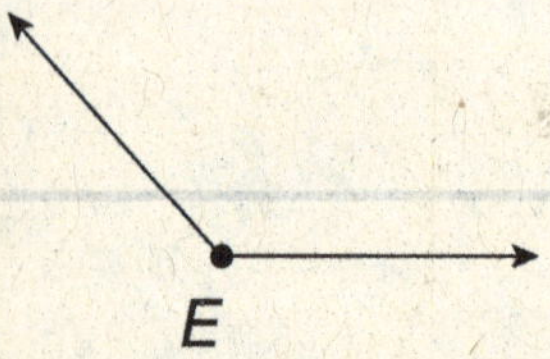

This angle is an obtuse angle because its measure is greater than 90°.

Classify each angle.

9.

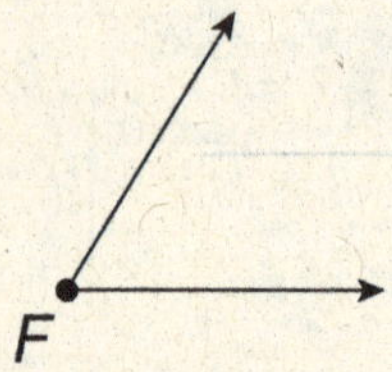

____

10.

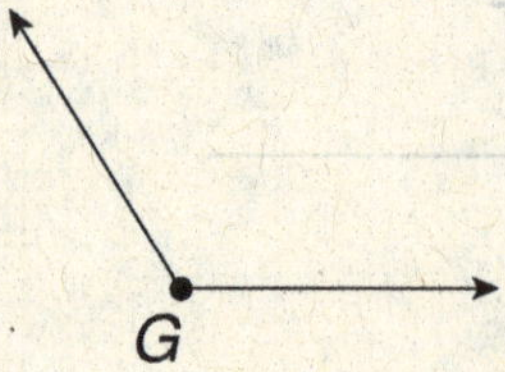

____

Name ______________________ Date ____________ Class ____________

LESSON 8B

Homework and Practice

Measuring and Classifying Angles

Use a protractor to measure each angle. Then classify the angle.

1.

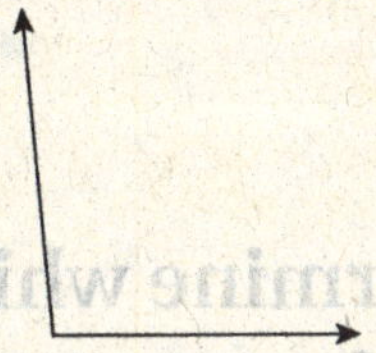

____

2.

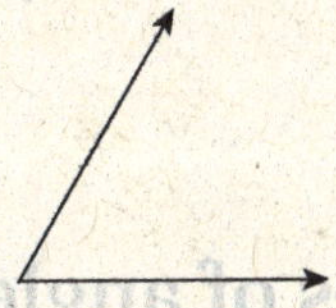

____

3.

____

Use a protractor to draw an angle with each given measure.

4. 90°

____

5. 20°

____

6. 125°

____

Estimate the measure of each angle, and then use a protractor to check the reasonableness of your estimate.

7.

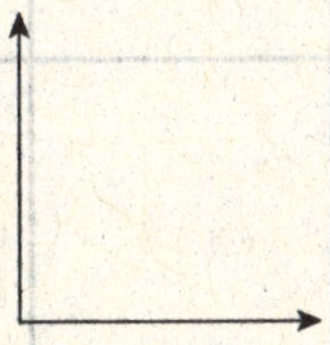

____

8.

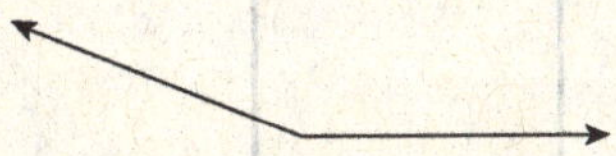

____

9.

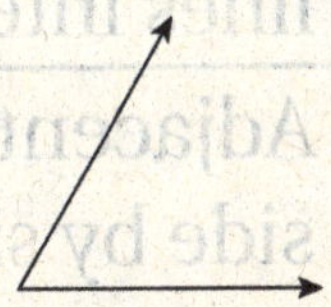

____

10. The screen of a computer has four angles that make it a quadrilateral shape. Describe each of these angles.

____

11. What kinds of angles are in each of the letters in this word?
LEFTY

____

____

____

EXPLORATION

Program of Studies and Core Content
MA-6-M-S-MPA2, MA-6-M-S-MPA4, MA-6-G-S-SR3, MA-06-3.1.1

8C Angle Relationships

The line segments in some letters, symbols, and numbers form angles.

Numbers 1–5 describe types of angle pairs. Determine which marked angle pairs in the figures apply to each description, and then check the appropriate boxes.

		Z	X	F	↗	4
1.	Congruent angles: same measure					
2.	Vertical angles: opposite each other when two lines intersect					
3.	Adjacent angles: side by side with a common vertex and ray					
4.	Complementary angles: sum equals 90°					
5.	Supplementary angles: sum equals 180°					

Think and Discuss

6. Discuss other examples of vertical angles in the real world.

Name ______________________________ Date ___________ Class ___________

LESSON 8C

Hands-On Lab

Folding Paper to Make Angle Bisectors

An *angle bisector* is a line segment that divides one angle into two smaller, congruent angles.

Activity

1. Using a ruler and a protractor, draw a right angle on a blank piece of paper with a dark marker.

2. Next, fold the paper onto itself so that the sides of the right angle lie directly on top of one another.

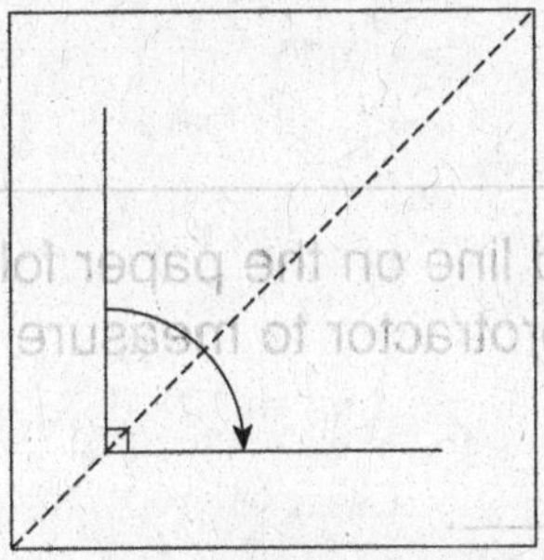

3. Now unfold the paper and draw a solid line on the paper fold. This line is the angle bisector. Label the new angles 1 and 2.

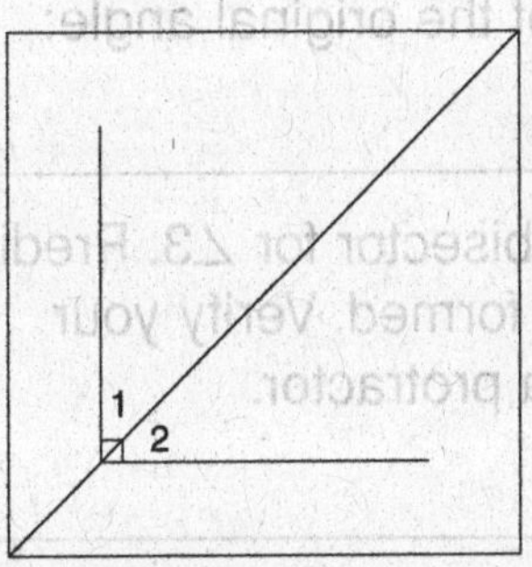

4. Use the definition of angle bisector to predict the measures of angles 1 and 2. Verify your predictions by measuring each angle.

$m\angle 1 =$ __________ $m\angle 2 =$ __________

Name ______________________ Date __________ Class __________

LESSON 8C

Hands-On Lab

Folding Paper to Make Angle Bisectors, continued

Think and Discuss

1. What geometric term could you use to describe the relationship between angles 1 and 2?

____

2. What other tools, besides paper-folding, could you use to construct and bisect angles?

____

Try This

1. Using only a ruler, draw an angle of any size on a blank piece of paper.

____

2. Next, find the angle bisector created by folding the paper so that the sides of the angle coincide.

____

3. Now, unfold the paper and draw a solid line on the paper fold. Label the new angles 3 and 4. Use a protractor to measure the angles.

4. $m\angle 3 =$ ______________.

5. $m\angle 4 =$ ______________.

6. Use the definition of angle bisector to determine the measure of the original angle. Write the measure of the original angle:

____

7. Fold the paper again to form an angle bisector for $\angle 3$. Predict the measure of the two smaller angles formed. Verify your answer by measuring the angles with a protractor.

____

Name ______________________________ Date ___________ Class ___________

LESSON **8C**

Technology Lab

Angle Relationships

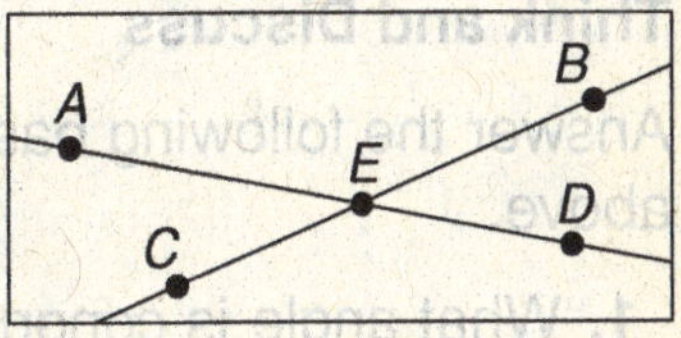

Angles are formed anywhere two lines or line segments intersect. If two lines intersect, four angles are formed.
∠*AEB* and ∠*CED* are vertical angles, as are ∠*BED* and ∠*AEC*.
∠*AEB* and ∠*BED* are adjacent angles.
The adjacent angles ∠*BED* and ∠*CED* are supplementary angles since the combined measure of their angles, m∠*BEC*, equal 180°.
If the combined measure of two angles equals 90° then the two angles are considered to be complementary angles.
If the measure of two angles is the same, then the angles are congruent. Vertical angles are always congruent.

Activity

Using geometry software, construct a right angle with two lines resembling the one in the first figure below. Label a point on one line *A*, one on the other line *B*, the intersection *C*, and calculate m∠*ACB*. Now draw an additional line through point *C* and label points on the new line on each side of *C* as points *D* and *E* as shown in the second figure. Show the measures of ∠*ACD* and ∠*DCB*.

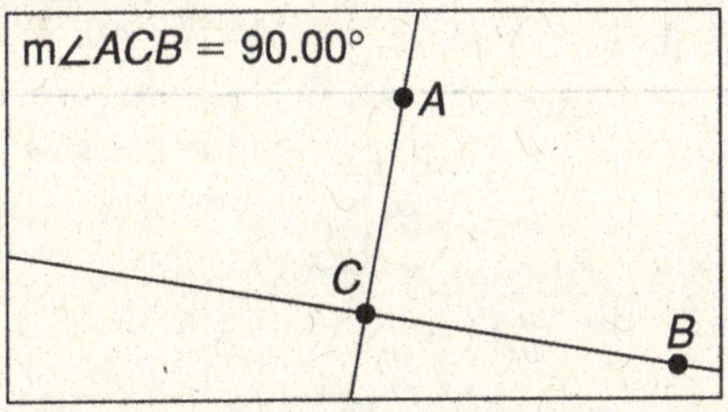

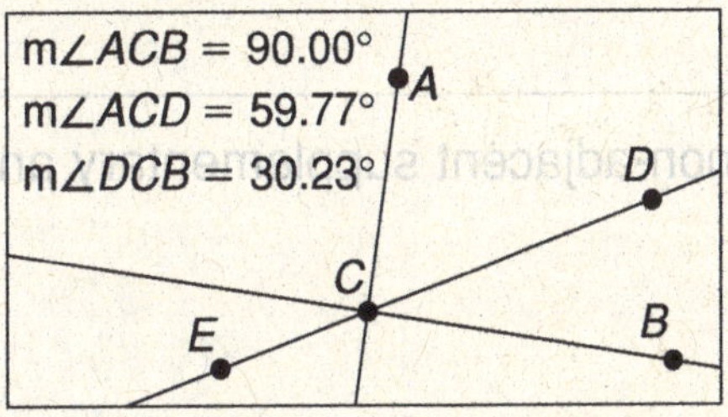

Now label additional points *F* and *G* as shown below on the left. Notice the relationships between the measures of the angles and the placement of the angles. Now rotate line *ED* around point *C* and notice how the measures of the angles change, as shown in the drawing below on the right.

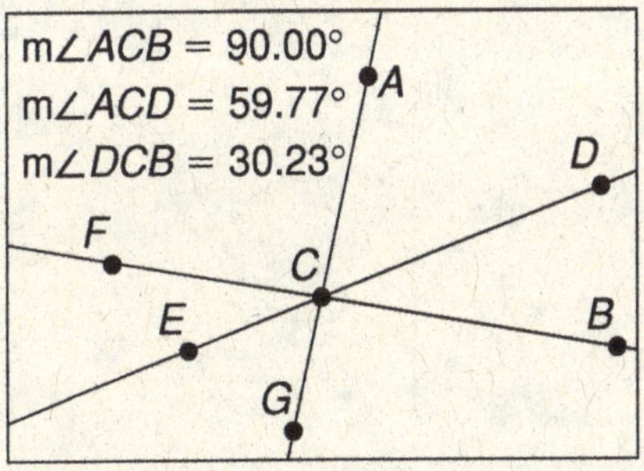

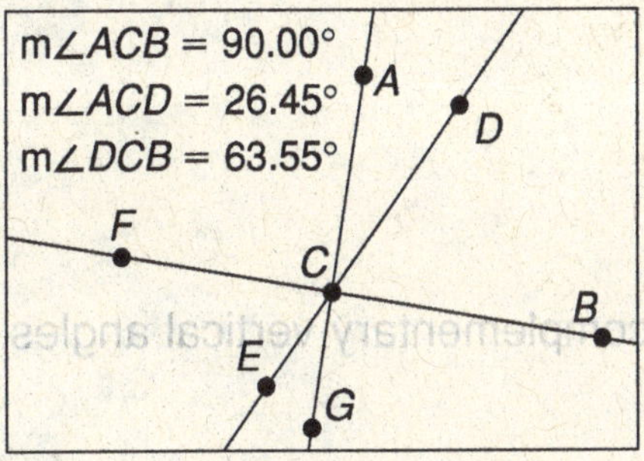

Name ______________________ Date __________ Class __________

LESSON 8C

Technology Lab

Angle Relationships, continued

Think and Discuss

Answer the following based on the final drawing in the activity above.

1. What angle is congruent to $\angle DCB$?

____

2. Are $\angle ACD$ and $\angle DCB$ complementary or supplementary?

____

3. What are the three supplemental angles of $\angle FCG$?

____

4. What angle is congruent to $\angle FCD$?

____

Try This

Use geometry software to construct the following scenarios within the space given.

____

1. Construct two non-adjacent supplementary angles.

2. Construct two non-supplementary adjacent angles.

3. Construct two complementary vertical angles.

Name ______________________ Date ____________ Class ____________

LESSON **8C**

Review for Mastery

Angle Relationships

When two lines intersect, two pairs of opposite angles, called vertical angles, are formed. Vertical angles are congruent.

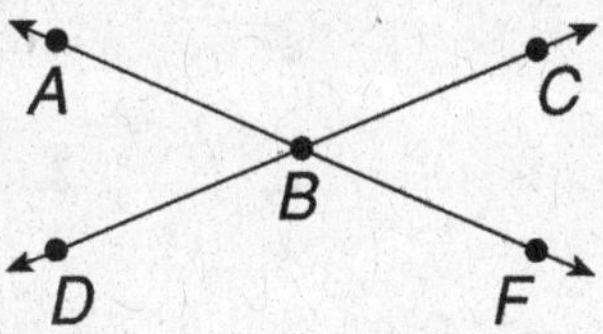

∠*ABC* and ∠*DBF* are vertical angles.
∠*ABD* and ∠*CBF* are vertical angles.

Adjacent angles have a common side and vertex.

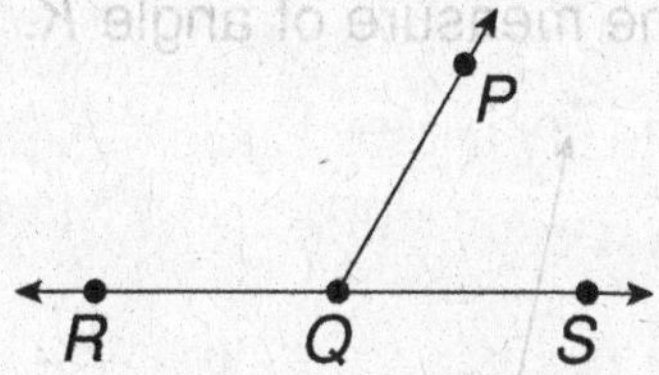

∠*PQR* and ∠*PQS* are adjacent angles.

Identify the type of each angle pair shown.

1.

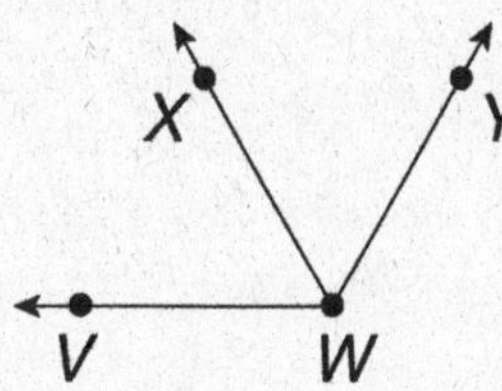

____

2.

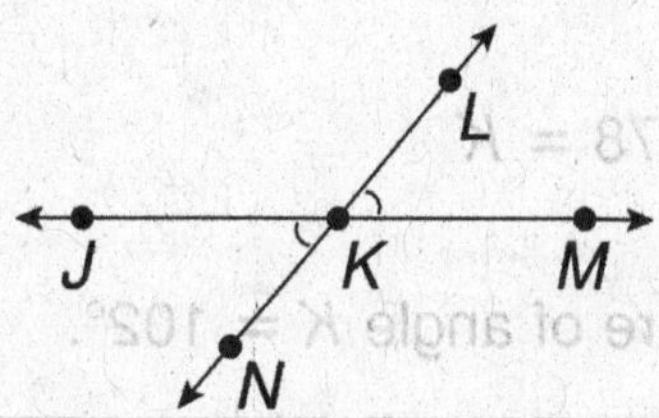

____

If the sum of the measures of two angles is 90°, then the angles are complementary angles.

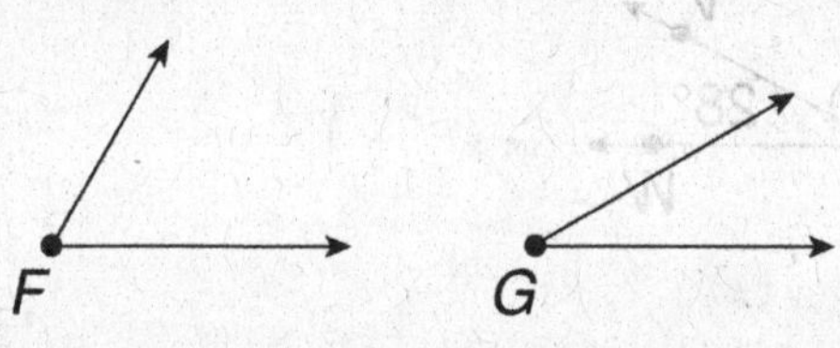

∠*F* and ∠*G* are complementary angles.

If the sum of the measures of two angles is 180°, then the angles are supplementary angles.

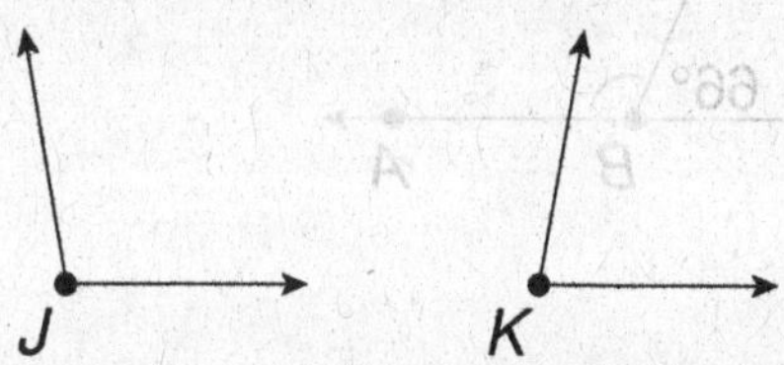

∠*J* and ∠*K* are supplementary angles.

Identify the type of each angle pair shown.

3.

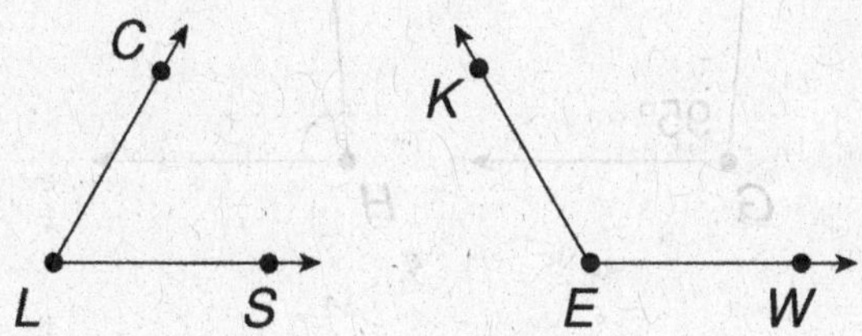

____

4.

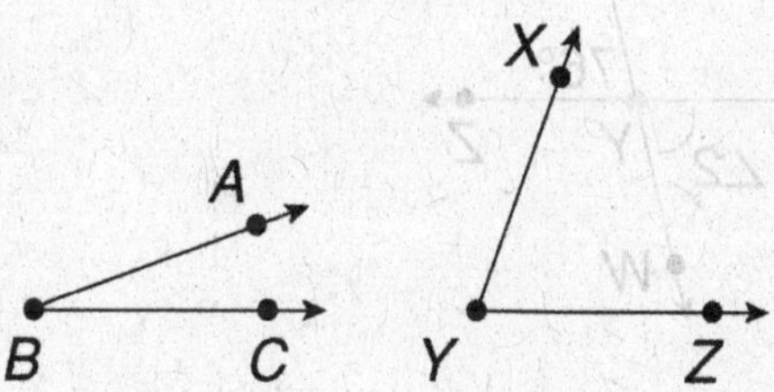

____

Name ______________________ Date ____________ Class ____________

LESSON 8C

Review for Mastery

Angle Relationships, continued

You can use related facts to find an unknown angle measure.

If you know that angles J and K are supplementary and $m\angle J = 78°$, you can find the measure of angle K.

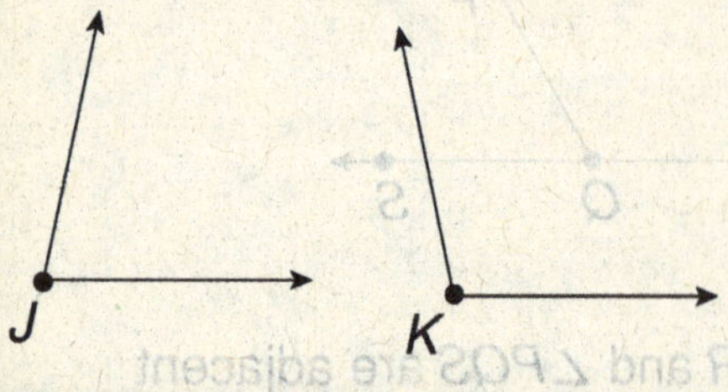

First write an equation.

$78 + K = 180$ Supplementary angles have a sum of 180°.

Then use related facts to solve the equation.

$78 + K = 180$
Think: $180 - 78 = K$
$K = 102$

So, the measure of angle $K = 102°$.

Find each unknown angle measure.

5. The angles are supplementary.

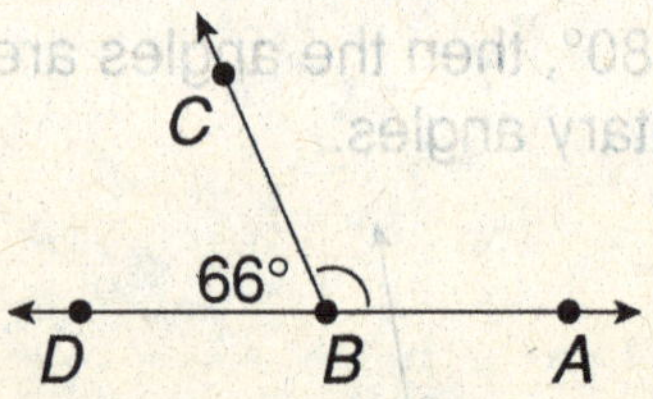

____

6. The angles are complementary.

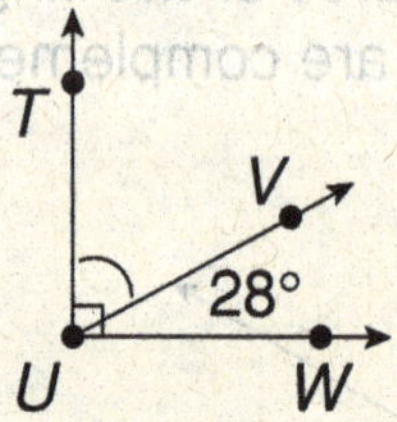

____

7. The angles are vertical angles.

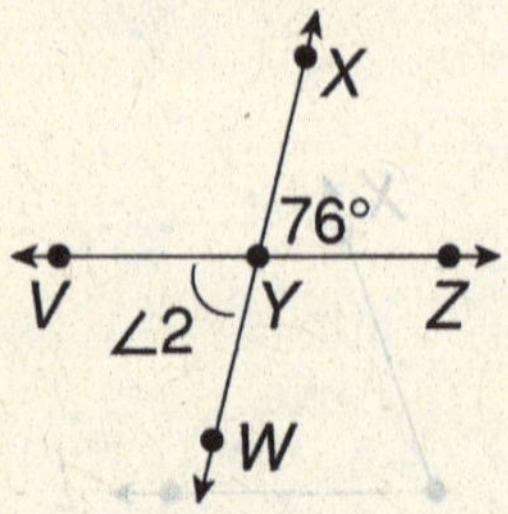

____

8. The angles are supplementary.

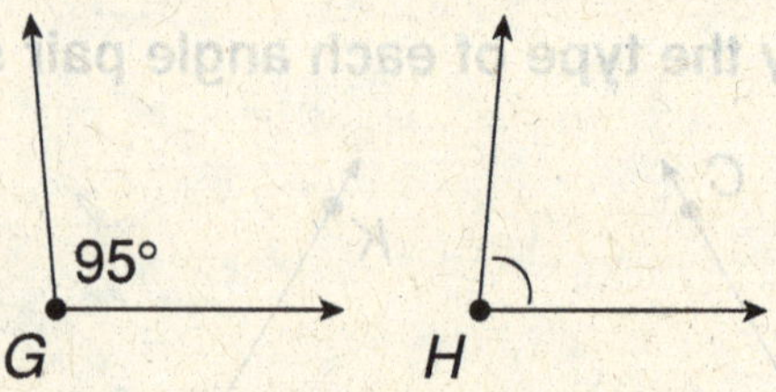

____

Name ______________________ Date __________ Class __________

LESSON 8C

Homework and Practice

Angle Relationships

Identify the type of each angle pair shown.

1.

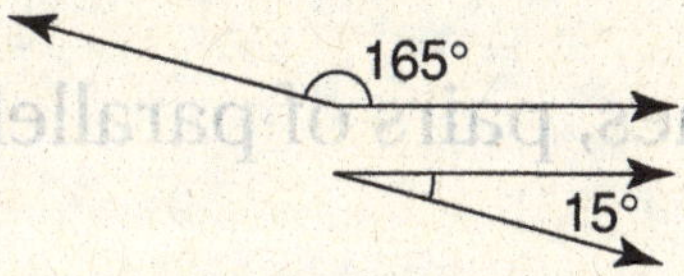

____

2.

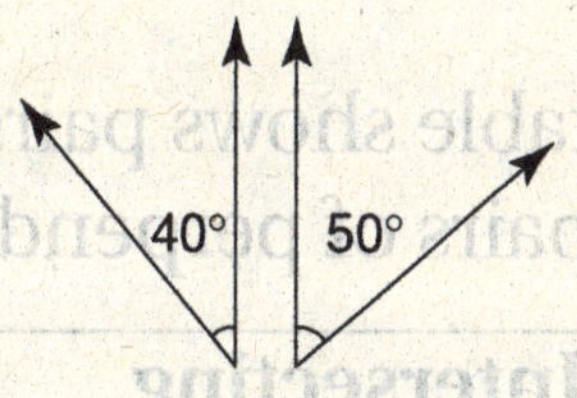

____

3.

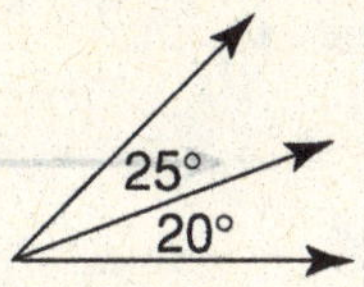

____

4.

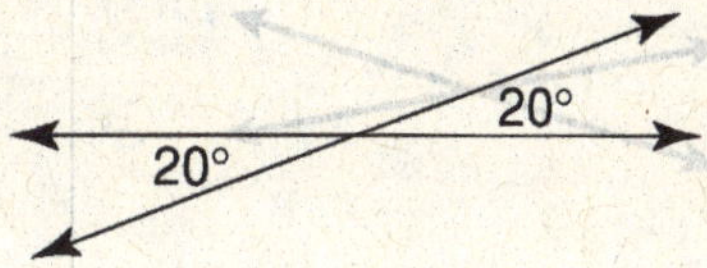

____

Find each unknown angle measure.

5. The angles are supplementary.

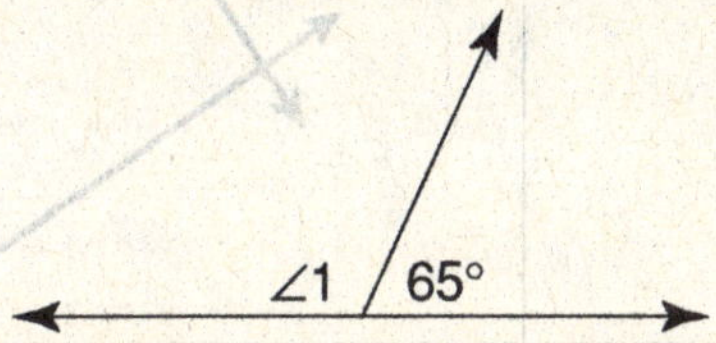

____

6. The angles are complementary.

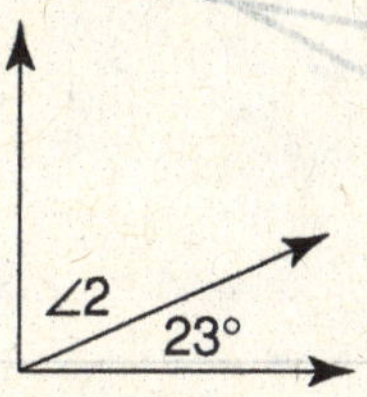

____

7. Frank says the letter X forms 2 pairs of vertical angles. Juan says it forms 2 pairs of congruent angles. Who is correct? Explain.

____

____

____

8. Is the following statement always true, sometimes true, or never true? Explain your reasoning. Two congruent angles that are supplementary both measure 95°.

____

____

EXPLORATION

Program of Studies and Core Content
MA-6-G-S-SR2, MA-06-3.1.1

8D Classifying Lines

The table shows pairs of intersecting lines, pairs of parallel lines, and pairs of perpendicular lines.

Intersecting Lines	Parallel Lines	Perpendicular Lines
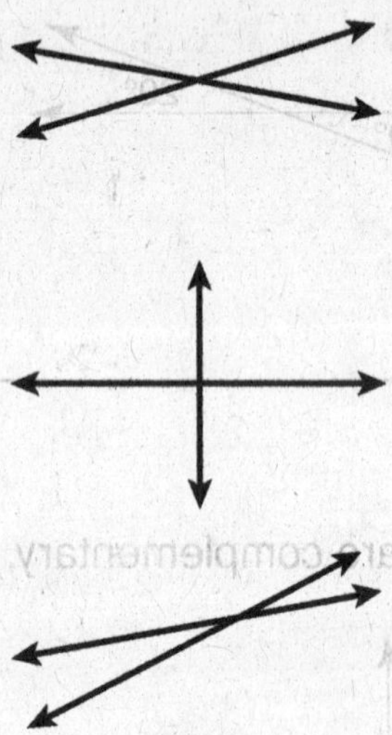	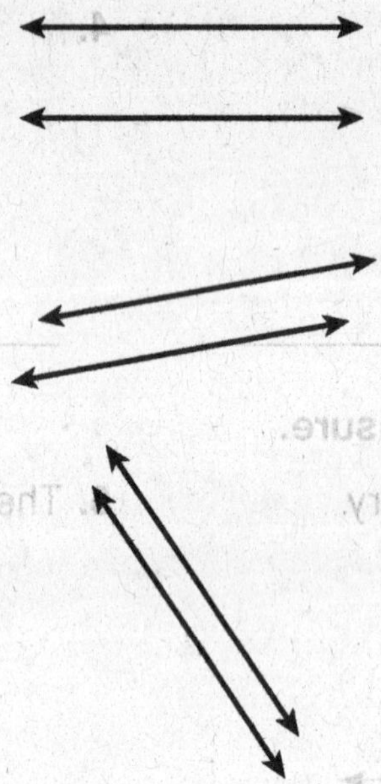	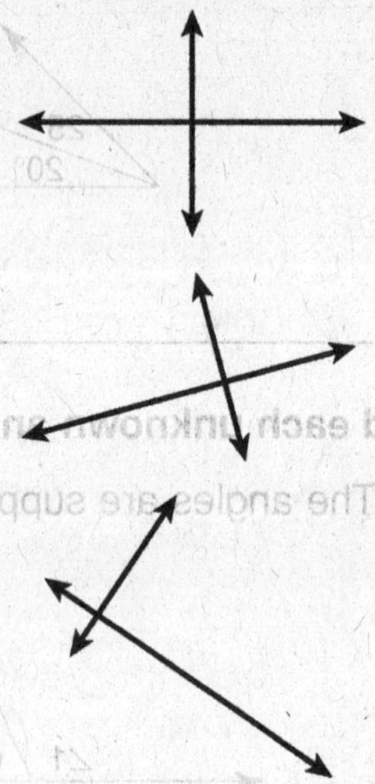

1. Draw your own examples of intersecting lines, parallel lines, and perpendicular lines.

Think and Discuss

2. **Describe** how you can tell if two lines are parallel.
3. **Discuss** what makes perpendicular lines different from other lines that intersect.

Name ______________________________ Date ____________ Class __________

LESSON 8D

Review for Mastery

Classifying Lines

Some lines have relationships.

Intersecting lines cross each other at one point.

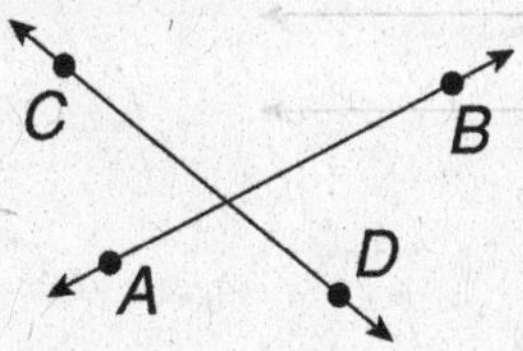

$\overleftrightarrow{AB}$ intersects $\overleftrightarrow{CD}$.

Perpendicular lines intersect to form right angles.

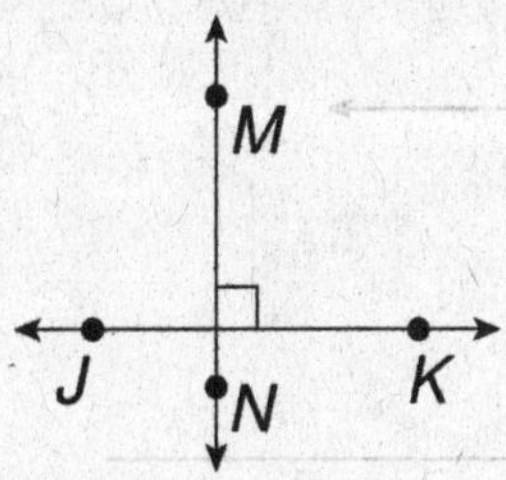

$\overleftrightarrow{JK} \perp \overleftrightarrow{MN}$

Parallel lines lie in the same plane but never intersect.

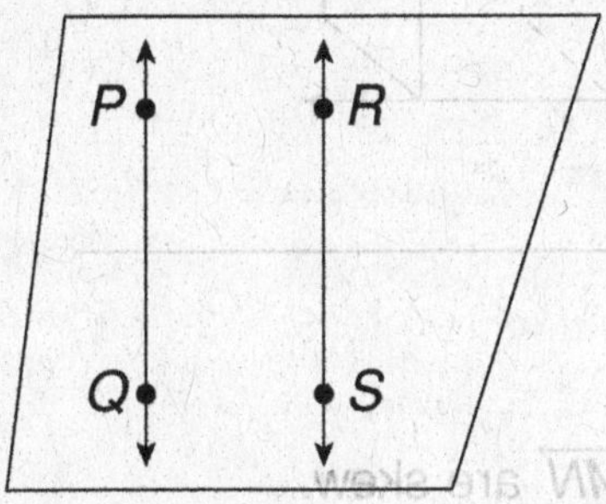

$\overleftrightarrow{PQ} \parallel \overleftrightarrow{RS}$

Skew lines lie in different planes and do not intersect.

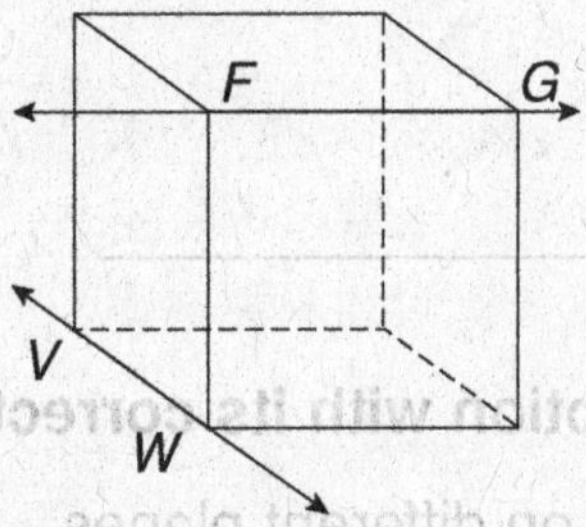

$\overleftrightarrow{FG}$ and $\overleftrightarrow{VW}$ are skew.

Classify each pair of lines.

1.

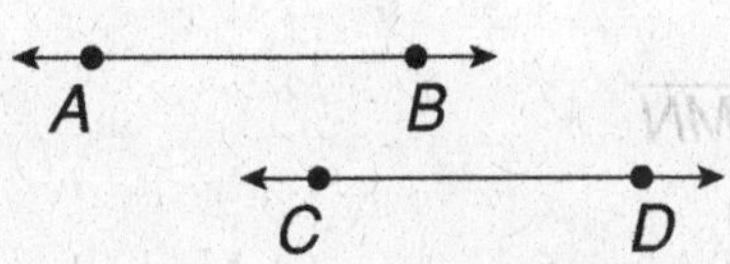

____

2.

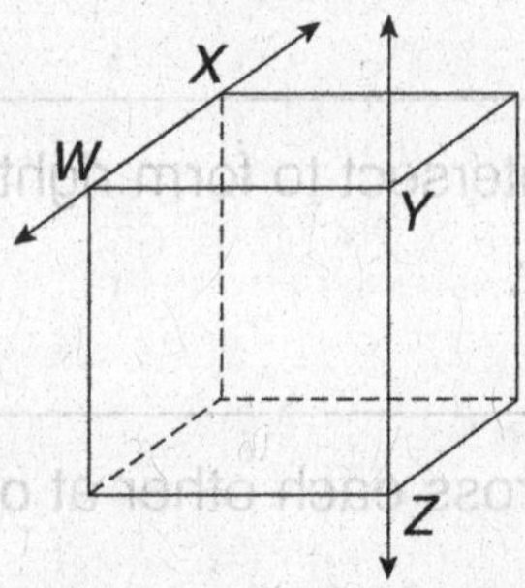

____

3.

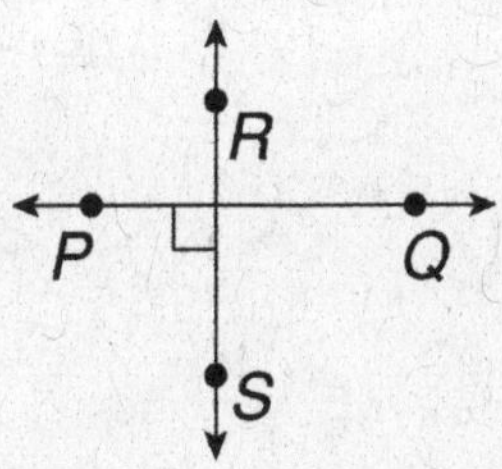

____

4.

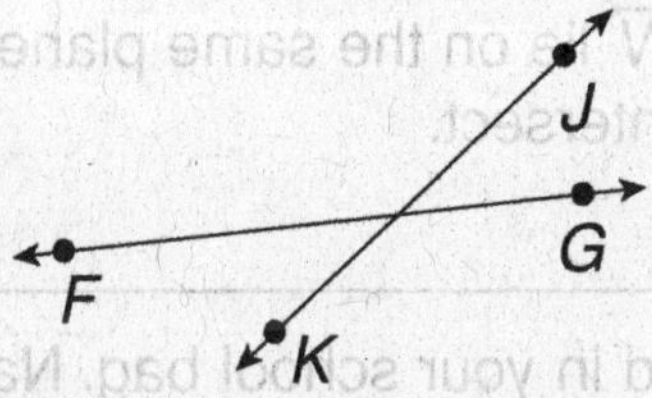

____

Name ______________________ Date ____________ Class ____________

CHAPTER **8D**

Homework and Practice

Classifying Lines

Classify each pair of lines.

1.

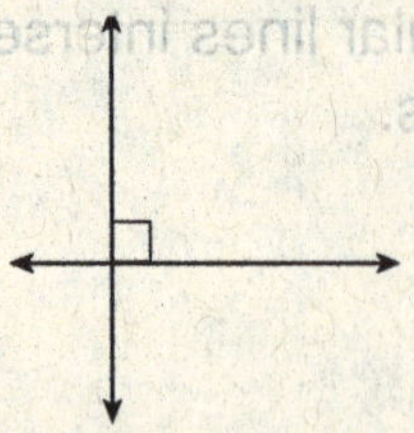

____

2.

____

3.

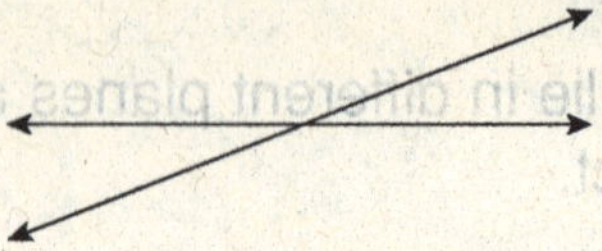

____

4.

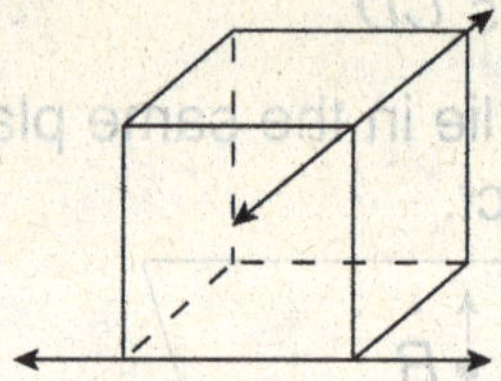

____

Match each description with its correct classification.

5. $\overleftrightarrow{XY}$ and $\overleftrightarrow{MN}$ lie on different planes and are neither parallel nor intersecting.

____

6. $\overleftrightarrow{XY}$ and $\overleftrightarrow{MN}$ intersect to form right angles.

____

7. $\overleftrightarrow{XY}$ and $\overleftrightarrow{MN}$ cross each other at one common point.

____

8. $\overleftrightarrow{XY}$ and $\overleftrightarrow{MN}$ lie on the same plane and never intersect.

____

A. $\overleftrightarrow{XY}$ and $\overleftrightarrow{MN}$ are skew.

B. $\overleftrightarrow{XY}$ intersects $\overleftrightarrow{MN}$

C. $\overleftrightarrow{XY} \parallel \overleftrightarrow{MN}$

D. $\overleftrightarrow{XY} \perp \overleftrightarrow{MN}$

9. Look around in your school bag. Name a pair of parallel lines that you see.

____

EXPLORATION

Program of Studies and Core Content
MA-6-G-S-SR1, MA-6-G-S-SR4, MA-06-3.1.2

8E Triangles

You can classify triangles by the measures of their angles.

Measure the angles of each triangle. Then check the box that gives the correct classification of the triangle.

	Acute Triangle: has only acute (less than 90°) angles	Obtuse Triangle: has one obtuse (greater than 90°) angle	Right Triangle: has one right (90°) angle
1. 			
2. 			
3. 			

Think and Discuss

4. Find the sum of the angle measures in each triangle.

5. Make a generalization about the sum of the angle measures in a triangle.

Name ______________________ Date __________ Class __________

LESSON 8E

Review for Mastery

Triangles

You can classify triangles by the measures of their angles.

An acute triangle has only acute angles.
A right triangle has one right angle.
An obtuse triangle has one obtuse angle.

To classify the triangle below by its angles, you first need to find the measure of all the angles.

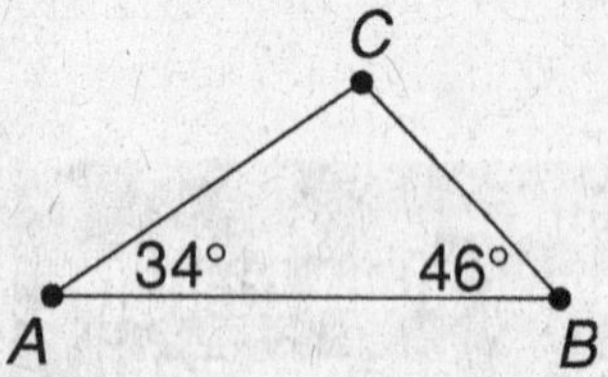

The sum of the angles of a triangle is 180°.

To find the unknown angle, first find the sum of the two known angles.

34 + 46 = 80

Then subtract the sum from 180.

180 − 80 = 100

The difference is the measure of the third angle.

The measures of the angles are 34°, 46°, and 100°. So the triangle is obtuse.

Find each unknown angle. Then classify each triangle.

1.

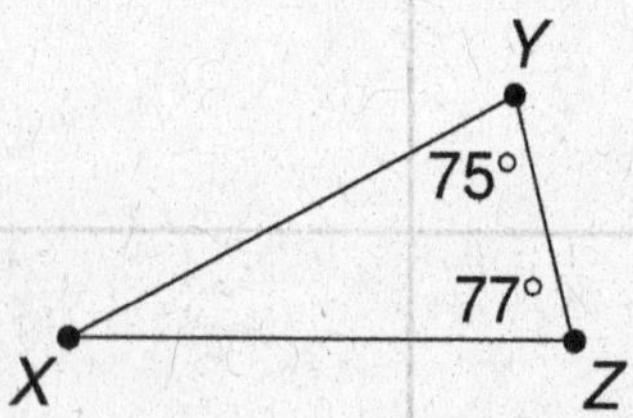

____

2.

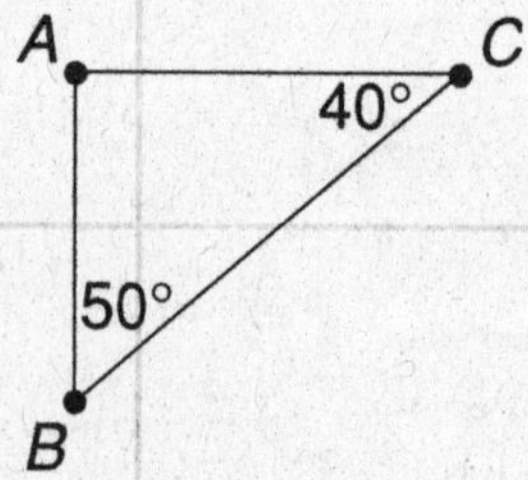

____

3.

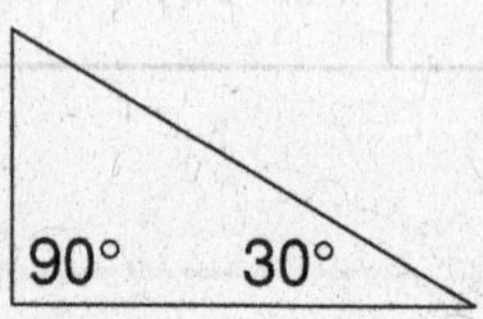

____

4.

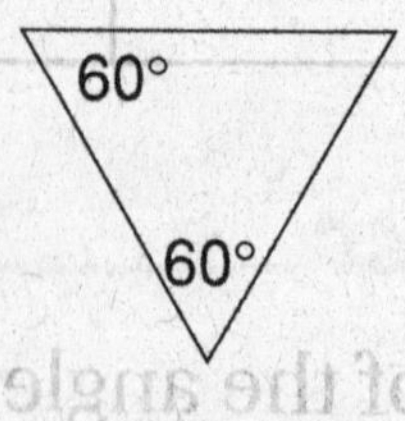

____

Name ______________________________ Date ____________ Class ____________

LESSON **8E**

Review for Mastery

Triangles, continued

You can also classify triangles by the lengths of their sides.

A scalene triangle has no congruent sides.
An isosceles triangle has at least 2 congruent sides.
An equilateral triangle has 3 congruent sides.

The perimeter of the triangle below is 18.6 cm. To classify the triangle below by its sides, you first need to find the length of the unknown side.

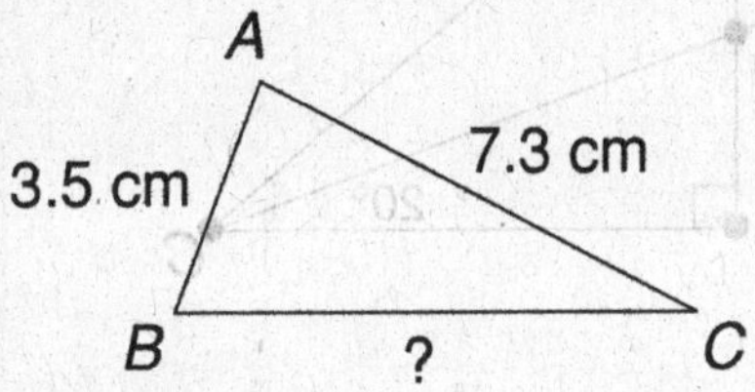

To find the unknown length, first find the sum of the two known lengths.

7.3 + 3.5 = 10.8

Then subtract this sum from the perimeter.

18.6 − 10.8 = 7.8

The length of the third side is 7.8 cm.

The lengths of the sides of the triangle are 3.5 cm, 7.3 cm, and 7.8 cm, so the triangle is scalene.

Find each missing length. Then classify each triangle.

5.

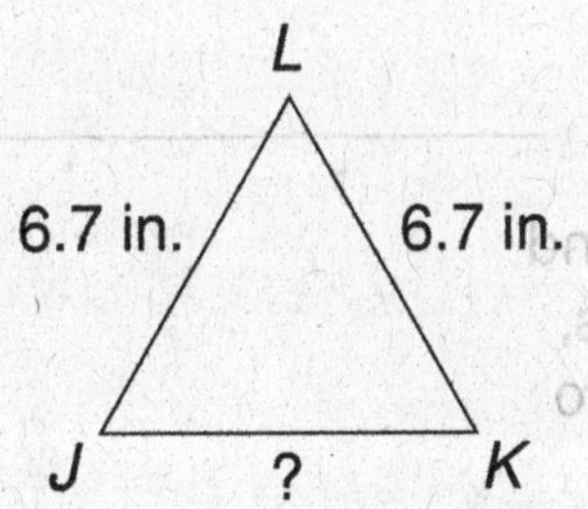

The perimeter of the triangle is 20.1 inches.

____

6.

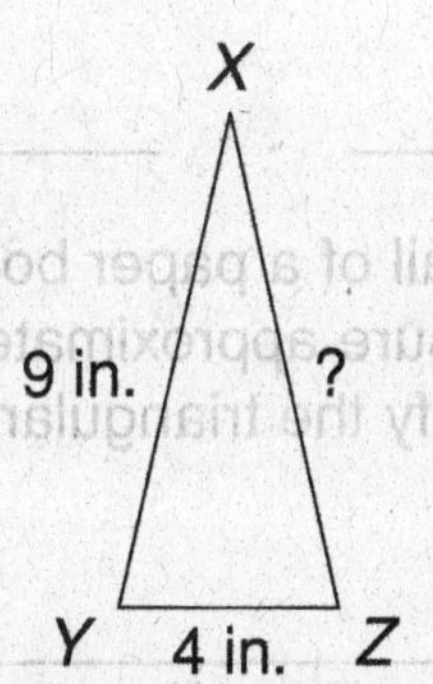

The perimeter of the triangle is 22 inches.

____

Name ______________________ Date ____________ Class ____________

LESSON 8E

Homework and Practice

Triangles

Use the diagram to find the measure of each indicated angle.

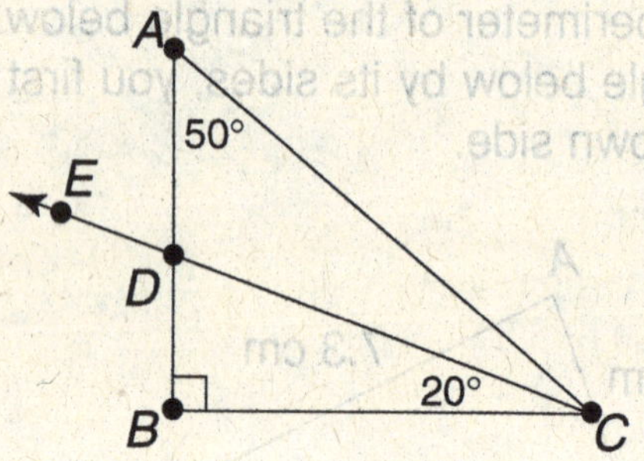

1. ∠*BDC*

____

2. ∠*BCA*

____

3. ∠*ADC*

____

4. ∠*ACD*

____

5. ∠*EDA*

____

Classify each triangle using the given information.

6. The sum of the lengths of the three sides is 15 cm.

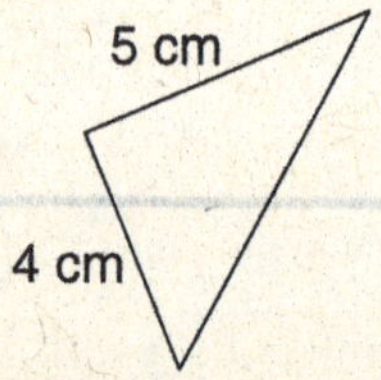

____

7. The sum of the lengths of the three sides is 21 in.

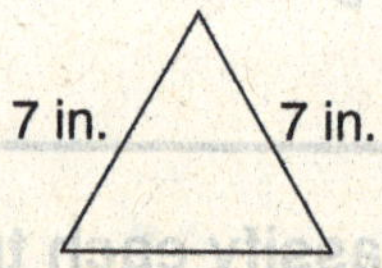

____

8. The sum of the lengths of the three sides is 16 ft.

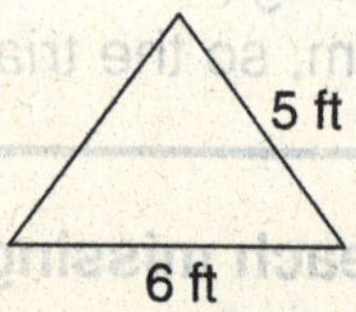

____

9. The angles of the sail of a paper boat measure 90°, 45°, and 45°. The sides measure approximately 2.8 inches, 2 inches, and 2 inches. Classify the triangular shape of the sail in two different ways.

____

10. One angle in one triangle is congruent to one angle in another triangle. What can you conclude about the other two angles in both triangles?

____

EXPLORATION

Program of Studies and Core Content

MA-6-G-S-SR4, MA-06-3.1.2

8F Quadrilaterals

Find a real-world example for each quadrilateral.

	Quadrilateral	Example
1.	 Parallelogram	
2.	 Rhombus	
3.	 Rectangle	
4.	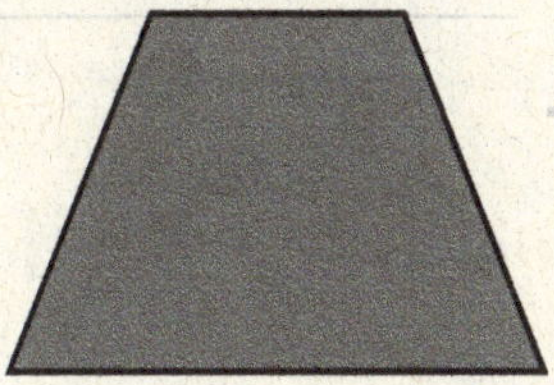 Trapezoid	

Think and Discuss

5. Discuss how all the quadrilaterals are alike.

6. Explain what is special about the rectangle.

Name ______________________ Date ____________ Class ____________

LESSON 8F

Review for Mastery

Quadrilaterals

A quadrilateral is a plane figure with four sides and four angles.

There are special types of quadrilaterals.

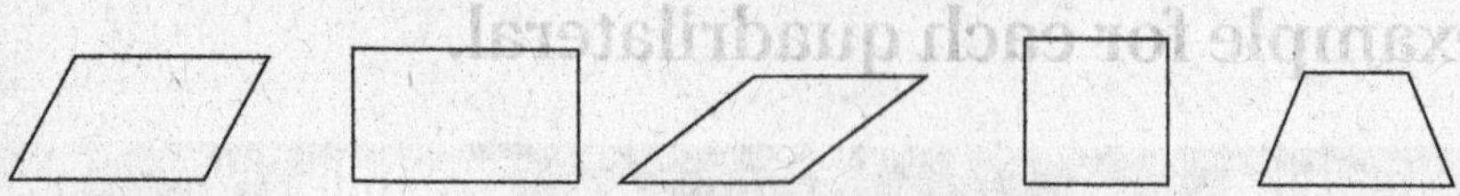

parallelogram rectangle rhombus square trapezoid

A parallelogram has opposite sides that are parallel and congruent. Opposite angles are also congruent.

A rectangle is a parallelogram that has four right angles.

A rhombus is a parallelogram with four congruent sides.

A square is a rectangle with four congruent sides.

A trapezoid is a quadrilateral with exactly two parallel sides.

The best description for the figure below is rectangle because it is a parallelogram with four right angles.

Give the best name for each figure.

1. 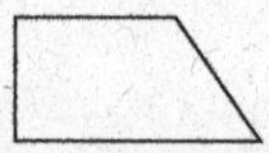

____

2. 

____

3. 

____

4. 

____

5. 

____

6. 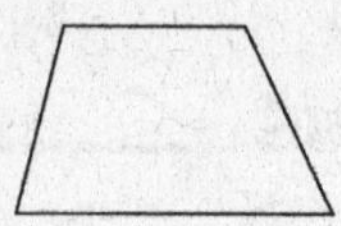

____

Name ______________________ Date __________ Class __________

LESSON 8F

Homework and Practice

Quadrilaterals

Give the most descriptive name for each figure.

1. 

____

2. 

____

3. 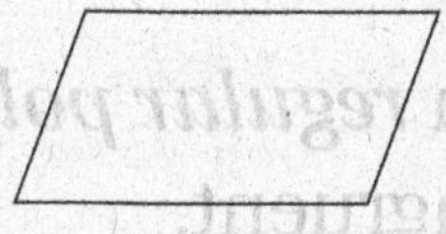

____

4. 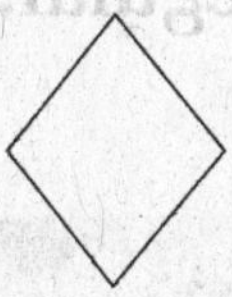

____

5. 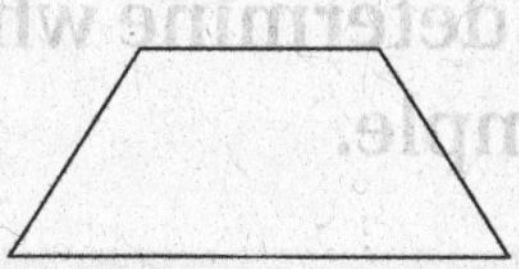

____

6. 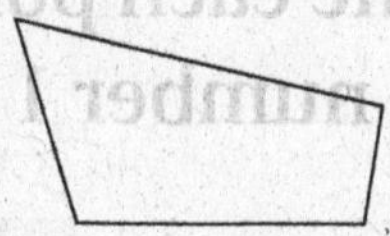

____

7. Are all rectangles also parallelograms? Explain.

____

8. Is a square always a rectangle? Explain.

____

9. Is a quadrilateral always a rhombus? Explain.

____

10. Is a quadrilateral any plane figure with four sides and four angles? Explain.

____

11. Barry made a table top in the shape of a quadrilateral. All of its angles measure 90°. It has two different lengths of sides. What is the shape of Barry's table top?

____

12. The perimeter of a rhombus is 84 inches. What is the length of each side of the rhombus? Explain.

____

EXPLORATION

Program of Studies and Core Content
MA-6-G-S-SR1,
MA-06-3.1.2

8G Polygons

In a ***regular polygon,*** all sides are congruent and all angles are congruent.

Name each polygon and determine whether it is regular. Use number 1 as an example.

	Polygon	Name	Regular?
1.	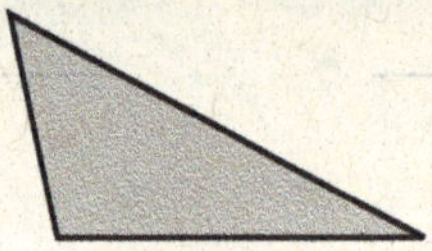	Triangle	no
2.	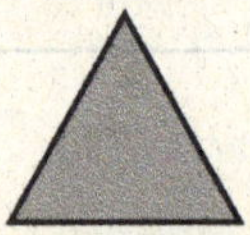		
3.			
4.			
5.			
6.			

Think and Discuss

7. Explain how you classified each polygon in numbers 2–6.

Name ______________________ Date __________ Class __________

LESSON 8G

Hands-On Lab 1

Determining Interior Angles

Activity

1. Tape two mirrors together to form a hinged mirror. Place the mirror pair on the figure below. What figure is formed by the line and its reflections?

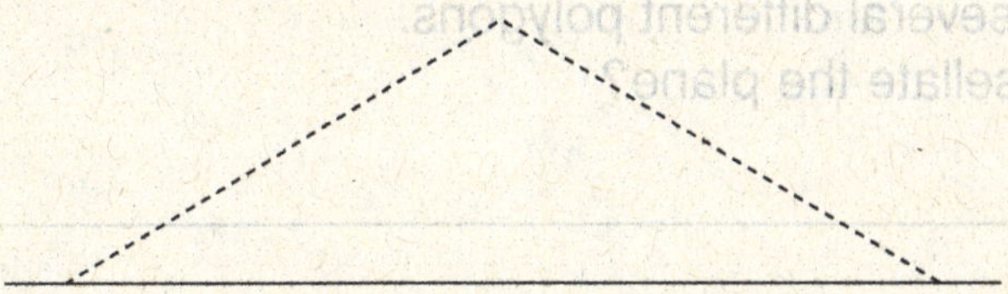

____

2. Pull the sides of the mirror together until you see a square. Measure the angle between the mirrors. This is the central angle of the square. Continue pulling the sides of the mirrors together until you see a pentagon, a hexagon, and an octogon. Measure the central angle each time.

Figure	Square	Pentagon	Hexagon	Octagon
Central angle				
Interior angle				

3. The interior angle of a polygon is the angle between two adjacent sides. Calculate the interior angle for each polygon and complete the table.

4. Place the corner of a polygon in the corner of the hinged mirror and pull the two mirrors against adjacent sides of the polygon. Sketch what you see.

____

5. Try other polygons. Which polygons will reflect a whole number of times?

____

Name ______________________________ Date ____________ Class ____________

LESSON 8G

Hands-On Lab 1

Determining Interior Angles, continued

Think and Discuss

1. Which polygons will tessellate the plane?

____

2. Divide 360° by the interior angles of several different polygons. What is true of polygons that will tessellate the plane?

____

____

Try This

1. Calculate the central angle for a heptagon.

____

2. What is the interior angle for a heptagon?

____

Name ______________________ Date ___________ Class ___________

LESSON 8G

Hands-On Lab 2

Patterns with Diagonals in Polygons

You can find a pattern for the number of diagonals in a polygon.

Activity

1. Draw all possible diagonals in the figure on the left. Then fill in the information in the chart.

Polygon	Number of Sides	Number of Diagonals
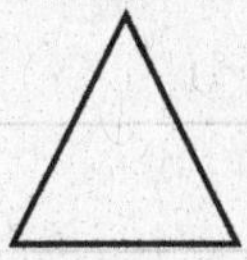	3	0
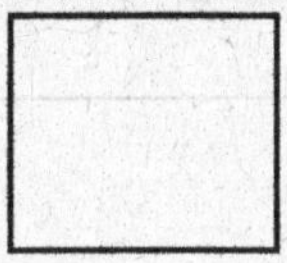	4	

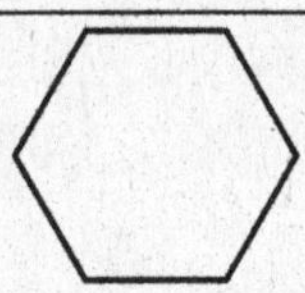		

2. What pattern do you see in the number of diagonals?

First list the diagonals in the order in the chart. Observe the difference between them.

0, 2, 5, 9,

+2 +3 +4

The difference increases by one for each additional side of the polygon.

Name ______________________ Date ____________ Class ____________

LESSON 8G

Hands-On Lab 2

Patterns with Diagonals in Polygons, continued

Think and Discuss

1. Will the number of diagonals be the same for any polygon with the same number of sides? Explain your answer. Use a quadrilateral as an example.

____

____

____

____

____

Try This

1. Use the pattern discovered in the Activity to determine the number of diagonals in an octagon (8-sided polygon).

____

2. Prove your answer to Exercise 1 by drawing the diagonals on the octagon below.

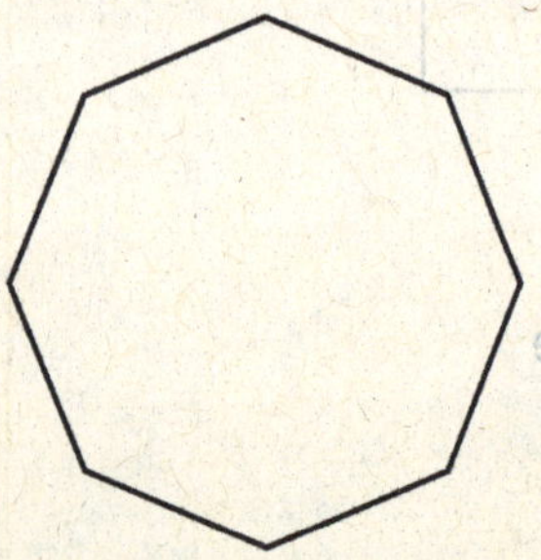

3. A polygon has 35 diagonals. What kind of polygon is it?

____

Name ______________________________ Date ____________ Class ____________

LESSON **8G**

Review for Mastery

Polygons

A polygon is a closed plane figure formed by three or more line segments.

Polygons are named by the number of their sides and angles.

Sides and Angles	**3**	**4**	**5**	**6**	**8**
Polygon	triangle	quadrilateral	pentagon	hexagon	octagon

A regular polygon has all congruent sides and angles.

Each angle is 90 degrees.
Each side is 3 centimeters long.

If the angles and the sides are not all congruent, then the polygon is not regular.

Name each polygon and tell whether it is regular or not regular.

1.

____

2.

____

To find the angle measures of a regular hexagon, first divide the polygon into triangles by drawing connecting segments from one vertex to all of its nonadjacent vertices.

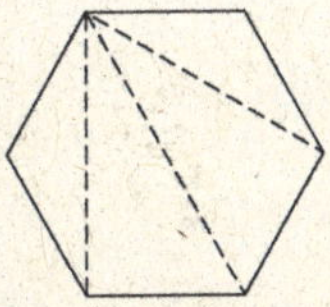

There are 4 triangles.

Next, multiply the number of triangles by 180.

$4 \cdot 180 = 720$ The sum of the angle measures of a triangle is 180°.

Divide the product by 6, the number of sides, to find the measure of each angle.

$\frac{720}{6} = 120$ All of the angles of a regular polygon are congruent.

Each angle is 120°.

Find the angle measure of each regular polygon.

3. regular quadrilateral ____________

4. regular pentagon ____________

Name ______________________________ Date ____________ Class ____________

LESSON 8G

Homework and Practice

Polygons

Name each polygon and tell whether it appears to be *regular* or *not regular*.

1.

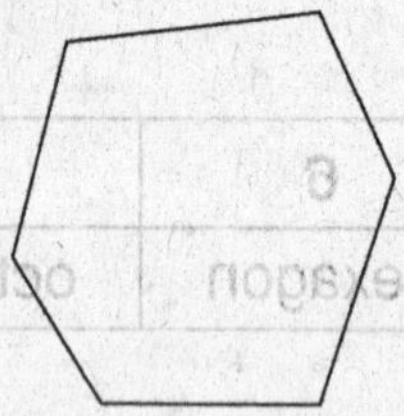

____

____

2.

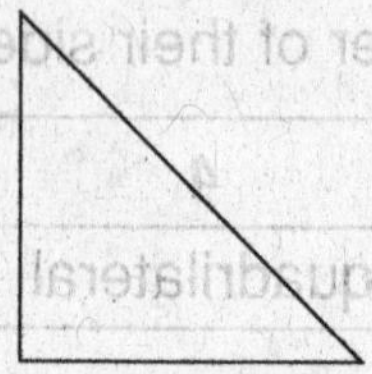

____

____

3.

____

____

4.

____

____

5.

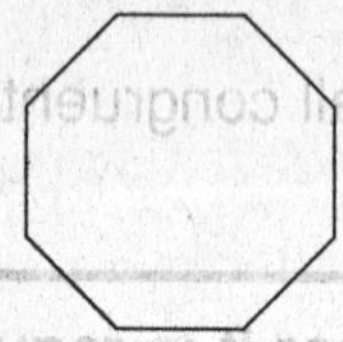

____

____

6.

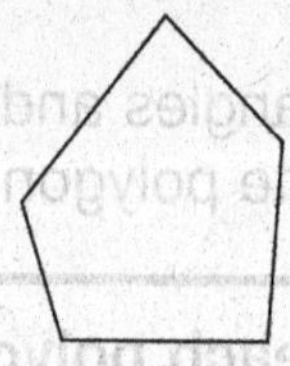

____

____

7. Tommy's large trampoline is in the shape of a regular pentagon. Each side of the trampoline measures 7 feet. What is the distance around the entire trampoline?

____

8. Draw a regular pentagon. Now draw one diagonal through that pentagon at any possible two points. Describe the two newly formed polygons.

____

EXPLORATION

Program of Studies and Core Content
MA-6-AT-S-PRF1, MA-06-5.1.1

8H Geometric Patterns

Look for a pattern, and draw the next three figures in the sequence.

1. 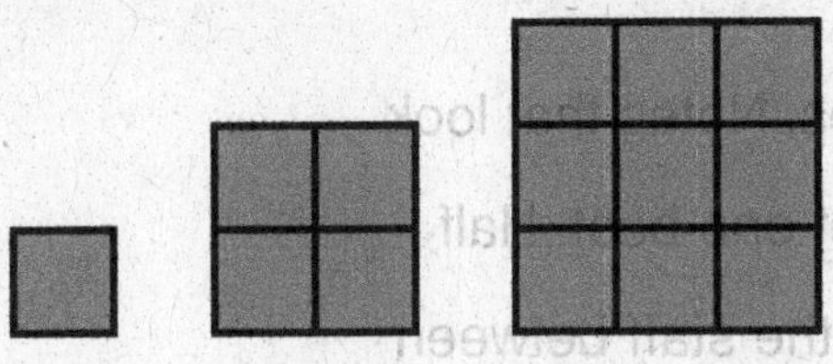

2. 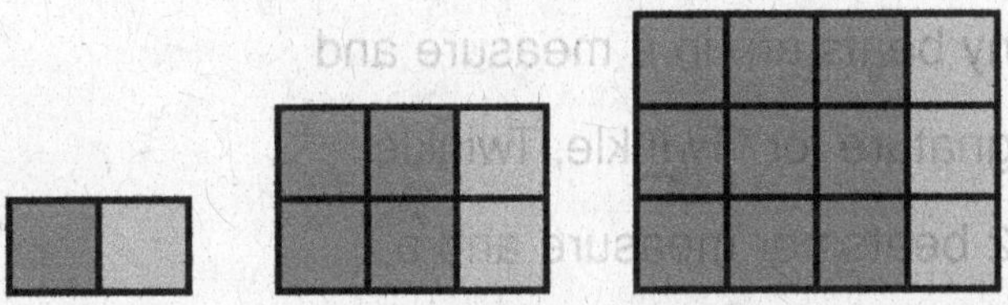

3. 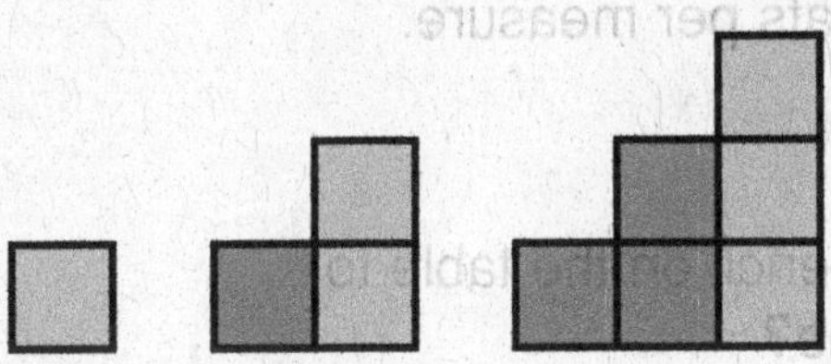

Think and Discuss

4. **Explain** how the sequence in number 2 is built on the sequence in number 1.
5. **Describe** in words the sequence in number 3.

Name ______________________________ Date ____________ Class ____________

LESSON **8H**

Hands-On Lab

Mathematics in Music

Twinkle Twinkle Little Star

Notes that look like this  are called quarter notes. Notes that look like this 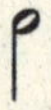 are called half notes. Quarter notes get one beat. Half notes get two beats. A measure is the portion of the staff between two vertical lines. This is one measure: . Most music has a time signature that indicates how many beats are in a measure and which note gets the beat. The time signature for "Twinkle, Twinkle Little Star" is $\frac{2}{4}$, indicating there are 2 beats per measure and a quarter note gets one beat. Other common time signatures are $\frac{3}{4}$ and $\frac{4}{4}$. In $\frac{3}{4}$ time (waltz time) there are three beats per measure.

Activity

1. Sing or hum the first line to the song. Tap a pencil on the table to keep time. How many beats are in the first line?

   ____

2. There are two phrases in the line. The two words that rhyme end the phrases. How many beats are in each phrase?

   ____

Name ______________________________ Date ____________ Class ____________

LESSON **8H**

Hands-On Lab

Mathematics in Music, continued

Think and Discuss

1. A famous hymn is written in $\frac{3}{2}$ time. How many beats are in one measure?

____

2. A folk dance is written in $\frac{7}{4}$ time. What note gets one beat?

____

Try This

1. A famous jazz composition by Dave Brubek is called "Take Five." Its time signature is $\frac{5}{4}$. How many beats are in each measure?

____

2. In "Take Five," what note gets one beat?

____

Name ______________________________ Date ____________ Class ____________

LESSON 8H

Review for Mastery

Geometric Patterns

Sometimes, patterns involve geometric shapes and figures. The pattern can be based on shape, color, size, position, or quantity.

Before you draw a missing figure in a pattern, you first have to look at the other figures to recognize a relationship among the figures.

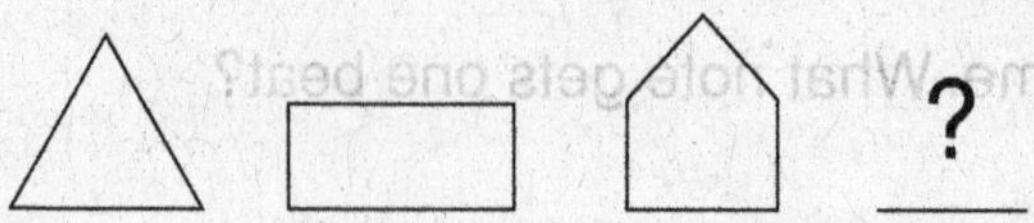

In the pattern above, each figure has one more side than the figure that proceeds it.

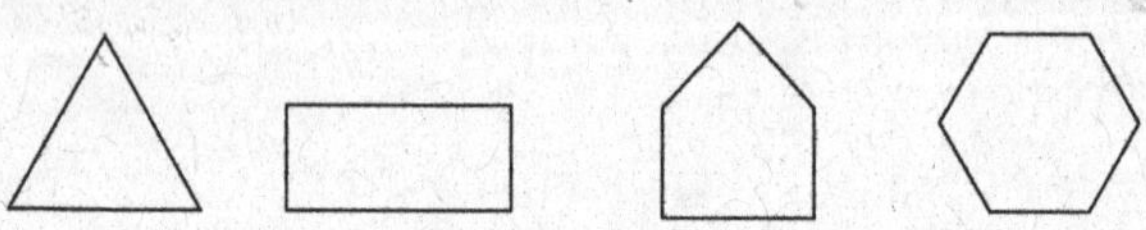

To continue the pattern, the next figure would be a hexagon.

Identify a pattern for each. Then draw the missing figure.

1.

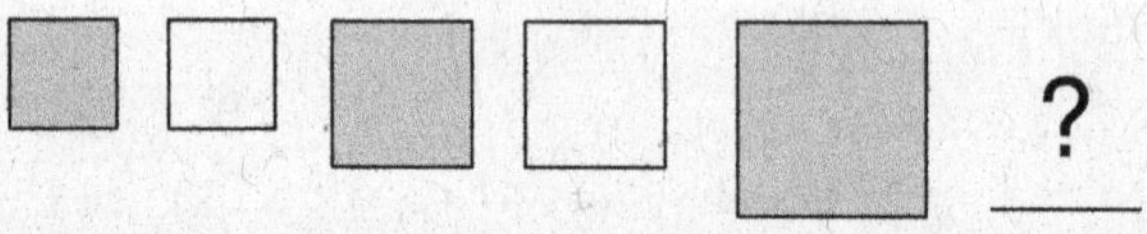

2.

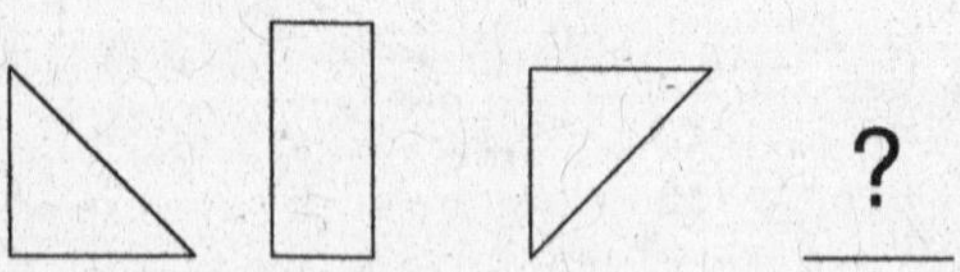

3.

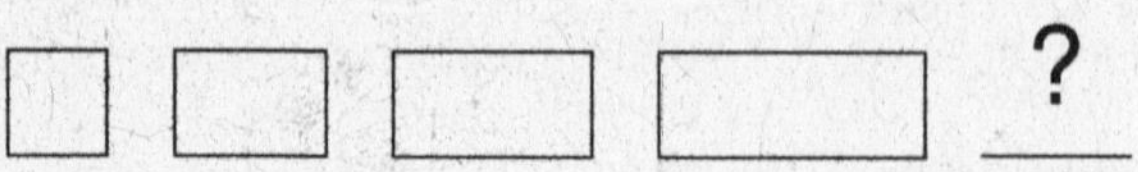

Name ______________________ Date __________ Class __________

LESSON 8H

Homework and Practice

Geometric Patterns

Identify a possible pattern. Use the pattern to draw the next figure.

1. 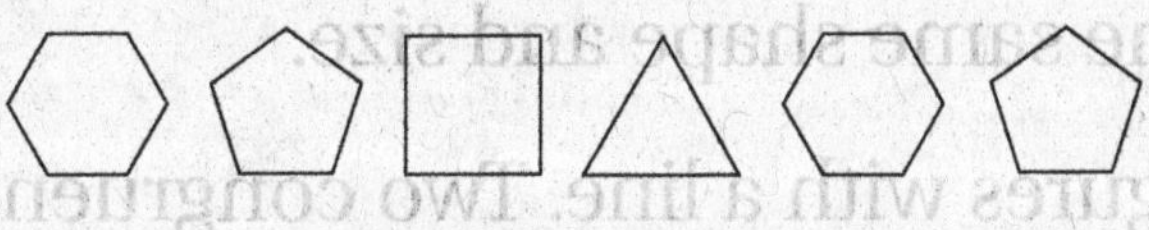 ____

____

____

2. **R Q R P R Q** ___

____

3. 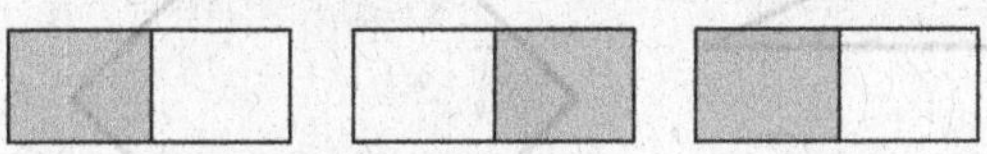 ______

____

____

4.  ____

____

5. Books are arranged in a pattern that is made up of 5 rows. The first row has two books. The second has 3 books on it. The third row has 5 books on it, and the fourth row has 8 books on it. How many books will be on the fifth row at the bottom of the stack.

____

6. Use polygons (more than one type) to create a geometric pattern. Describe you pattern.

____

____

EXPLORATION

Program of Studies and Core Content
MA-6-G-S-SR6, MA-06-3.1.4

81 Congruence

Congruent figures are exactly the same shape and size.

1. Connect two congruent figures with a line. Two congruent rectangles have been connected for you to use as a guide.

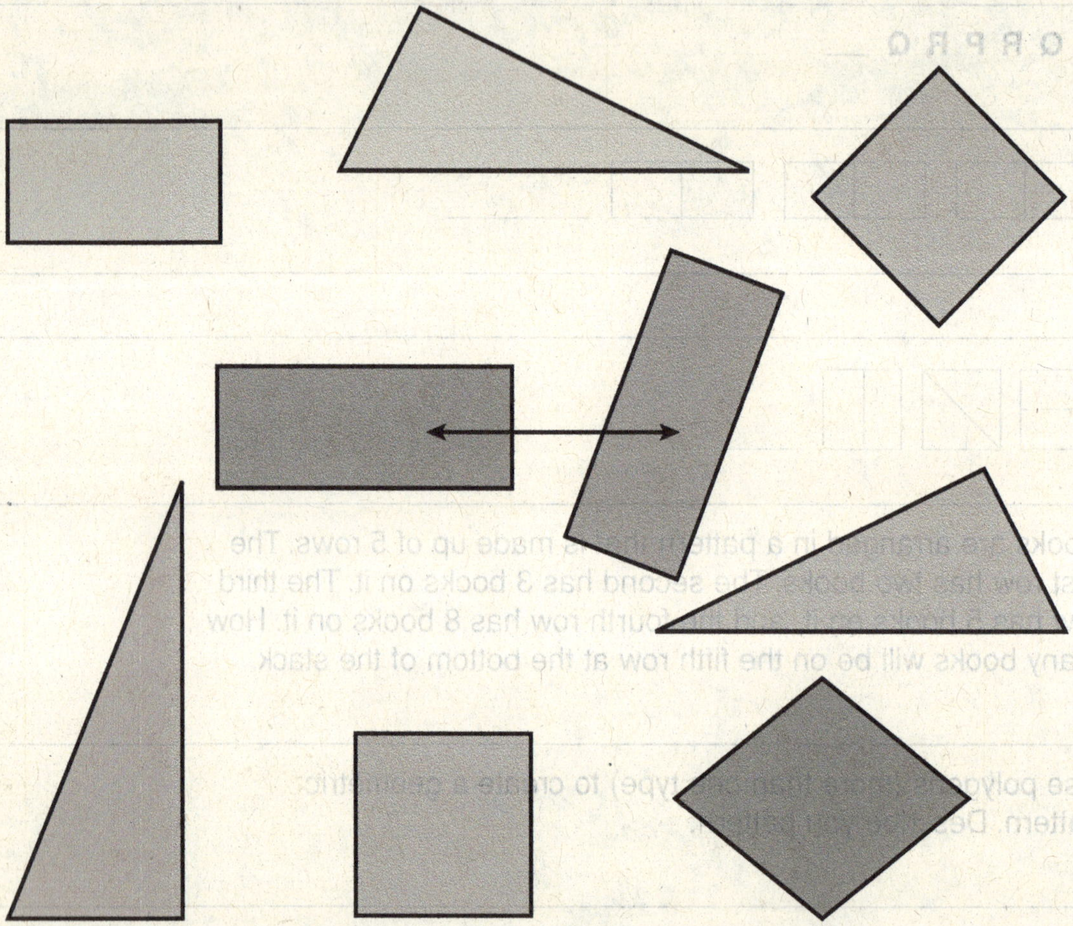

2. Measure the sides and angles of each pair of connected figures to be sure that they are congruent.

Think and Discuss

3. **Give examples** of congruent figures that occur in the real world.

Name ____________________ Date ____________ Class ____________

LESSON **8I**

Review for Mastery

Congruence

Two figures are congruent if they have the same size and shape.

Look at the two figures.

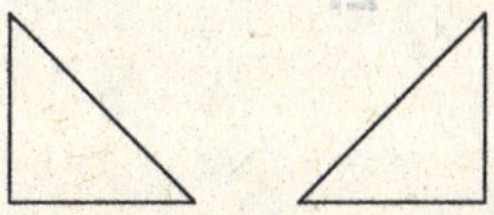

They are congruent because they are both triangles and they are the same size.

Now look at these two figures.

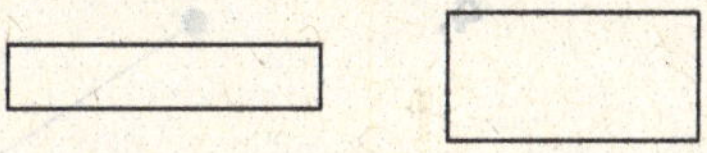

These figures are not congruent. They are both rectangles but they are not the same size.

Decide whether the figures in each pair are congruent. If not, explain.

1.

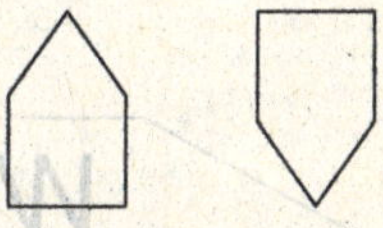

____

2.

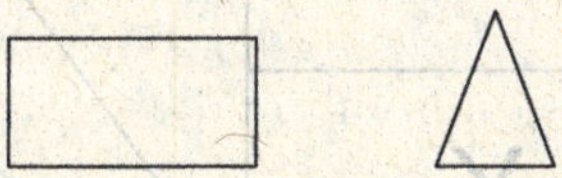

____

3.

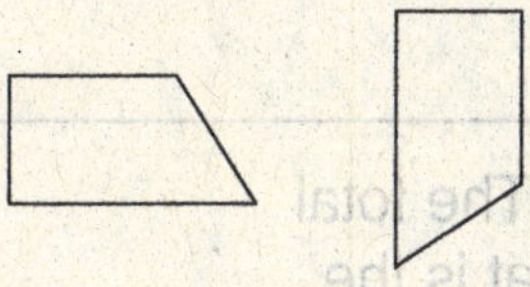

____

Name ______________________________ Date __________ Class __________

LESSON 8I

Homework and Practice

Congruence

Decide whether the figures in each pair are congruent. If not, explain.

1.

____

____

2.

____

____

3.

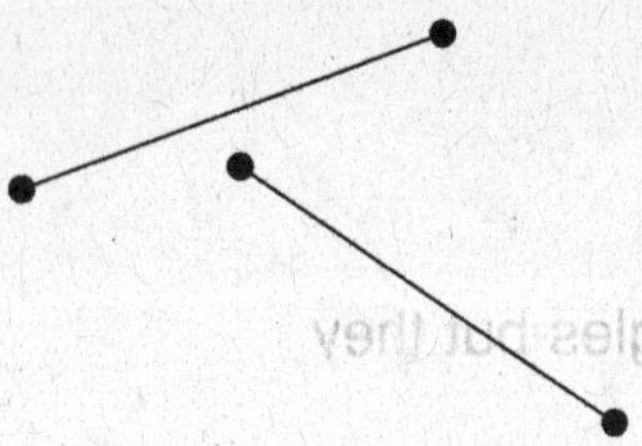

____

____

4.

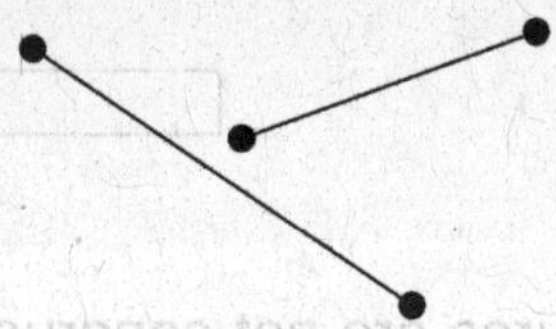

____

____

Use the diagram for Exercises 5-7.

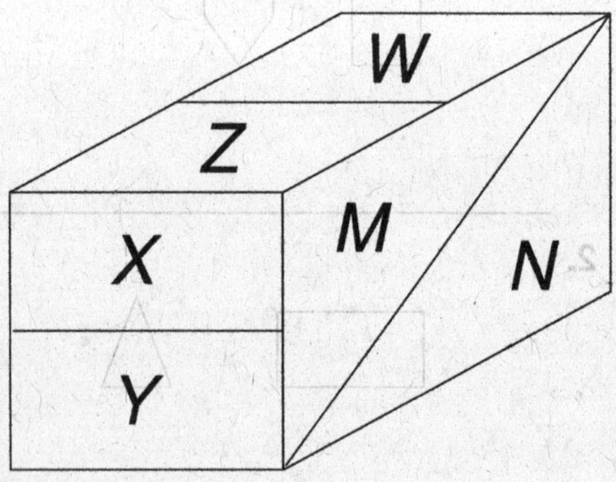

5. Which part of the figure is congruent to *N*?

____

6. Which part of the figure is congruent to *W*?

____

7. Which part of the figure is congruent to *X*?

____

8. Name two parts of a car that appear to be congruent.

____

9. Hexagon *ABCDEF* is congruent to hexagon *GHJKLM*. The total length of the sides of hexagon *ABCDEF* is 36 feet. What is the perimeter of hexagon *GHJKLM*?

____

CHAPTER 8

Chapter Summary

Answer these questions to summarize the important concepts from Chapter 8 in your own words.

1. Explain how to measure an angle with a protractor.

2. Explain why the angle measures 55° and 35° are complementary angles.

3. Two angle measures in a triangle are 85° and 62°. Explain how to find the measure of the third angle in the triangle.

4. Explain how to find the sum of the angle measures in a polygon if you only know the number of sides of the polygon.

For more review of Chapter 8:

- Complete the Chapter 8 Study Guide and Review on pages 474–476 of your textbook.
- Complete the Ready to Go On quizzes on pages 434, 454, and 470 of your textbook.

Name ______________________ Date __________ Class __________

CHAPTER 8

KCCT Prep

Multiple Choice

1. Damon drew an obtuse angle. He found the measure of the angle and then multiplied the angle measure by 2. Which of the following must be true about his result?

- (A) It is less than 90°.
- (B) It is between 90° and 180°.
- (C) It is between 180° and 360°.
- (D) It is greater than 360°.

2. Michelle wants to draw a rectangle by graphing four points on a coordinate plane and then connecting the points. She graphs the points (1, 1), (6, 1), and (6, 4). What is the last point that she should graph?

- (A) (1, 4)
- (B) (4, 1)
- (C) (4, 6)
- (D) (1, 6)

3. Which fraction is closest to 24%?

- (A) $\frac{1}{24}$
- (B) $\frac{7}{25}$
- (C) $\frac{9}{25}$
- (D) $\frac{11}{33}$

4. The triangles shown here are congruent. What is the sum of the side lengths of ΔMNP?

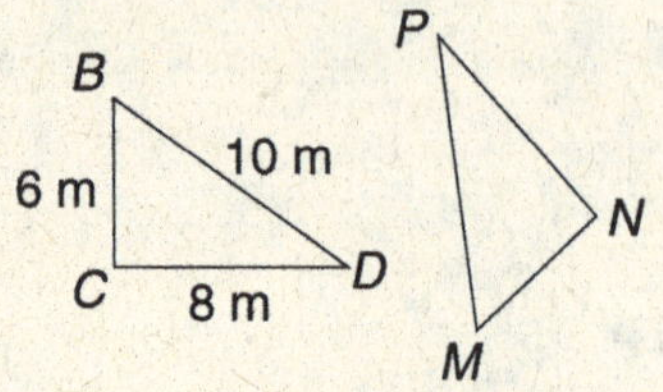

- (A) 14 m
- (B) 18 m
- (C) 20 m
- (D) 24 m

5. The trapezoid is translated so that point *A* moves to (6, 7). What is the final location of point *D*?

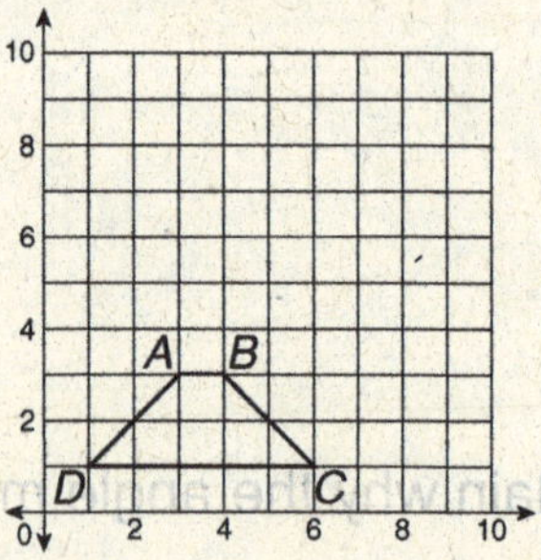

- (A) (4, 5)
- (B) (5, 4)
- (C) (4, 4)
- (D) (5, 5)

6. On both Monday and Tuesday, Franco ran $4\frac{1}{4}$ miles and Nick ran $2\frac{2}{3}$ miles. How much farther did Franco run than Nick over the course of the two days?

- (A) $2\frac{5}{6}$ miles
- (B) $3\frac{1}{6}$ miles
- (C) $4\frac{5}{6}$ miles
- (D) $5\frac{1}{6}$ miles

7. Julie correctly wrote the prime factorization of 200, but part of her work got covered by a drop of ink, as shown below. Which of these is the part of her work that got covered?

$200 = \bullet \cdot 5^2$

- (A) 2^2
- (B) 2^3
- (C) 3^2
- (D) 3^3

8. Which of these is closest to 15%?

- (A) 0.17
- (B) $\frac{1}{16}$
- (C) 0.014
- (D) $\frac{3}{50}$

9. Chantal drew the square shown below. Then she translated it 3 units down. Which ordered pair gives the final location of point A?

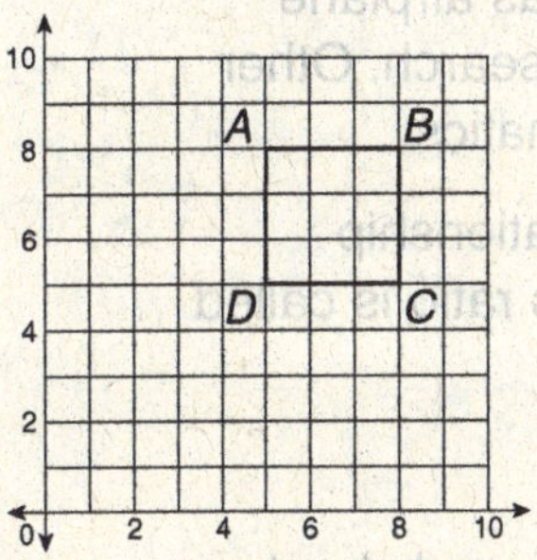

Ⓐ (5, 2)   Ⓒ (5, 8)
Ⓑ (5, 5)   Ⓓ (8, 5)

10. Luis reads 49 pages of a book in 16.2 minutes. Which is the best estimate of his unit rate?

Ⓐ $\frac{1}{3}$ page per minute
Ⓑ 3 pages per minute
Ⓒ 16 pages per minute
Ⓓ 24 pages per minute

11. Two towns are 10 inches apart on a map. The scale of the map is 1 in:20 mi. If Lorena drives at 50 mi/h, how long will it take her to drive from one town to the other?

Ⓐ 2 h   Ⓒ 5 h
Ⓑ 4 h   Ⓓ 10 h

12. Trail mix costs \$2.65 per pound. Tim buys 3.2 pounds of the trail mix. Which is the best estimate of the total cost of the trail mix?

Ⓐ \$5.80   Ⓒ \$7.00
Ⓑ \$6.50   Ⓓ \$8.50

13. Benito drew 5 hexagons. How many vertices did he draw altogether?

Ⓐ 20   Ⓒ 30
Ⓑ 25   Ⓓ 45

Open Response

14. Ming graphs the points $A(2, 5)$, $B(6, 5)$, and $C(6, 3)$ on a coordinate plane. Then he graphs point D so that the four points can be connected to form a rectangle.

a. Give the ordered pair for point D.

b. Graph and label points A, B, C, and D. Then connect them to form a rectangle.

c. Ming reflects the rectangle over the line through points C and D. Draw the rectangle in its new position.

15. The stem-and-leaf plot gives the prices of 12 computer keyboards.

Stems	Leaves
2	4 4 4 8
3	3 4 6
4	2 5 8
5	9 9

Key: 2|4 means \$24

a. Find the mean, median, and mode of the data set.

b. Rachel needs a keyboard. She wants to convince her mother that it would not be too expensive to buy a new one. Explain whether she should describe the data set using the mean, median, or mode.

16. Josh has two coupons for a clothing store. One coupon offers $\frac{1}{5}$ off the regular price of any item. The other coupon offers 15% off the regular price of any item.

a. Josh is buying a pair of jeans. He can only use one coupon. Which coupon should he use? Why?

b. The jeans cost \$26. What is the amount of the discount if Josh uses the coupon from Part a?

c. What is the final price of the jeans before sales tax?

CHAPTER 9

Project

Measurement and Geometry: Circular Reasoning

Career: Mathematician

Some mathematicians apply their knowledge in areas such as airplane scheduling, medical safety, and automobile and industrial research. Other mathematicians prefer to study the concepts behind mathematics.

For hundreds of years, mathematicians have studied the relationship between the circumference and the diameter of a circle. This ratio is called *pi* and is represented by the Greek letter π.

Before you start the project, work through the Discovering Pi worksheet on the next page.

In this chapter project . . .

You calculate your own approximation of π.

What to do:

- Learn that π is the ratio of the circumference of a circle to its diameter.
- Measure the circumferences and diameters of several circular objects.
- Use the measurements you collected to approximate π. The Project Worksheet asks you to measure four circular objects, but try measuring several more and find out if this improves your approximation. Compare your approximation to your classmates' approximations to see who can find the most accurate value.

What to present:

- A chart showing your measurement data and approximations.

Where's the math?

- Choose appropriate units to measure your circles.
- Calculate the mean of your ratios of circumference to diameter.
- Compare your estimate for π to the value given by your calculator.

go.hrw.com
Chapter Project Online
KEYWORD: MR7 Ch9

Name ______________________________ Date __________ Class __________

CHAPTER 9

Project Recording Sheet

Measurement and Geometry

Discovering Pi

Pi is the mathematical constant that describes the ratio of the circumference of a circle to its diameter. That constant, pi (π), that we have all heard about since elementary school is a repeating decimal fraction. Below are the first 399 decimal places of π.

3.1415926535 8979323846 2643383279 5028841971 693993751
05820974944 5923078164 0628620899 8628034825 342117067
98214808651 3282306647 0938446095 5058223172 535940812
84811174502 8410270193 8521105559 6446229489 549303819
64428810975 6659334461 2847564823 3786783165 271201909
14564856692 3460348610 4543266482 1339360726 024914127
37245870066 0631558817 4881520920 9628292540 917153643
67892590360 0113305305 4882046652 1384146951 941511609

This number has caused many people over the centuries to wonder why they couldn't just round π off to 3.00. It would make the calculations so much easier!

Is π the same for every circle? Check it out.

Gather four different-sized cans and other circular objects. Measure each diameter and circumference. Complete the chart below.

Object	**Diameter (*D*)**	**Circumference (*C*)**	$\frac{C}{D}$
			Mean:

What do you notice about the results in the $\frac{C}{D}$ column? Find the average (mean) in the $\frac{C}{D}$ column. Do you see where π comes from and why it can't be three? Why doesn't each experimental result equal 3.1416?

- Use your calculator. What happens to the area of a circle when you use π with more decimal points?

Extension: Over the years several states have attempted to legislate a simpler number for π. Find out which states and why they attempted to do this.

9A Understanding Customary Units of Measure

Program of Studies and Core Content
MA-6-M-S-MPA2, MA-6-M-S-MPA4, MA-6-M-S-MPA5, MA-6-M-S-SM1, MA-6-M-S-SM2, MA-06-2.1.2

Different units of measure are used for measuring length, weight, and capacity. Capacity is the amount that a container can hold.

1. Decide whether each of the units of measure below is used to measure length, weight, or capacity. Write the name of the unit in the appropriate column of the table.

Pint	Pound	Inch	Ton
Foot	Quart	Mile	Gallon
Yard	Ounce	Cup	Fluid Ounce

Length	Weight	Capacity

Think and Discuss

2. Describe an object that could be measured using yards.

3. Explain how you would decide whether to measure an object using inches or feet.

Name ______________________________ Date ____________ Class ____________

LESSON **9A**

Hands-On Lab

Measuring Standard Lengths

You can use a standard ruler to find the length of an object to the nearest $\frac{1}{2}$, $\frac{1}{4}$, $\frac{1}{8}$, and $\frac{1}{16}$ inch. For example, if your thumb is $1\frac{5}{16}$ inches long, it is $1\frac{1}{2}$ inches long rounded to the nearest $\frac{1}{2}$ inch, $1\frac{1}{4}$ inches rounded to the nearest $\frac{1}{4}$ inch, $1\frac{3}{8}$ inches rounded to the nearest $\frac{1}{8}$ inch, and $1\frac{5}{16}$ inch rounded to the nearest $\frac{1}{16}$ inch.

Activity

Measure the length of the following figure.

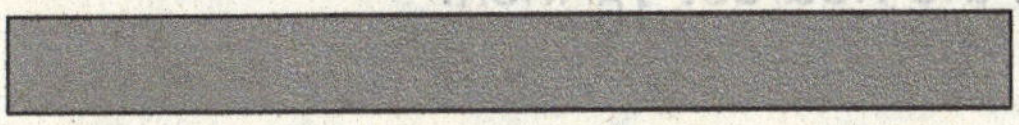

1. What is its length to the nearest $\frac{1}{2}$ inch?

____

2. What is its length to the nearest $\frac{1}{4}$ inch?

____

3. What is its length to the nearest $\frac{1}{8}$ of an inch?

____

4. What is its length to the nearest $\frac{1}{16}$ of an inch?

____

Name ____________________ Date ____________ Class ____________

LESSON 9A

Hands-On Lab

Measuring Standard Lengths, continued

Think and Discuss

1. When the length of an object is rounded to the nearest $\frac{1}{2}$, $\frac{1}{4}$, $\frac{1}{8}$, and $\frac{1}{16}$ inch, is it possible for these lengths to be equivalent? Justify your answer with an example.

____

____

2. Describe a situation when you might decide to round off a measurement to the nearest $\frac{1}{2}$ inch rather than the nearest $\frac{1}{16}$ inch.

____

____

Try This

1. Measure the length of this object to the nearest $\frac{1}{4}$ inch.

____

2. Measure the length of this object to the nearest $\frac{1}{2}$ inch.

____

3. Measure the length and width of a book to the nearest $\frac{1}{8}$ of an inch.

____

Name ________________________________ Date ___________ Class ___________

LESSON 9A

Review for Mastery

Understanding Customary Units of Measure

The customary system of measurement uses inches, feet, yards, and miles to measure length, width, distance, and height.

- Use inches to measure objects that are a few widths of your thumb.
- Use feet to measure objects that are a few times the distance from your shoulder to your elbow.
- Use yards to measure objects that are a few widths of a classroom door.
- Use miles to measure distances that are a few times the lengths of 18 football fields.

Write *inches*, *feet*, *yards*, or *miles*.

1. Your hand is about 3 ________ wide.

2. A grown man is about 2 ________ tall.

The customary system of measurement uses ounces, pounds, and tons to measure weight.

- Use ounces to measure objects that weigh as much as a few slices of bread.
- Use pounds to measure objects that weigh as much as a few loaves of bread.
- Use tons to measure objects that weigh as much as a few small cars.

Write *ounces*, *pounds*, or *tons*.

3. A feather weighs about 2 ________.

4. A hippo weighs about 4 ________.

The customary system of measurement uses fluid ounces, cups, pints, quarts, and gallons to measure capacity.

- Use fluid ounces to measure objects that have a capacity of a few spoonfuls.
- Use cups to measure objects that have a capacity of a few glasses of juice.
- Use pints to measure objects that have a capacity of a few small bottles of salad dressing.
- Use quarts to measure objects that have a capacity of a few small containers of paint.
- Use gallons to measure objects that have a capacity of a few large containers of milk.

Write *fluid ounces*, *cups*, *pints*, *quarts*, or *gallons*.

5. Two teaspoons hold about 2 ________.

6. Five mugs hold about 5 ________.

Name ______________________ Date __________ Class __________

LESSON **9A**

Homework and Practice

Understanding Customary Units of Measure

Explain why each situation does or does not use the proper customary unit. If it does not, write the proper customary unit.

1. The distance between cities is given in inches.

2. The length of a plane is measured in miles.

3. The weight of a computer monitor is in pounds.

4. The volume of a swimming pool is in tons.

5. The circumference of a basketball is given in inches.

6. What is the length of the pencil? Write your answer to the nearest half, fourth, or eighth of an inch.

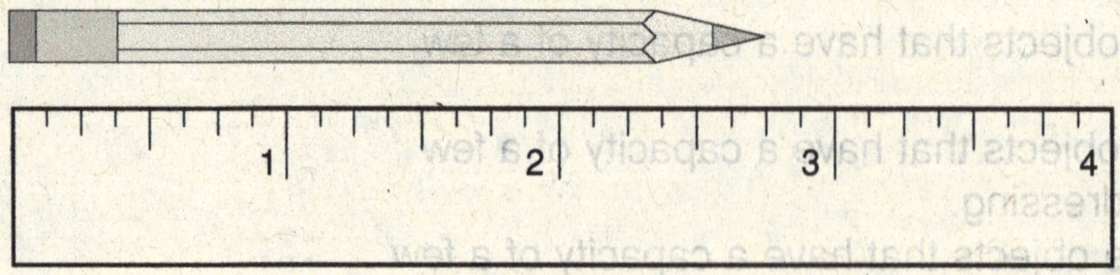

Name ______________________ Date __________ Class __________

LESSON **9A**

Homework and Practice

Understanding Customary Units of Measure, continued

Write the unit of measure that provides the best estimate.

7. A refrigerator weighs about 0.3 ____________.

8. The distance across a parking lot is about 50 ____________.

9. A piece of notebook paper is just about 1 ____________ wide.

10. A salad is topped with approximately 3 ____________ of dressing.

11. A gallon of milk weighs about 9 ____________.

12. The gas tank of a car holds about 12 ____________ of fuel.

Use benchmarks to estimate each measure.

13. the width of a parking space

____

14. the capacity of a trash barrel

____

15. the weight of pair of scissors

____

16. the circumference of a bagel

____

Are the following statements true or false? If false, explain.

17. A fluid ounce of any two liquids each have the same weight.

____

____

18. An ounce of feathers and an ounce of lead weigh the same.

____

____

EXPLORATION

Program of Studies and Core Content

MA-6-M-S-MPA2, MA-6-M-S-MPA4, MA-6-M-S-MPA5, MA-6-M-S-SM1, MA-6-M-S-SM2, MA-06-2.1.2

9B Understanding Metric Units of Measure

In the metric system, the basic unit of length is the meter. One meter (1 m) is about the width of a classroom doorway. Other metric units of length are based on the meter.

Unit	Fraction of a Meter	Decimal Part of a Meter
Millimeter	$\frac{1}{1{,}000}$ of a meter	0.001 m
Centimeter	$\frac{1}{100}$ of a meter	0.01 m
Decimeter	$\frac{1}{10}$ of a meter	0.1 m

1. Which of the units listed is the smallest? How do you know?

2. Which of the units listed is the largest? How do you know?

3. In the metric system, the basic unit of capacity is the liter (L). The prefixes *milli-*, *centi-*, and *deci-* are used in the same way to create other units of capacity. Complete the table.

Unit	Fraction of a Liter	Decimal Part of a Liter
Milliliter	___?___ of a liter	___?___
Centiliter	___?___ of a liter	___?___
Deciliter	___?___ of a liter	___?___

Think and Discuss

4. Explain which is longer, a stick that is 5 centimeters long or a stick that is 5 decimeters long.

5. Explain which holds more, a container with a capacity of 7 milliliters or a container with a capacity of 7 centiliters.

Name ______________________ Date ____________ Class ________

LESSON 9B

Hands-On Lab 1

Choosing Measuring Tools

Activity

In order to get the most precise measurement, the correct measurement tool must be used. The smallest reasonable measuring tool and unit should always be used for accurate measurements.

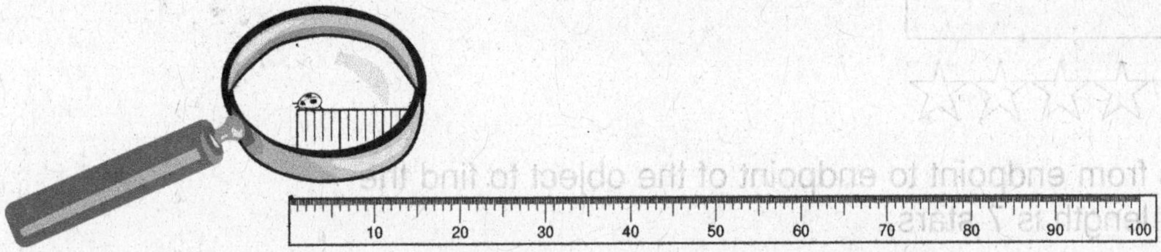

The measurement of the bug is more accurate with the cm ruler because the measurement marks are smaller. On the meter stick, the accuracy of measurement is poor because the measurement marks are too large for the bug.

The greatest possible error (GPE) of a measurement is half (0.5) of the measurement unit being used, therefore, the smaller the measurement being used, the less the error in measurement.

Think and Discuss

1. Why is it better to measure a classroom in feet rather than millimeters?

____

____

____

____

Try This

1. Use an appropriate measuring tool to find the length of your desk.

   What is the length of your desk? ____________________

   What is the GPE of your measurement? ____________________

Name ______________________________ Date ____________ Class __________

LESSON 9B

Hands-On Lab 2

Length

Key

 = 1 unit

You can use nonstandard units of measurement to determine length.

Activity

Determine the length of this object in the nonstandard unit shown above.

Place stars from endpoint to endpoint of the object to find the length. The length is 7 stars.

A ruler can also be used to measure length.

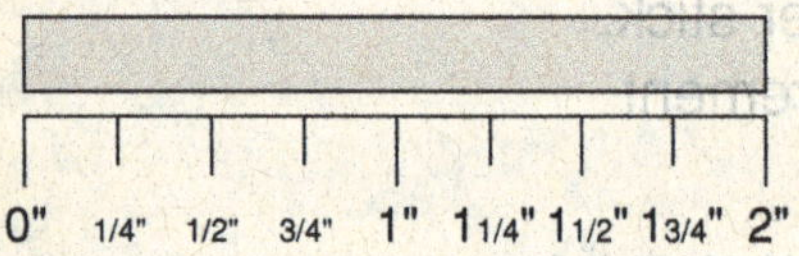

Line up your ruler at the endpoint of the object. Then, read the length of the object on the ruler. The object is 2 inches.

Think and Discuss

1. Give an example of how you might use a nonstandard unit of measurement to find the length of an object.

____

____

Name ______________________________ Date ____________ Class ____________

LESSON **9B**

Hands-On Lab 2

Length, continued

Try This

Use a centimeter ruler to measure the length of each object. Give your answer to the nearest centimeter.

1.  ____________

2.  ____________

3.  ____________

4.  ____________

5.  ____________

6.  ____________

7.  ____________

Use an inch ruler to measure the length of each object. Give your answer to the nearest $\frac{1}{8}$ in.

8.  ____________

9.  ____________

10.  ____________

11.  ____________

12.  ____________

13.  ____________

14.  ____________

Name ______________________ Date ____________ Class ____________

LESSON **9B**

Hands-On Lab 3

Weight and Mass

Key

□ = 1 gram **▯ = 10 grams** **☐ = 100 grams**

Activity

Determine the mass of this object.

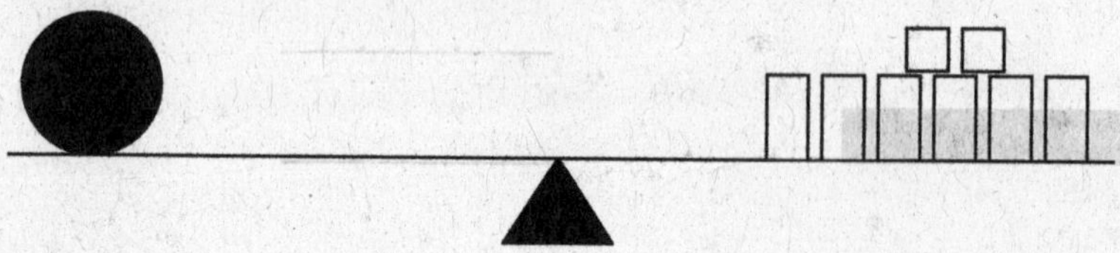

Place weights on the balance scale until the scale is even. Then, count the total mass of the weights on the scale. This determines the weight, 62 grams.

Use the balance or the scale provided by your teacher. Place 62 grams on one side of the balance, place an object in the room on the other side of the balance. Try different objects to find one that is close to weighing 62 grams.

Think and Discuss

1. A coaster is placed in a tray on one side of a balance. Sam uses 22 quarters on the other tray in order to balance the coaster. If Sam places 8 coasters on one side of the tray, how many quarters will he need in order to balance the coasters?

____

Name ______________________ Date __________ Class __________

LESSON **9B**

Hands-On Lab 3

Weight and Mass, continued

Try This

Use the balance or the scale provided by your teacher. Determine the mass of each object using a standard unit of measure.

1. a stapler ______________________

2. a pen ______________________

3. a shoe ______________________

4. a tape dispenser ______________________

5. a small book ______________________

6. a chalkboard eraser ______________________

7. a necklace ______________________

8. a box of tissue ______________________

Determine the mass of each object using the nonstandard unit of measure described.

9. How many pencils does a small book weigh? ______________________

10. How many socks does one watch weigh? ______________________

11. How many cups of water does one book weigh? ______________________

12. Estimate how many cups of water three books weigh. ______________________

13. How many cups of water does three books weigh? ______________________

14. How close was your estimate in Exercise 12 to your answer in Exercise 13?

____

15. How many pennies does a quarter weigh?

____

16. Estimate how many pennies five quarters weigh. ______________________

17. How many pennies does five books weigh? ______________________

18. How close was your estimate in Exercise 16 to your answer in Exercise 17?

____

Name ______________________ Date __________ Class __________

LESSON **9B**

Hands-On Lab 4

Volume

Activity

Find the capacity of this container.

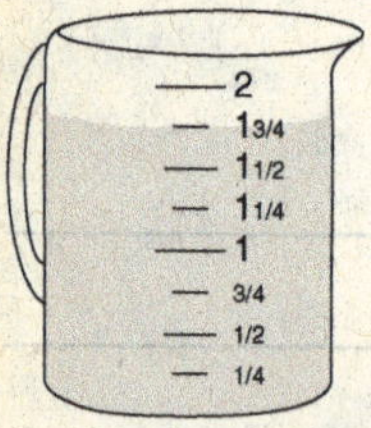

To find the capacity, look at the top of the fluid in the container. Then, read the measurement on the measuring cup.

$1\frac{3}{4}$ cups

Think and Discuss

1. Examine the measuring devices you have in your classroom (measuring cup, beaker, graduated cylinder). Why is a graduated cylinder a better tool to measure milliliter capacity than a measuring cup? Explain.

Try This

1. Estimate how many cups of water will fit into one liter. Then using your measuring cup, fill a one-liter beaker with water. Keep track of how many cups it actually takes to fill the beaker. Compare your actual results with your estimate.

2. Estimate how many milliliters of water are needed to fill a 2-cup container. Then using your graduated cylinder, fill a 2-cup container with water. Keep track of how many milliliters it actually takes to fill the container. Compare your actual results with your estimate.

3. Fill an empty water bottle with sand. Then pour the sand from the bottle into an empty 2-liter bottle. Approximately how many water bottles of sand does it take to fill one 2-liter bottle? Estimate how many water bottles of sand it would take to fill 100 2-liter bottles.

Name ______________________________ Date ____________ Class ____________

LESSON 9B

Review for Mastery

Understanding Metric Units of Measure

The metric system of measurement uses millimeters, centimeters, decimeters, meters, and kilometers to measure length, width, distance, and height.

- Use millimeters to measure objects that are a few thicknesses of a dime.
- Use centimeters to measure objects that are a few widths of a fingernail.
- Use decimeters to measure objects that are a few widths of a CD case.
- Use meters to measure objects that are a few widths of a single bed.
- Use kilometers to measure distances that are a few distances around a city block.

Write *millimeters*, *centimeters*, *decimeters*, *meters*, or *kilometers*.

1. Your arm is about 3 ____________ long.

2. An ant is about 5 ____________ long.

The metric system of measurement uses milligrams, grams, and kilograms to measure mass.

- Use milligrams to measure objects that have a mass of a few very small insects.
- Use grams to measure objects that have a mass of a few large paper clips.
- Use kilograms to measure objects that have a mass of a few textbooks.

Write *milligrams*, *grams*, or *kilograms*.

3. A ruler has a mass of about 5 ____________.

4. A kitten has a mass of about 2 ____________.

The metric system of measurement uses milliliters and liters to measure capacity.

- Use milliliters to measure objects that have a capacity of a few drops of water.
- Use liters to measure objects that have a capacity of a few blender containers.

Write *milliliters* or *liters*.

5. A fish bowl has a capacity of about 7 ____________.

6. A soup spoon has a capacity of about 20 ____________.

Name ______________________ Date ____________ Class ____________

LESSON **9B**

Homework and Practice

Understanding Metric Units of Measure

Explain why each situation does or does not use the proper metric unit. If it does not, write the proper metric unit.

1. The mass of a pony is given in milligrams.

____

____

2. The amount of ink in a pen is about 2 milliliters.

____

____

3. The capacity of a swimming pool is in kilograms.

____

____

4. The mass of a strawberry is in grams.

____

____

5. The approximate radius of the earth is in meters.

____

____

6. What is the length of the pencil? Write your answers below.

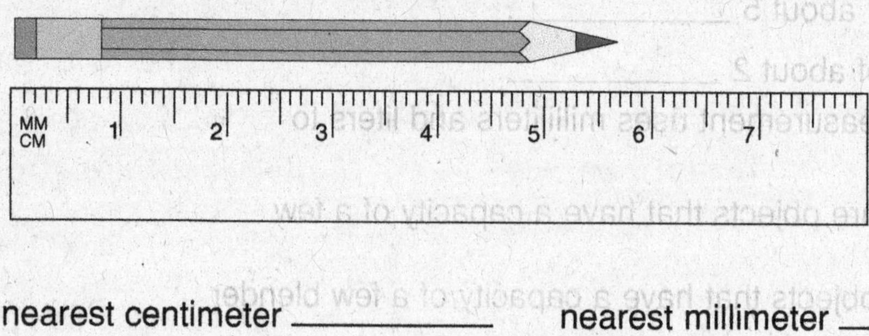

nearest centimeter ____________ nearest millimeter ____________

Name ______________________ Date __________ Class __________

LESSON 9B

Homework and Practice

Understanding Metric Units of Measure, continued

Write the unit of measure that provides the best estimate.

7. A car tire has a mass of about 10 ______________.

8. The width of a movie theater screen is about 7 ______________.

9. A test tube holds about 0.01 ______________ of liquid.

10. The mass of a sugar cube is about 1 ______________.

11. A pitcher of lemonade weighs about 1 ______________.

12. A peanut (in its shell) is approximately 3 ______________ long.

Use benchmarks to estimate each measure.

13. the height of a kitchen counter ______________

14. the capacity of a tea bag ______________

15. the mass of a surfboard ______________

16. the perimeter of a standard envelope ______________

Are the following statements true or false? If false, explain.

17. A liter of any two liquids each have the same mass.

____

18. A kilogram is a metric unit of capacity.

____

EXPLORATION

Program of Studies and Core Content
MA-6-M-S-SM1, MA-6-M-S-SM2, MA-06-2.2.1

9C Converting Customary Units

One foot is equal to 12 inches. You can use this fact to explore the relationship between feet and inches.

1. Complete the table.

Feet	1	2	3	4	5	6	7	8
Inches	12	24						

2. Complete the table.

Feet								
Inches	108	120	132	144	156	168	180	192

3. If you are given a length in feet, how can you convert the length to inches?

4. If you are given a length in inches, how can you convert the length to feet?

Think and Discuss

5. Describe what you would do to convert 324 inches to feet.

6. Explain how you could use the above tables to convert 3.5 feet to inches.

Name ______________________________ Date ____________ Class ____________

LESSON 9C

Hands-On Lab

Converting Customary Units

In this lab, you will convert customary units into smaller or larger units using proportions.

Activity

1. Fill in the correct numbers:

1 yard = _____ feet   1 foot = ______ inches   1 yard = _____ inches

Since 1 yd = 3 ft, the proportions of yards to feet is always $\frac{1}{3}$.
That means that any fraction that forms a proportion with $\frac{1}{3}$ is also a ratio of yards to feet. For instance, $\frac{2}{6}$ forms a proportion with $\frac{1}{3}$, and 2 yards equal 6 feet.

2. Fill in each of the following fractions so that it forms a proportion with $\frac{1}{3}$. Show how each fraction relates a number of yards to a number of feet.

a. $\frac{?}{12}$ ______

b. $\frac{18}{?}$ ______

c. $\frac{7}{?}$ ______

d. $\frac{?}{15}$ ______

Name ______________________________ Date ____________ Class ____________

LESSON 9C

Hands-On Lab

Converting Customary Units, continued

Think and Discuss

1. Helga says: "Since a yard is 3 times as long as a foot, to convert feet to yards, I need to multiply by 3." Is Helga right? If the same length is measured in feet and yards, which number would be larger?

____

____

____

2. If you took the reciprocal of each fraction you formed in Activity 1, would all the fractions still form a proportion with each other? Would they still show a relation between feet and yards?

____

____

____

____

Try This

1. Find three fractions that form a proportion with $\frac{1}{12}$. Find the cross-products to check that all the fractions really have the same proportion.

____

____

____

2. What customary units could you convert using fractions in this proportion?

____

____

Name ______________________________ Date ___________ Class ___________

LESSON 9C

Review for Mastery

Converting Customary Units

You can use the table below to convert customary units.

Length	Weight	Capacity
1 foot = 12 inches 1 yard = 36 inches 1 yard = 3 feet 1 mile = 5,280 feet 1 mile = 1,760 yards	1 pound = 16 ounces 1 ton = 2,000 pounds	1 cup = 8 fluid ounces 1 pint = 2 cups 1 quart = 2 pints 1 quart = 4 cups 1 gallon = 4 quarts 1 gallon = 128 fluid ounces

To figure out how many pounds are in 32 ounces, set up a proportion where the first ratio uses 16 ounces is 1 pound, and the second ratio uses a variable for the value you are trying to find.

$$\frac{16 \text{ ounces}}{1 \text{ pound}} = \frac{32 \text{ ounces}}{x \text{ pounds}}$$

Then solve the proportion.

$$\frac{16 \text{ ounces}}{1 \text{ pound}} = \frac{32 \text{ ounces}}{x \text{ pounds}}$$ First, find the cross products.

$16x = 32$

Think: $32 \div 16 = x$ Then, use a related math sentence to solve the equation.

$x = 2$

So, there are 2 pounds in 32 ounces.

Use the table above to set up a proportion. Then find each of the values.

1. the number of pounds in 80 ounces

$\frac{1}{16} =$ ____________

____

2. the number of quarts in 6 gallons

$\frac{4}{1} =$ ____________

____

3. the number of yards in 5 miles

____

4. the number of cups in 20 pints

____

Name ______________________ Date ____________ Class ____________

LESSON **9C**

Homework and Practice

Converting Customary Units

Convert.

1. 12 feet = __________ yards

2. __________ cups = 16 fluid ounces

3. __________ ounces = 6 pounds

4. 12 inches = __________ yards

5. 24 pints = __________ quarts

6. __________ pounds = 1.5 tons

7. 6,160 yards = __________ miles

8. 2 gallons = __________ cups

9. __________ yards = 108 inches

10. __________ quarts = 8 pints

Compare. Write <, >, or =.

11. 2 miles □ 10,000 feet

12. 4.5 gallons □ 72 cups

13. 2 pounds □ 36 ounces

14. 32 quarts □ 8 cups

15. 33 inches □ 3 feet

16. $\frac{1}{2}$ ton □ 16,000 ounces

17. 64 fluid ounces □ 2 quarts

18. 9 yards □ 30 feet

19. Long Island is about 110 miles long. Manny estimates that this is about 20,000 yards. Is he correct? Explain.

____

____

Name ______________________________ Date ____________ Class ____________

LESSON **9C**

Homework and Practice

Converting Customary Units, continued

Complete the conversion tables. In the last row of each table, write an expression using the variable indicated.

20.

Feet	Inches
1	
2	
3	
4	
f	

21.

Gallons	Cups
1	
2	
3	
4	
g	

22.

Pints	Fluid Ounces
1	
2	
3	
4	
p	

23.

Pounds	Ounces
1	
2	
3	
4	
g	

24.

Quarts	Cups
1	
2	
3	
4	
q	

25.

Yards	Inches
1	
2	
3	
4	
p	

A *furlong* is a distance often used in horse racing. One furlong is equal to 220 yards. Use this information to complete the conversions below.

26. 3 furlongs = ____________ yards

27. ____________ furlongs = 110 yards

28. ____________ miles = 24 furlongs

29. 660 feet = ____________ furlongs

30. Jasna has a 10-foot piece of ribbon. She cuts the ribbon into 15 pieces of equal length. How long is each piece of ribbon?

____

EXPLORATION

Program of Studies and Core Content

MA-6-M-S-SM1, MA-6-M-S-SM2, MA-06-2.2.1

9D Converting Metric Units

Because the metric table is based on powers of 10, it is important to be able to multiply and divide quickly by powers of 10.

1. Complete the table by multiplying by 10, by 100, and by 1,000. You may use a calculator or any other method you wish. Look for patterns as you work.

Number	× 10	× 100	× 1,000
42			
3.8			
0.97			
0.065			

2. Complete the table by dividing by 10, by 100, and by 1,000. You may use a calculator or any other method you wish. Look for patterns as you work.

Number	÷ 10	÷ 100	÷ 1,000
512			
63.9			
4.05			
0.772			

Think and Discuss

3. **Describe** shortcuts for multiplying by 10, by 100, and by 1,000.

4. **Explain** shortcuts for dividing by 10, by 100, and by 1,000.

Name ______________________________ Date ____________ Class ____________

LESSON 9D

Hands-On Lab

Converting Metric Units

In this lab, you will convert metric units into other metric units using proportions.

milli-meter (mm): means *one-thousandth of one meter*
centi-meter (cm): means *one-hundredth of one meter*
deci-meter (dm): means *one-tenth of one meter*
deka-meter (dkm): means *ten times one meter*
hecto-meter (hm): means *one hundred times one meter*
kilo-meter (km): means *one thousand times one meter*

Activity

1. Fill in the blanks: 1 meter = ______ mm; 1 meter = ______ cm; 1 meter = 10 decimeters; 1 meter = 0.1 dekameters; 1 meter = 0.01 hectameters; 1 m = ______ km

2. If you know that the proportion of meters to centimeters is 1 to 100 for any given length, you can convert meters to centimeters in the form of a proportion.

   To convert 24 meters to centimeters, use the proportion 24 : x = 1 : 100. Solve by finding the cross-products. What is the result? ________________

   If you converted 24 meters to centimeters by moving the decimal point, which way would you move it, and how many places? ________________

3. Now convert 24 meters to kilometers, using the proportion 24 : x = 1000 : 1. Solve by finding the cross-products. What is the result? ________________

   If you converted 24 meters to kilometers by moving the decimal point, which way would you move it, and how many places?

   ____

4. Summarize your findings by filling in the following table:

	Converting to larger or smaller unit?	Proportion used	Decimal point moved right or left?	Number of places decimal point moved
m → cm				
m → km				

Name ______________________________ Date ____________ Class ____________

LESSON **9D**

Hands-On Lab

Converting Metric Units, continued

Think and Discuss

1. When you converted to a smaller unit, you moved the decimal point to the ______. Justify your response.

____

____

2. When you converted to a larger unit, you moved the decimal point to the ______. Justify your response.

____

____

3. When you converted by a factor of 100, you moved the decimal point 2 places. When you converted by a factor of 1000, you moved the decimal point 3 places. Why did you do this? ______

Try This

kilometer	hectometer	dekameter	meter	decimeter	centimeter	millimeter

You can convert between any of the units in the table above by moving the decimal point in the same direction and by as many places as you need to move from place to place in the table. To move from centimeters to meters in the table, for instance, you need to move two places to the left, and to convert centimeters to meters, you likewise need to move the decimal point two places to the left.

1. Explain why this works.

____

____

____

2. Convert 2.1 meters to millimeters using the table.

____

3. Convert 2.1 meters to kilometers using the table.

____

Name ______________________ Date __________ Class __________

LESSON **9D**

Review for Mastery

Converting Metric Units

There are patterns in powers of ten.

10 • 1 = 10   10 •10 = 100   10 •10 •10 = 1,000

The number of times 10 is a factor equals the number of zeros in the power of ten.

You can use these patterns to multiply and divide by powers of ten. The number of zeros or the number of tens tells you how many places to move the decimal point.

If you are dividing by a power of ten, move the decimal point left.

7,345 ÷ 100 = 73.45

two zeros, two places

73.45
2 1

If you are multiplying by a power of ten, move the decimal point right.

23.4 • 10 •10 •10 = 23,400

three tens, three places

23,400
1 2 3

Multiply or divide.

1. 4.25 • 10 •10 •10 •10

____

2. 1,347.8 ÷ 100

____

3. 9.4 ÷ 1,000

____

4. 18.05 • 100

____

The metric system uses powers of ten.

Each place value is 10 times as large as the place value to the right.						
1,000	100	10	1	0.1	0.01	0.001
thousands	hundreds	tens	ones	tenths	hundredths	thousandths
km	hm	dam	m	dm	cm	mm

2 m = ___ cm The centimeter unit is 2 places to the right of the meter.

2 m = 200 cm Move the decimal point 2 places to the right.

Use the chart to convert each measure.

5. 3.4 km = ______________ mm

6. 7 dm = ______________ hm

7. 4.32 dam = ______________ m

8. 34.8 cm = ______________ dm

Name ______________________ Date ____________ Class ____________

LESSON 9D

Homework and Practice

Converting Metric Units

Convert.

1. 500 cm = __________ m

2. __________ L = 10,000 mL

3. __________ g = 6 kg

4. 855 mg = __________ g

5. 3,650 mL = __________ dL

6. __________ mm = 77 cm

7. 1.75 g = __________ mg

8. 450 m = __________ km

9. __________ cm = 0.55 km

10. __________ kg = 230,500 mg

Compare. Write <, >, or =.

11. 70 mm ☐ 7 cm

12. 3,300 kg ☐ 3.3 g

13. 1,800 mL ☐ 180 L

14. 16 km ☐ 16,000 cm

15. 90.1 m ☐ 91,000 mm

16. $\frac{1}{2}$ kg ☐ 500 g

17. 3 L ☐ 300 mL

18. 900 mm ☐ 0.09 km

19. Tanya wants to convert a measurement of 44.50 milliliters to liters. What is the conversion factor? Be sure to include the units.

____

Name ______________________ Date __________ Class __________

LESSON 9D

Homework and Practice

Converting Metric Units, continued

Fill in the missing information for each conversion. Be sure to include the units if any are missing.

20. 5,600 mL • $\frac{1\text{L}}{\square}$ = __________ L

21. 66 g • $\frac{\square}{1\text{g}}$ = __________ mg

22. __________ • $\frac{\square}{1{,}000\text{ mm}}$ = 645.5 m

23. In table tennis, an official table must be 2.74 meters long and 1.525 meters wide. What are these dimensions in centimeters?

____

24. Official table tennis balls are spherical, with a diameter of 0.04 centimeters. What is the diameter in millimeters?

____

25. The official mass of a table tennis ball is 2.7 grams. What is the mass of the ball in centigrams?

____

A *picometer* is a very small metric unit of length. It is equal to one trillionth of a meter! Complete each conversion.

26. 1 meter = __________ picometers

27. __________ millimeters = 50,000 picometers

28. __________ picometers = 0.0004 centimeters

29. Abdul has a 3-meter piece of wood. He cuts the wood into 8 pieces of equal length. To the nearest centimeter, how long is each piece of wood?

____

EXPLORATION

Program of Studies and Core Content
MA-6-M-S-MPA1, MA-6-M-S-MPA5, MA-06-2.1.1

9E Perimeter

Felicia is working with her dad to design a deck for their yard. The grid below represents the floor space of the deck with the side length of each square representing 2 feet.

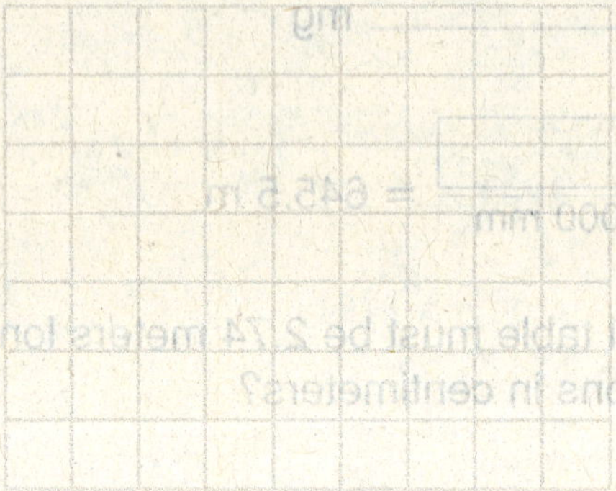

1. Label the dimensions in feet of the deck.
2. The final task in building the deck is to nail a trim piece all the way around the deck. Find the distance around the deck.
3. The distance around the deck is called the ***perimeter.*** What are three other real-world situations in which you might want to find the perimeter?

Think and Discuss

4. **Discuss** your method for finding the perimeter of the deck.
5. **Explain** how you could write a formula for perimeter of a rectangle using ℓ for length and w for width.

Name ______________________________ Date ____________ Class ____________

LESSON 9E

Hands-On Lab

Perimeter

The perimeter of an object is the distance around the object. Being able to find the perimeter of a polygon is handy for anyone who wants to build a fence, plant a border for a garden, buy gutters for a house, and many other applications.

Activity

1. Three objects are shown below: a stop sign, a yield sign, and a school crossing sign. The stop sign is an octagon, the yield sign is a triangle, and the school crossing sign is a pentagon.

2. Estimate the perimeter of each object in centimeters. Then determine the perimeter of each object by measuring the length of each side and adding the lengths together. Complete the following table.

Sign	Number of sides	Estimated Perimeter	Actual Perimeter
Stop sign			
Yield sign			
School crossing sign			

3. Now estimate the perimeter of your desktop. Then measure the actual length of each side. Add the lengths to find the perimeter of your desktop. Are any of the sides equal to any of the other sides?

____

____

Name ______________________ Date __________ Class __________

LESSON 9E

Hands-On Lab

Perimeter, continued

4. You can use a shortcut to determine the perimeter of a rectangle. The opposite sides of a rectangle are equal, so you can measure any two sides that touch, add their lengths, and then double the answer. Estimate the perimeter of this desktop.

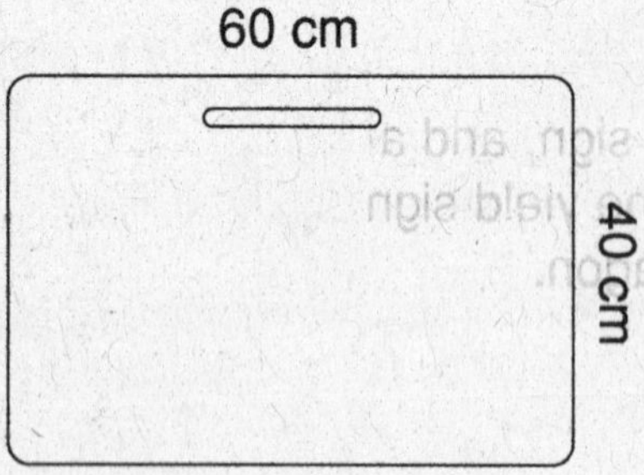

Think and Discuss

1. If a polygon is a regular polygon, you can determine the perimeter by measuring only one side and then multiplying by the number of sides. This works for regular polygons because all of the sides are the same length. Which of the signs from the Activity are regular polygons?

____

2. Would this procedure work for the school crossing sign?

____

Try This

1. A rectangle has a length of 8 cm and a width of 3 cm. First, estimate the perimeter. Then calculate the actual perimeter.

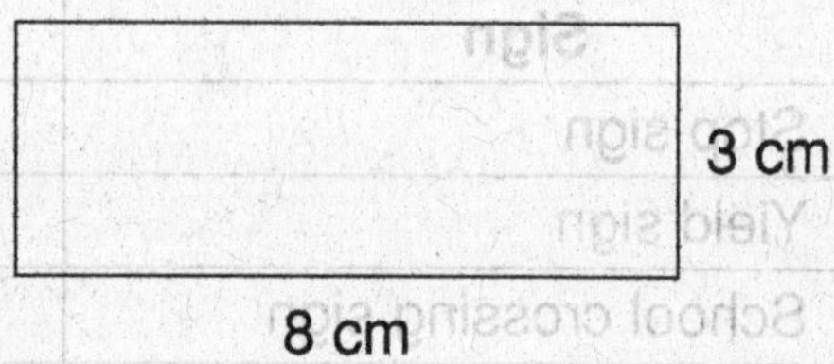

____

2. One side of a regular pentagon is 3 in. What is its perimeter?

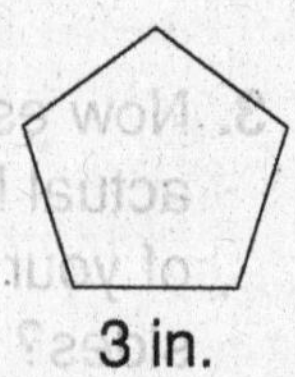

____

3. A gardener is designing a flowerbed. The bed will have the shape shown. Estimate how many meters of edging will be necessary to surround the bed?

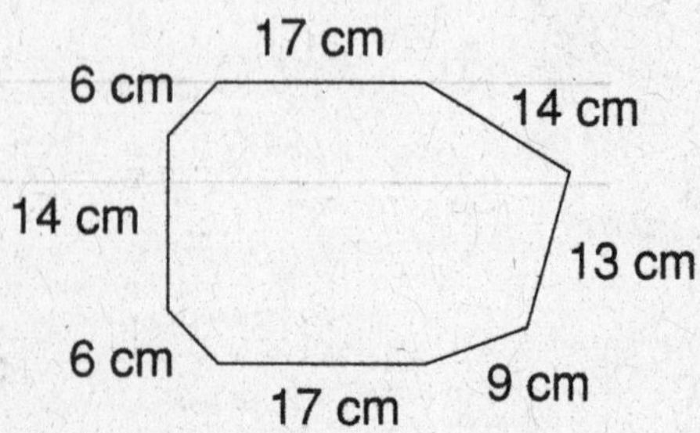

____

Name ______________________ Date __________ Class ________

LESSON 9E

Technology Lab

Finding Perimeter

The perimeter of a polygon is the sum of the lengths of all its sides.

Activity

1. Use geometry or drawing software to draw a regular octagon with $\ell = 3$ cm. Find the perimeter.

Draw the first side at the top and measure it with the ruler tool.

Then copy the length of that first side and copy it into the remaining sides of the octagon.

$\ell = 3$

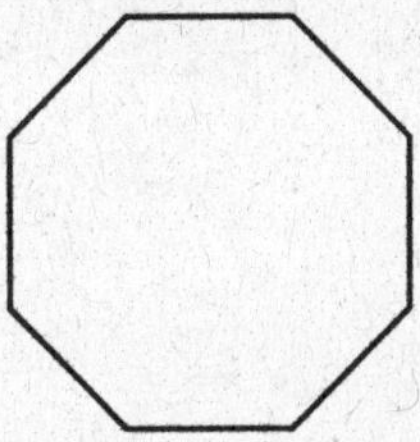

$P_{OCT} = 8\ell = 24$

The perimeter is 24 cm.

2. Use the same steps to draw and find the perimeter of a regular pentagon with $\ell = 4$ in.

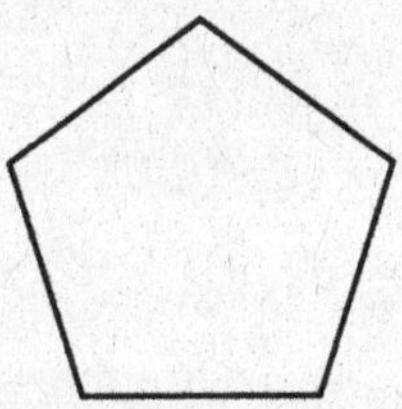

$\ell = 4$

$P_{PEN} = 5\ell = 20$

The perimeter is 20 in.

3. Find the length of a side of an equilateral triangle with a perimeter $P_{TRI} = 21$ in.

Draw the triangle.

$\ell = ?$

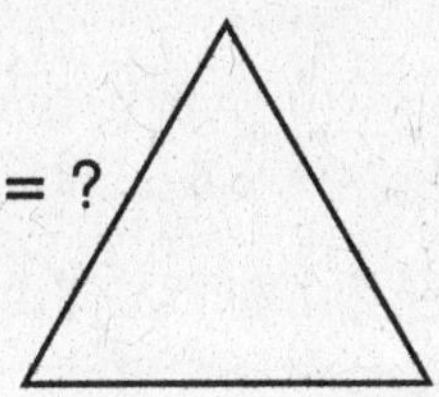

$3\ell = P_{TRI} : 3\ell = 21 : \ell = 7$

The length of the side is 7 in.

Name ______________________ Date __________ Class __________

LESSON 9E

Technology Lab

Finding Perimeter, continued

Think and Discuss

1. If you double the length of the sides of the triangle in Exercise 3, does the perimeter double? Why or why not?

____

____

____

Try This

1. Draw and find the perimeter of a regular hexagon with $\ell = 5$ cm.

____

2. Draw a parallelogram with $s = 5$ in. and $\ell = 10$ in. What is its perimeter?

____

Name ______________________ Date __________ Class __________

LESSON **9E**

Review for Mastery

Perimeter

The perimeter of a figure is the distance around it. To find the perimeter of a figure, find the sum of the lengths of its sides.

Look at the figure below.

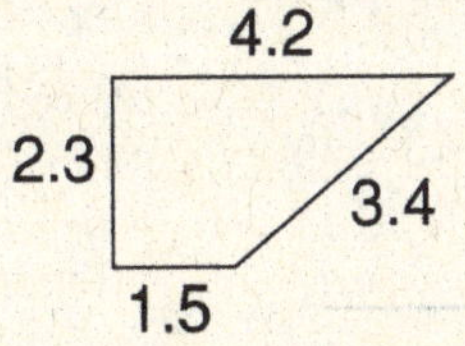

The perimeter is the sum of the lengths of its sides.

$P = 2.3 + 1.5 + 3.4 + 4.2$
$= 11.4$

So, the perimeter of the figure is 11.4 units.

Find the perimeter of each figure.

1.

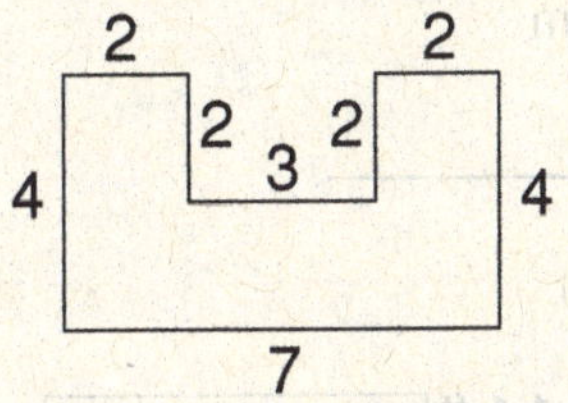

____

2.

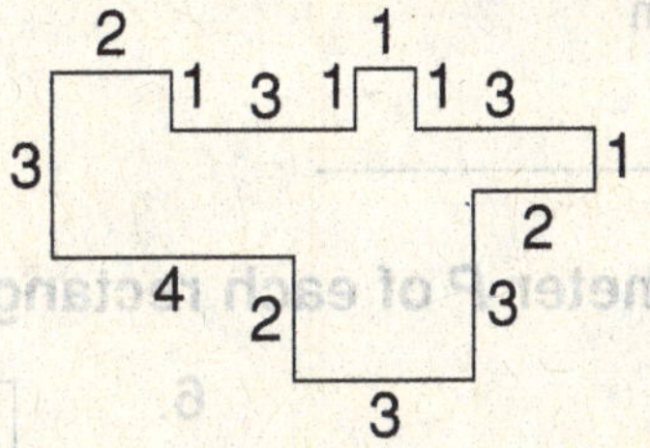

____

If you know the perimeter of a figure, you can find the unknown length of a side.

The perimeter of the figure below is 94 units.

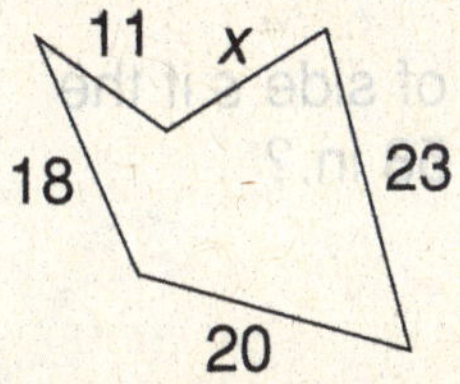

To find the missing length *x* first find the sum of the known lengths.

$23 + 18 + 20 + 11 = 72$

Then subtract the sum from the perimeter.

$94 - 72 = 22$ So, side *x* is 22 units long.

Find the missing length for each figure.

3.

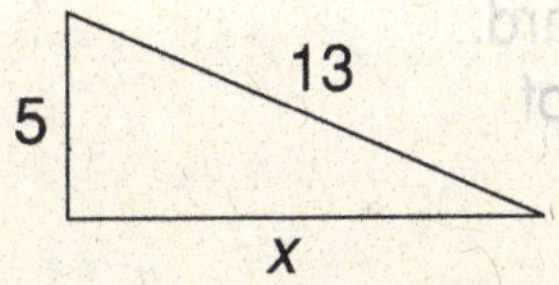

Perimeter = 30 units

____

4.

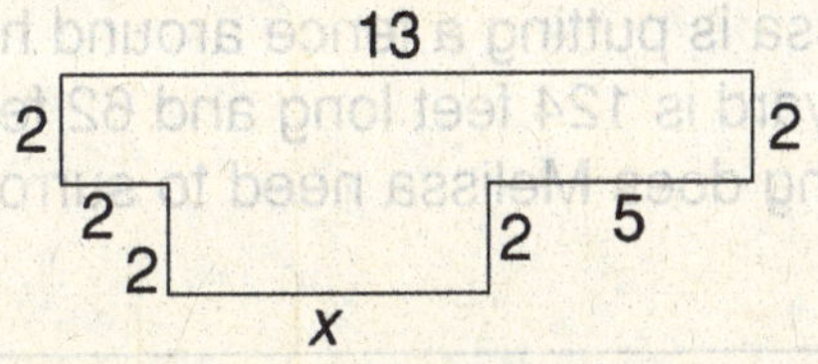

Perimeter = 34 units

____

Name ________________ Date ________ Class ________

LESSON 9E

Homework and Practice

Perimeter

Find the perimeter of each figure.

1.

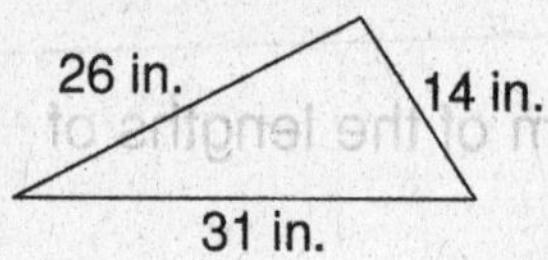

2.

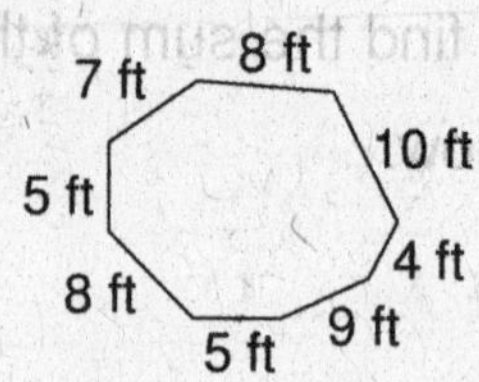

3.

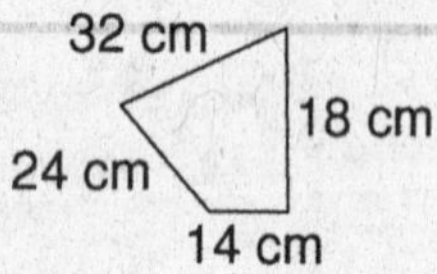

4.

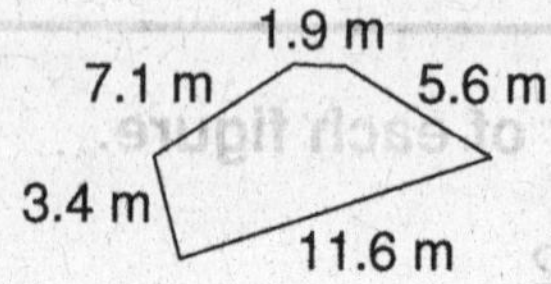

Find the perimeter *P* of each rectangle.

5.

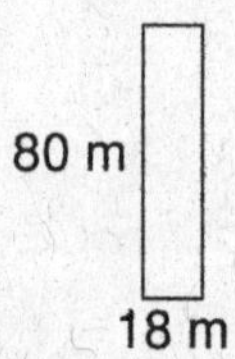

6.

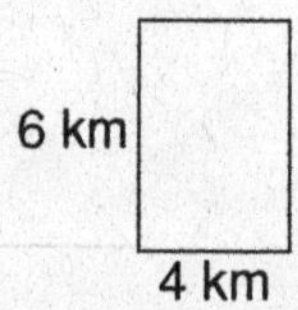

7.

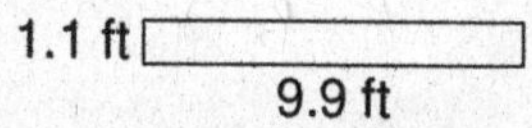

Find the unknown measure.

8. What is the length of side *b* if the perimeter equals 52 cm?

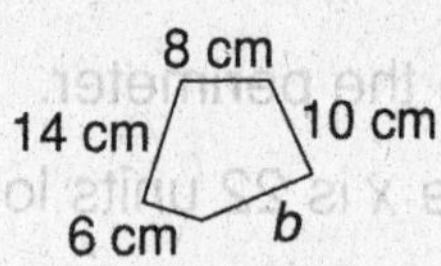

9. What is the length of side *s* if the perimeter equals 158 in.?

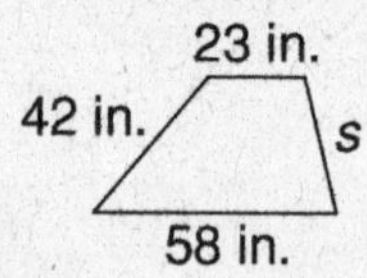

10. Melissa is putting a fence around her rectangular-shaped yard. The yard is 124 feet long and 62 feet wide. How many feet of fencing does Melissa need to surround her entire yard?

11. Jeremy is building a pentagon-shaped pen whose sides measure 6 ft, 8 ft, 5 ft, 10 ft, and 4 ft. How many feet of fence should Jeremy purchase?

9F Circles and Circumference

Program of Studies and Core Content
MA-6-G-S-SR3, MA-6-G-S-SR4, MA-6-AT-S-EI2, MA-06-3.1.1, MA-06-3.1.2

People in ancient civilizations learned to estimate the distance around a circle (*circumference*) by multiplying the distance across (*diameter*) by three.

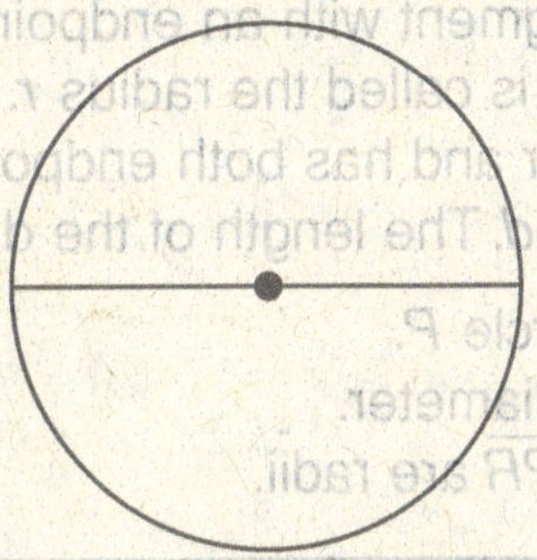

1. Does the distance around the circle above look like three times the diameter?

2. The table below contains actual measurements, to the nearest tenth of a centimeter, of two cans. Use a calculator to find the ratio $\frac{\text{circumference}}{\text{diameter}}$.

Object	Diameter (cm)	Circumference (cm)	$\frac{\text{Circumference}}{\text{Diameter}}$
Juice can	5.2	16.3	
Coffee can	15.7	49.3	

3. Are the $\frac{\text{circumference}}{\text{diameter}}$ ratios you found in number 2 reasonably close to 3?

Think and Discuss

4. **Discuss** whether the ratio $\frac{\text{circumference}}{\text{diameter}}$ changes according to the size of the circle.

Name ____________________ Date __________ Class __________

LESSON **9F**

Review for Mastery

Circles and Circumference

A circle is the set of all points in a plane that are the same distance from a given point, called the center.

A line segment with an endpoint at the center and an endpoint on the circle is called the radius *r*. A line segment that passes through the center and has both endpoints on the circle is called the diameter *d*. The length of the diameter is twice the radius.

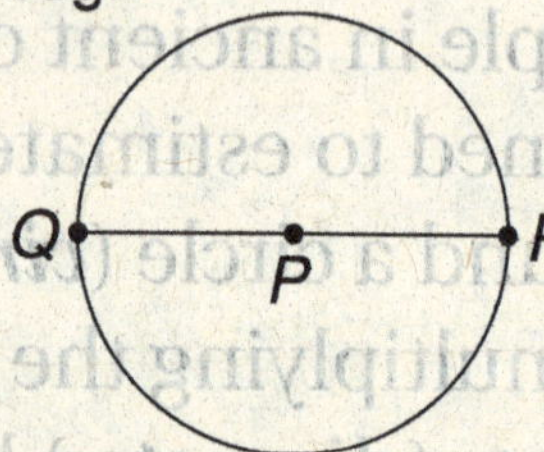

This is circle *P*.
$\overline{QR}$ is a diameter.
$\overline{PQ}$ and $\overline{PR}$ are radii.

Identify each part of the circle below.

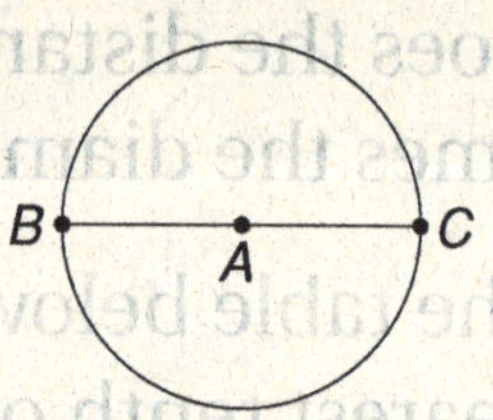

1. center ______________

2. diameter ______________

3. two radii ______________

The distance around a circle is called the circumference. The ratio of the circumference of a circle to its diameter is always the same. This relationship is called *pi* and is represented by the Greek letter π.

Pi is a decimal number that has no end and never repeats. 3.14 or $\frac{22}{7}$ is used to estimate *pi*.

To find the circumference *C* of a circle, use a formula.

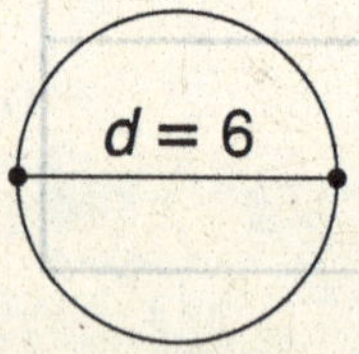

$C = \pi d$
$\approx 3.14 \cdot 6$
≈ 18.84
The circumference is about 18.84 units.

Find the circumference. Use 3.14 for π.

4.

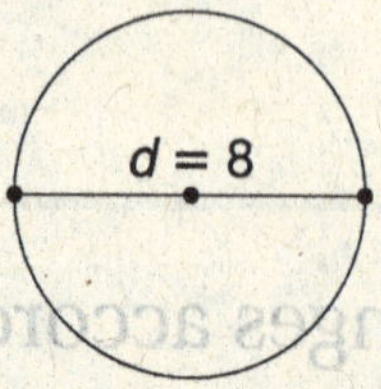

____

5.

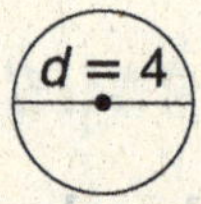

____

Name ______________________ Date ____________ Class ____________

LESSON 9F

Homework and Practice

Circles and Circumference

Use the circle to answer each question.

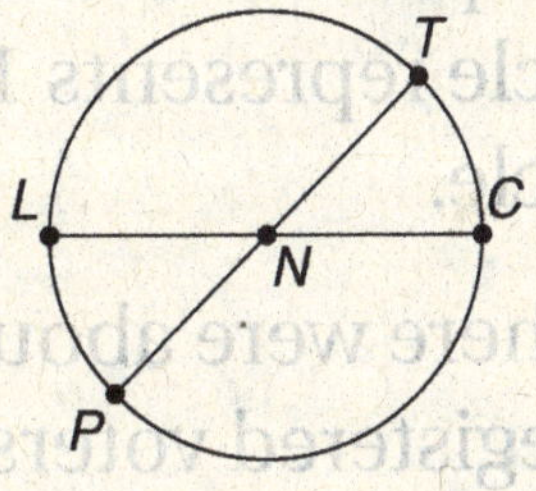

1. Name the circle. ______________
2. Name two diameters.

____

3. Name the four radii.

____

Find the circumference. Use 3.14 for pi.

4.

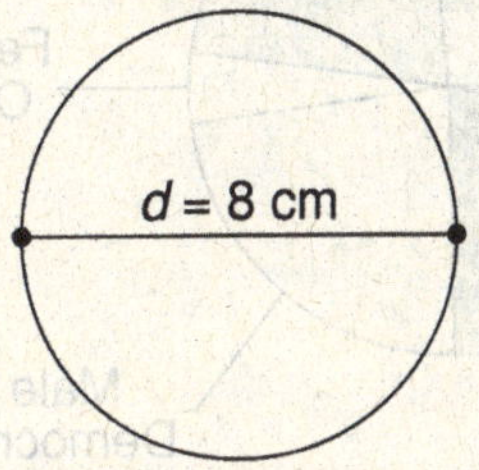

C = ? ______________

5.

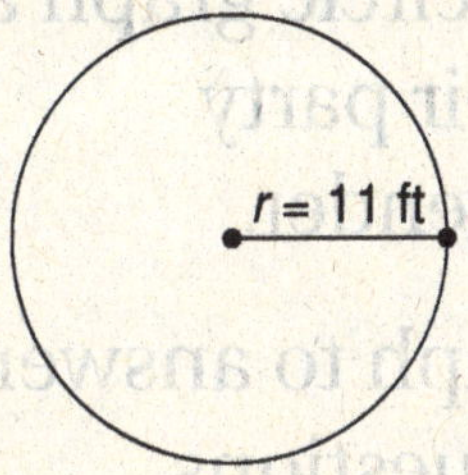

C = ? ______________

Find the circumference. Use $\frac{22}{7}$ for pi.

6.

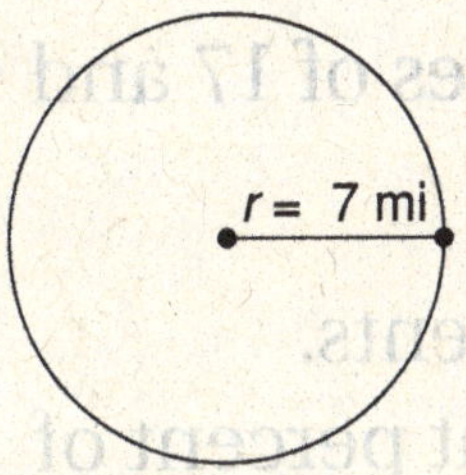

C = ? ______________

7.

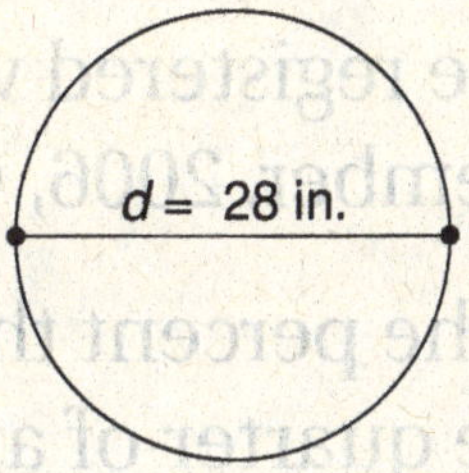

C = ? ______________

8. The first Ferris wheel was built in 1893 in Chicago. Its diameter was 250 feet. What was the actual circumference on the Ferris wheel? Use 3.14 for pi.

____

9. Stonehenge, a circle of large carved stones in England, was built more than 1,000 years ago. The circle of stones has a diameter of 108 feet. What is the circumference of Stonehenge in feet? Use 3.14 for pi.

____

EXPLORATION

Program of Studies and Core Content

MA-06-4.1.1

9G Circle Graphs

A circle graph shows a circle divided into *sectors*, or wedges. The whole circle represents 100%, and each sector is a percent of the whole circle.

In 2006, there were about 280,000 registered voters in Kentucky between the ages of 17 and 24. The circle graph at right shows their party affiliations by gender.

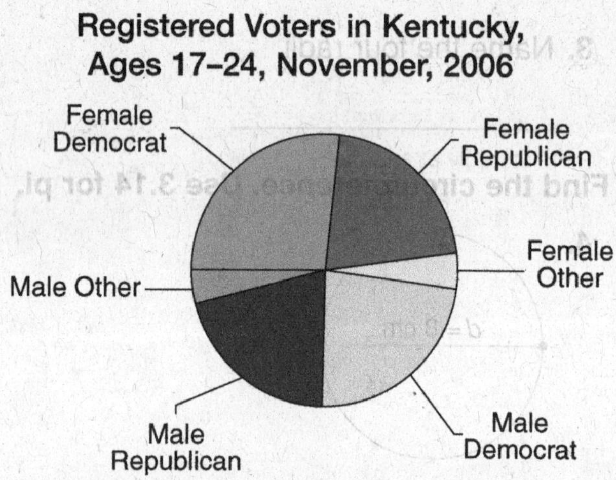

Analyze the graph to answer the following questions.

1. Were there more women or more men between the ages of 17 and 24 registered as Democrats in November, 2006?

2. Most of the registered voters between the ages of 17 and 24 in November, 2006, were what gender?

3. Estimate the percent that each sector represents. (Hint: One quarter of a circle represents what percent of the whole circle?)

Think and Discuss

4. **Describe** a data set that may be represented using a circle graph.

5. **Discuss** how you can estimate the number of voters each sector represents.

Name ______________________________ Date ____________ Class ____________

LESSON **9G**

Review for Mastery

Reading and Interpreting Circle Graphs

A **circle graph** shows how the parts of a complete set of data are related. The circle shows 100% of the data.

Choose a circle graph to show what percent (or part) of the total sales is represented by each type of pizza.

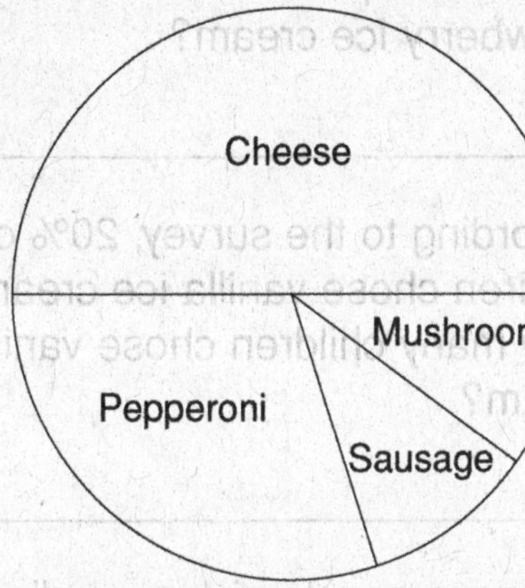

Choose a bar graph to show the number of each type of pizza sold.

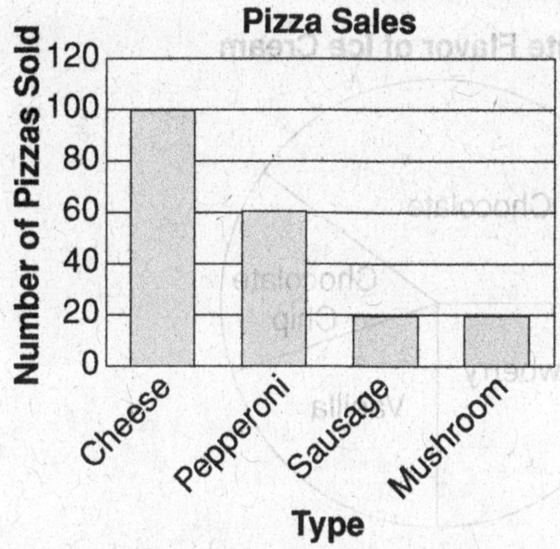

The circle graph shows that about one-half, or 50%, of the total sales was cheese pizzas. If 200 pizzas were sold, then about half, or 100, were cheese pizzas.

The circle graph shows the results of a survey of 100 teens who were asked about their favorite winter sport. Use the graph to answer each question.

Favorite Winter Sport

1. Did more teens pick skiing or snowboarding? ____________

2. About what percent of teens picked skiing? ____________

3. How many teens chose skiing? ____________

Decide whether a bar graph or a circle graph would best display the information. Explain your answer.

4. the number of hurricanes in each state during 1 year

____

____

____

5. the number of students who play soccer compared with the total number of students in a school

____

____

____

Name ____________________________ Date ____________ Class ____________

LESSON 9G

Homework and Practice

Reading and Interpreting Circle Graphs

The circle graph directly below shows the results of a survey of 120 children who were asked about their favorite flavor of ice cream. Use the graph for Exercises 1–3.

Favorite Flavor of Ice Cream

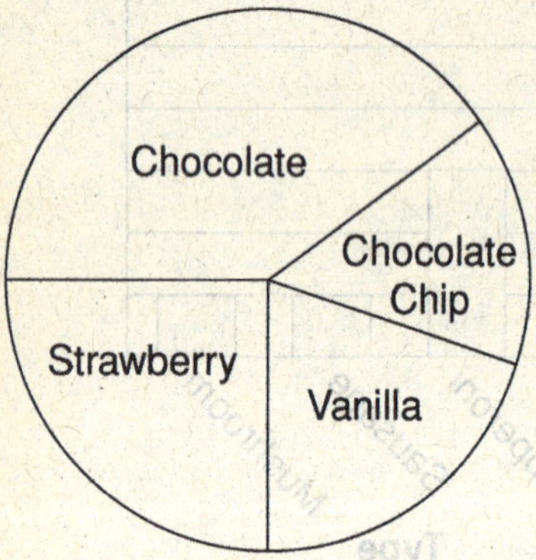

1. Did more children choose chocolate chip or vanilla as their favorite flavor of ice cream?

____

2. About what percent of children chose strawberry ice cream?

____

3. According to the survey, 20% of the children chose vanilla ice cream. How many children chose vanilla ice cream?

____

The circle graph below shows the results of a survey of 100 people who were asked about their favorite type of movie. Use the graph for Exercises 4–6.

Favorite Type of Movie

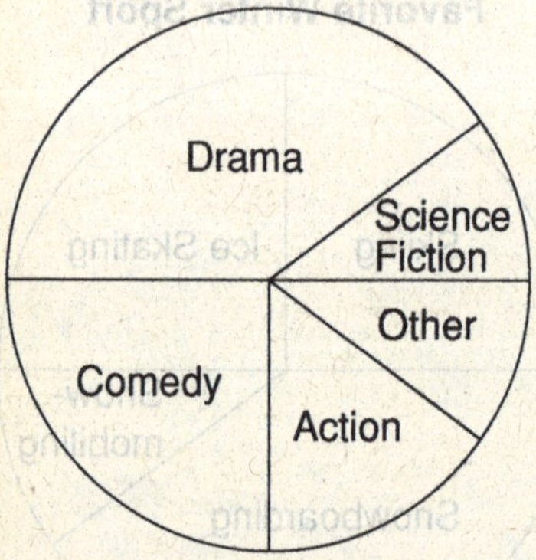

4. Did more people pick comedies or science fiction movies?

____

5. About what percent of people picked action movies?

____

6. According to the survey, 40% of the people chose drama. How many people chose drama?

____

Decide whether a bar graph or a circle graph would best display the information. Explain.

7. The number of hurricanes in the United States during the past ten years.

____

____

Chapter Summary

Answer these questions to summarize the important concepts from Chapter 9 in your own words.

1. Explain how to convert 108 inches to feet.

2. Explain how to convert 5 weeks to seconds.

3. Explain how to find the perimeter of a rectangle with length 4 ft and width 7 ft.

4. Explain how to find the circumference of a circle with radius 6 inches.

For more review of Chapter 9:

- Complete the Chapter 9 Study Guide and Review on pages 530–532 of your textbook.
- Complete the Ready to Go On quizzes on pages 508 and 526 of your textbook.

Name ______________________ Date __________ Class __________

CHAPTER 9

KCCT Prep

Multiple Choice

1. Which is the best estimate of the perimeter of the rectangle?

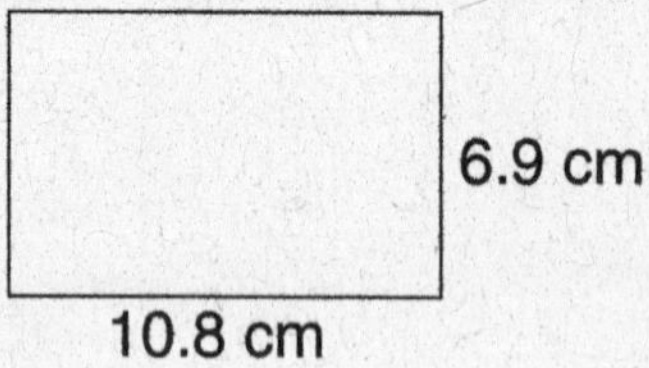

Ⓐ 18 cm Ⓒ 45 cm
Ⓑ 36 cm Ⓓ 77 cm

2. What is the greatest common factor of 45, 90, and 30?

Ⓐ 1 Ⓒ 5
Ⓑ 3 Ⓓ 15

3. Tracie solves the equation $\frac{y}{12} = 4$ to find the number of crackers y in a box. How many crackers are there in 3 boxes?

Ⓐ 16 Ⓒ 96
Ⓑ 48 Ⓓ 144

4. The ratio of pens to pencils in a drawer is 3:4. There are 21 pens and pencils altogether. How many pens are there?

Ⓐ 3 Ⓒ 12
Ⓑ 9 Ⓓ 15

5. Which of these is an example of the Commutative Property?

Ⓐ $7 \times (2 \times 4) = (7 \times 2) \times 4$
Ⓑ $5 + 10 = 10 + 5$
Ⓒ $33 \times 1 = 33$
Ⓓ $(6 + 4) + 1 = 6 + (4 + 1)$

6. The perimeter of an equilateral triangle is 19.2 in. Which is closest to the length of a side of the triangle?

Ⓐ 3 in. Ⓒ 6 in.
Ⓑ 4 in. Ⓓ 9 in.

7. Which is the best estimate for the perimeter of the figure?

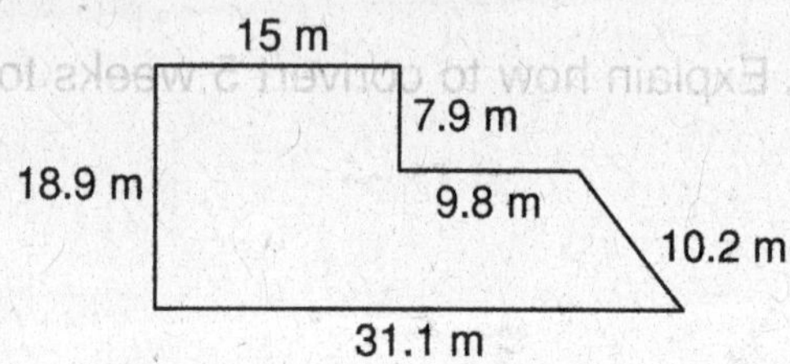

Ⓐ 65 m Ⓒ 95 m
Ⓑ 80 m Ⓓ 105 m

8. The radius of a circle is 18 in. A snail crawls along the diameter of the circle at a rate of 3 inches per minute. How long does it take the snail to cross the circle?

Ⓐ 3 min Ⓒ 9 min
Ⓑ 6 min Ⓓ 12 min

9. The least common multiple of 16 and another number is 80. Which of these could be the other number?

Ⓐ 4 Ⓒ 20
Ⓑ 12 Ⓓ 25

10. A map has a scale of 1 cm:3 km. The distance between Brisbane and Winslow on the map is 5.8 cm. Which is the best estimate of the actual distance between the towns?

Ⓐ 2 km Ⓒ 9 km
Ⓑ 6 km Ⓓ 18 km

11. Which fraction is closest to 15%?

Ⓐ $\frac{1}{14}$ Ⓒ $\frac{17}{100}$

Ⓑ $\frac{1}{5}$ Ⓓ $\frac{3}{10}$

12. What is the perimeter of the shape shown below?

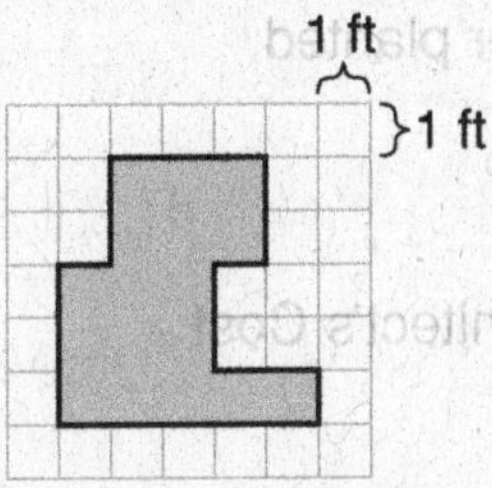

Ⓐ 17 ft Ⓒ 22 ft

Ⓑ 20 ft Ⓓ 25 ft

13. A circle has a radius of 10 meters. Which is closest to the circumference of the circle?

Ⓐ 15 m Ⓒ 60 m

Ⓑ 30 m Ⓓ 300 m

14. A rope is 17 ft long. Natasha cuts the rope into pieces that are each $1\frac{3}{8}$ ft long. What is the maximum number of such pieces that she can cut from the rope?

Ⓐ 9 Ⓒ 13

Ⓑ 12 Ⓓ 15

15. Simon wants to find the circumference of this circle using $\frac{22}{7}$ as an estimate for π. What answer should he get?

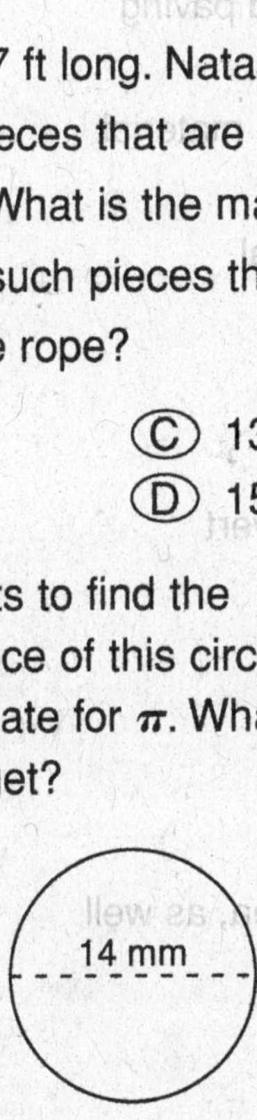

Ⓐ 22 mm Ⓒ 44 mm

Ⓑ 42 mm Ⓓ 88 mm

Open Response

16. Dan is making bouquets using roses and tulips. He has 36 roses and 48 tulips. He wants to use all the flowers to make the bouquets.

a. Dan wants each bouquet to have the same number of roses and the same number of tulips. What is the greatest number of bouquets he can make? Explain.

b. How many roses and how many tulips will be in each bouquet?

17. A city planner is designing a new playground. The playground has the shape of a square with an equilateral triangle on each end, as shown below.

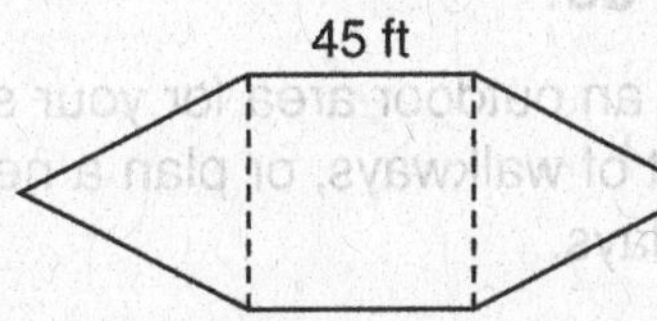

a. The city planner wants to place a fence around the playground. What is the total length of the fence?

b. The fence costs \$9.89 per foot. The budget for the fence is \$2,900. Use estimation to explain whether the city planner will be able to buy the fence and stay within the budget.

18. Flora is making a scale drawing of the buildings in her neighborhood. She uses a scale of 1 in:15 ft.

a. A grocery store is 75 feet long. How long should the store be in Flora's scale drawing? Explain.

b. A theater in the neighborhood is in the shape of a circle. On Flora's drawing, the diameter of the theater is 8 inches. What is the actual circumference of the theater? Show your work.

CHAPTER 10

Project

Measurement: The Architecture of Flowers

Career: Landscape Architect

Almost every shopping center has walkways, lawns, and trees. A landscape architect designs these so that they are appealing and well suited to the environment. Landscape architects also plan the layouts of residential areas, office parks, and school campuses. To calculate costs, landscape architects must estimate how much material is needed to cover planted areas.

Before you start the project, work through the Landscape Architect's Cost Estimate worksheet on the next page.

In this chapter project . . .

You design a new outdoor area for your school.

What to do:

- Design an outdoor area for your school. You could choose a playground or a set of walkways, or plan a new outdoor cafeteria for use during warm days.
- Make a scale drawing of your plans and include a list of materials needed, including ground cover, flowers, outdoor furniture, and paving material.
- Estimate the total area to be covered, and the amount of each material needed for your design.
- Research the cost of the various materials you chose at a local hardware store or on the Web.

What to present:

- An accurate scale drawing of your plans. Include a key to convert lengths on your scale drawing to actual lengths.
- A cost estimate for your proposed renovation.

Where's the math?

- Use formulas to find or estimate areas of your outdoor area.
- Calculate the amount of materials needed for your outdoor area, as well as the total cost based on these amounts.

go.hrw.com
Chapter Project Online
KEYWORD: MR7 Ch9

Name ______________________ Date __________ Class __________

CHAPTER 10

Project Recording Sheet

Measurement: Cover an Area

Landscape Architect's Cost Estimate

A landscape architect is designing three planted areas outside a new museum. Use the coverage chart to help you complete the table below. For each planted area, indicate the least number of bags needed.

Coverage for 1 Bag of Wood Chips	
Type of Chips	**Area Covered (ft^2)**
Cedar	150
Pine	120
Redwood	160
Rubberized	60

	Shape	Dimensions (ft)	Area (ft^2)	Type of Wood Chips	Bags of Wood Chips Needed
A	Rectangle	Length: 80 Width: 40		Redwood	
B	Circle	Diameter: 30		Cedar	
C	Triangle	Base: 50 Height: 20		Rubberized	

Use your completed table to answer the following questions.

1. How many bags of wood chips will be needed altogether? ____________

2. The cost of a bag of wood chips is as follows: Cedar $22, Pine $15, Redwood $30, and Rubberized $50. What will be the total cost of the wood chips for this project? ____________

3. The landscape architect needs to lower the total cost of the wood chips. Describe one way to change the wood chips in one or more of the planted areas so that the total cost is less than $1000. Assume the museum wants you to use 3 different types of wood chips.

____

____

____

Extension: Estimate the dimensions of a planted area near your school. Explain how to calculate its area.

____

____

EXPLORATION

Program of Studies and Core Content
MA-6-M-S-MPA3, MA-06-2.1.1

10A Estimating and Finding Area

Mr. and Mrs. Domínguez want to have the bottom of their pool refinished. A sketch of the pool is shown below with the side length of each square representing 1 yard. Before they begin the refinishing project, they have to estimate the area in square yards of the bottom of the pool.

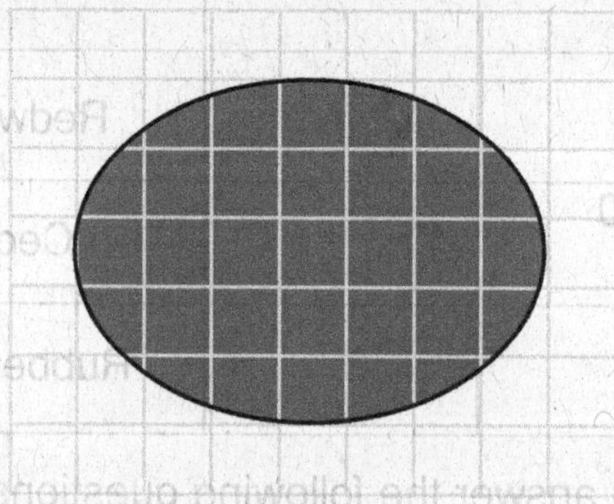

1. Estimate the area, in square yards, of the bottom of the pool.
2. Compare your estimate with the estimates of others in your class, and then average your estimates.
3. What are three other real-world situations in which you might want to estimate area?

Think and Discuss

4. **Discuss** the strategies you used for estimating the area of the bottom of the pool.
5. **Explain** how you could use squares to help you estimate the areas of irregular shapes.

Name ____________________ Date __________ Class __________

LESSON **10A**

Hands-On Lab

Estimating and Finding Area

The area of a figure is the amount of surface it covers. Being able to measure area is important to engineers, gardeners, painters, architects, and many others. The area of many figures can be calculated from the figure's dimensions. For irregular or oddly shaped figures, the area can be estimated.

Activity

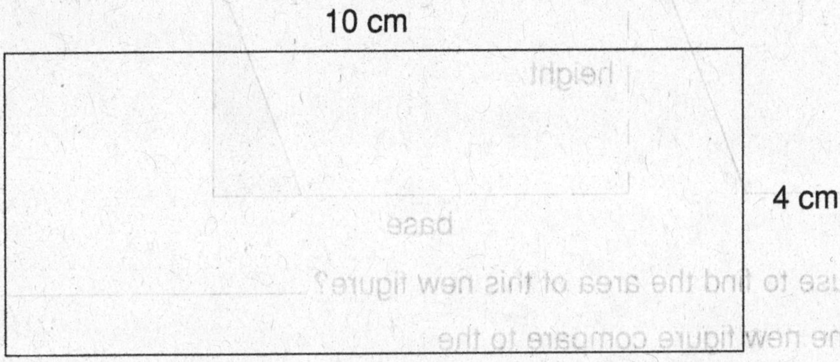

1. The rectangle above is 10 cm wide and 4 cm high. Cut several squares of paper 1 cm by 1 cm. Lay the squares of paper out on the rectangle and determine the number of squares it would take to cover the rectangle.

____

2. Multiply the width by the height of the rectangle. Compare the product with the result you got in the first step. How do they compare?

____

3. For all rectangles, the area can be calculated by multiplying the length by the width. The formula is $A = \ell \times w$. Use this formula to measure the area of your desktop. The result will be in square centimeters, written this way: cm^2.

____

Name ______________________ Date ____________ Class ____________

LESSON 10A

Hands-On Lab

Estimating and Finding Area, continued

4. There are simple methods for calculating the area of many other geometric figures. Copy this parallelogram onto a separate sheet of paper. Using scissors, carefully cut the small triangle from the left side of the parallelogram and move it to the right side as shown. What new geometric figure is formed? ____________

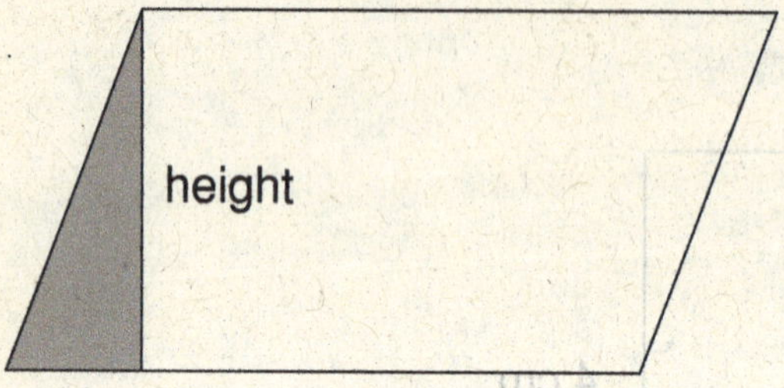

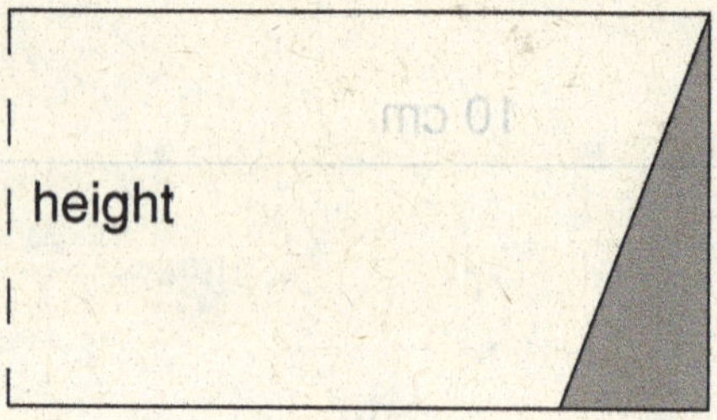

What formula can you use to find the area of this new figure? ____________

How does the area of the new figure compare to the original area? ____________

Calculate the area of a parallelogram with a height of 5 m and a base of 11 m. ____________

5. Copy this triangle onto a sheet of paper two times. Cut out one of the triangles and turn it over. Can you put the two triangles together to form a parallelogram? ____________

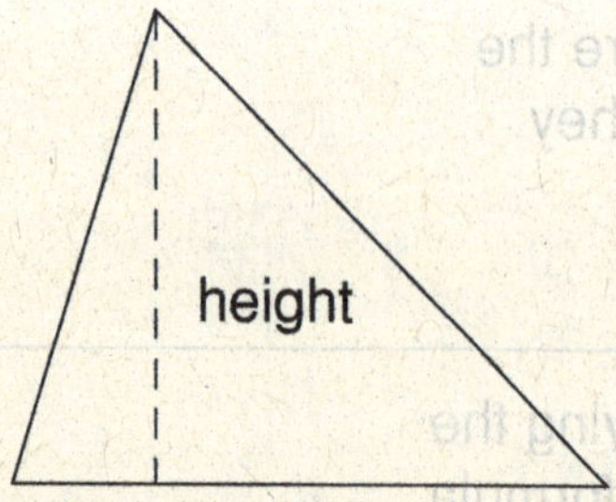

What formula can you use to find the area of the parallelogram formed by putting the two triangles together? ____________

How does the area of the parallelogram formed by the two triangles compare to the area of the original triangle? ____________

____

This shows that area of a triangle is $A = \frac{1}{2} \cdot b \cdot h$, where b is the base and h is the height. Calculate the area of a triangle with a base of 6 in. and a height of 3 inches. ____________

Name ______________________________ Date ____________ Class ____________

LESSON **10A**

Hands-On Lab

Estimating and Finding Area, continued

6. The area of irregularly shaped areas can be estimated. Trace the figure below onto centimeter grid paper. Count the number of complete squares within the figure.

   ____

   Is this value larger or smaller than the actual area?

   ____

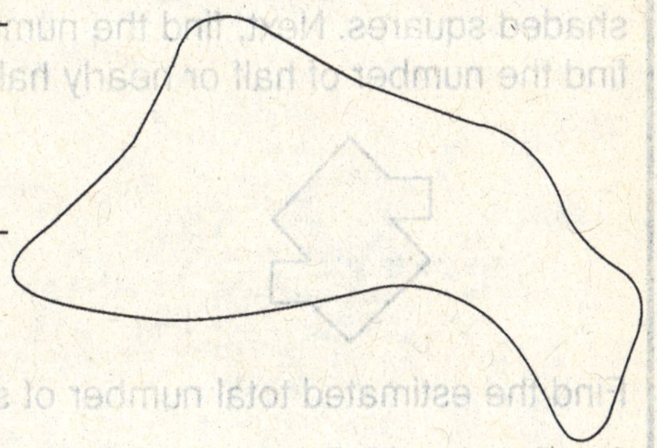

Think and Discuss

1. Can you think of a way to improve the accuracy of your estimate of the area of an irregular figure?

   ____

   ____

   ____

Try This

1. A rectangle has a length of 8 cm and a width of 3 cm. Find its area.

   ____

2. A triangle has the dimensions shown. What is its area?

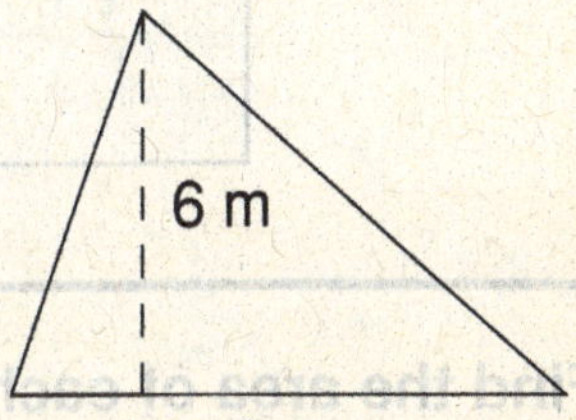

   ____

Name ______________________ Date ____________ Class ____________

LESSON 10A

Review for Mastery

Estimating and Finding Area

You can use grid paper to estimate and find area of figures. The area of a figure is the amount of surface it covers. Area is measured in square units.

To estimate the area of the figure below, first find the number of fully shaded squares. Next, find the number of nearly full squares. Then, find the number of half or nearly half-shaded squares.

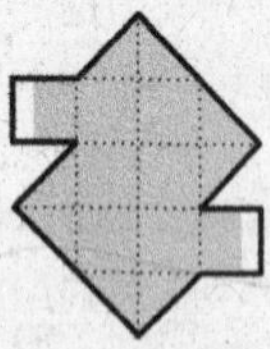

There are 6 fully shaded squares.
There are 2 nearly full squares.
There are 8 half-shaded squares.
$(8 \cdot \frac{1}{2} = 4)$

Find the estimated total number of squares shaded.

$6 + 2 + 4 = 12$. The area of the figure is about 12 square units.

Estimate the area of each figure.

1.

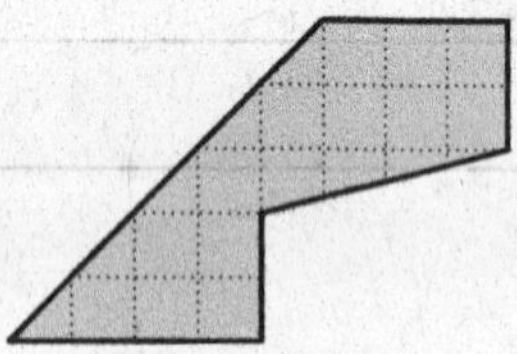

____

2.

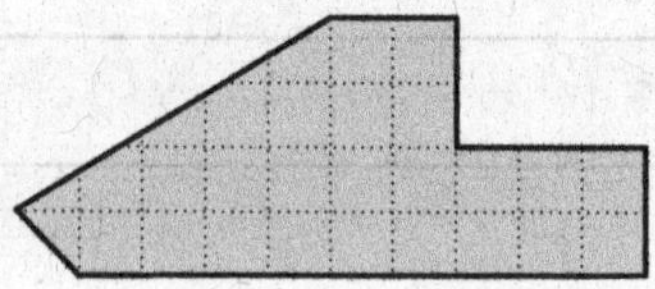

____

To find the area of a rectangle, find the total number of square units.

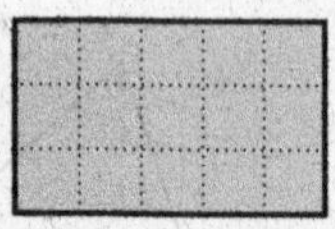

There are 3 rows of 5 squares.
$3 \cdot 5 = 15$
So, the area of the rectangle is 15 square units.

Find the area of each rectangle.

3.

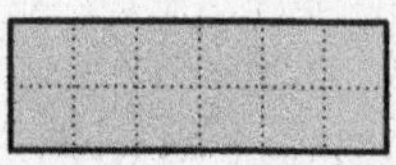

____

4.

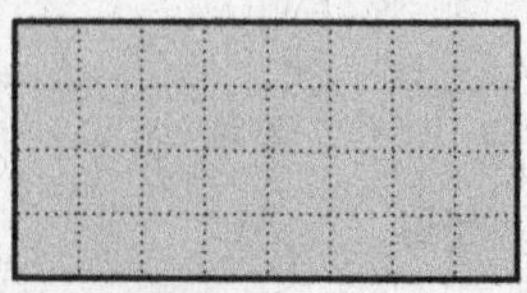

____

Name ______________________ Date ____________ Class ____________

LESSON 10A

Homework and Practice

Estimating and Finding Area

Estimate the area of each figure.

1.

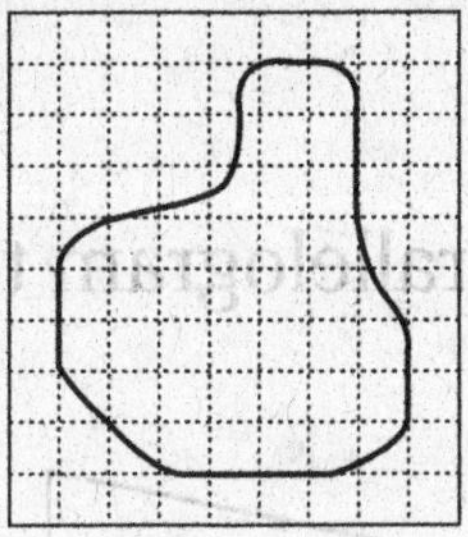

□ = 1 cm^2

____

2.

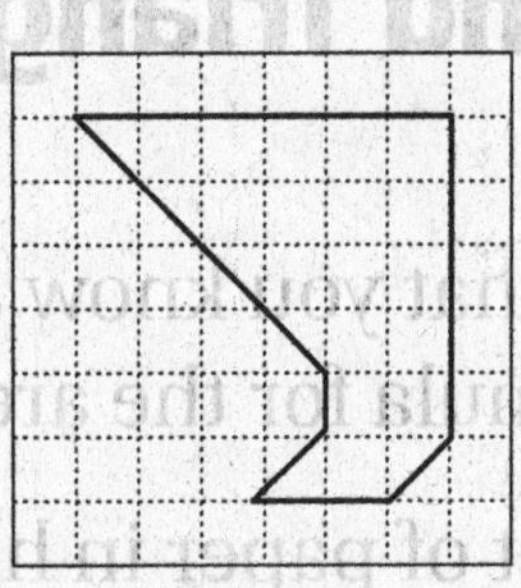

□ = 1 ft^2

____

Find the area of each rectangle.

3.

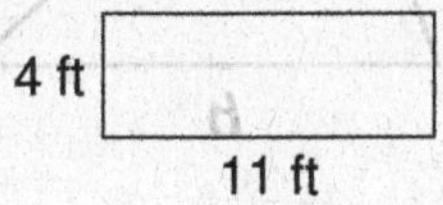

____

4.

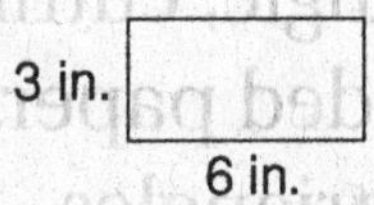

____

Find the area of each parallelogram.

5.

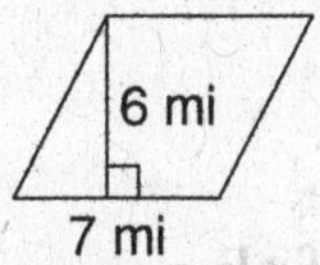

____

6.

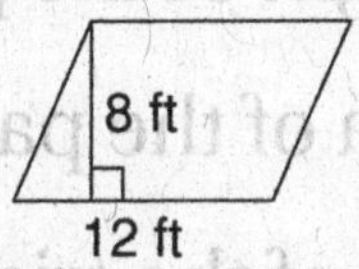

____

Find the area of each triangle.

7.

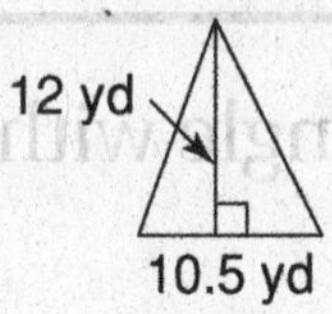

____

8.

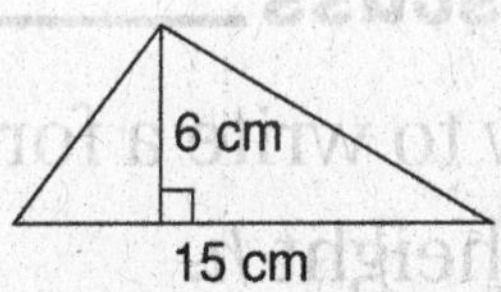

____

9. A painting at an art gallery is shaped like a parallelogram. Its base is 13 inches and its height is 8.5 inches. How much canvas was used to paint the picture?

____

EXPLORATION

Program of Studies and Core Content

MA-6-M-S-MPA3, MA-6-AT-S-EI2, MA-06-2.1.1, MA-06-2.1.3

10B Area of Rectangles and Triangles

You can use what you know about the area of a parallelogram to develop a formula for the area of a triangle.

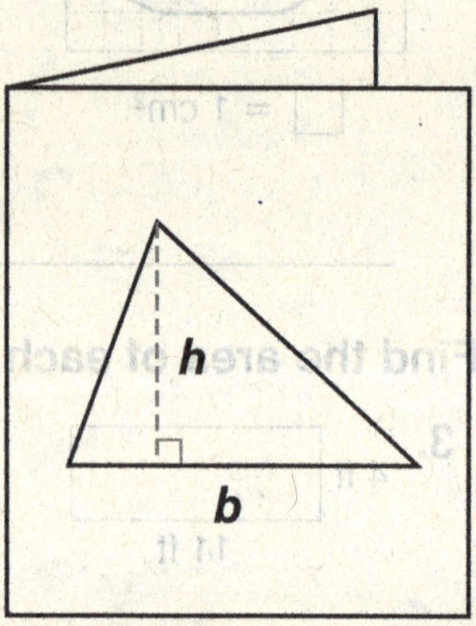

1. Fold a sheet of paper in half. Use a ruler to draw a triangle on one side of the folded sheet. Label the base and height of the triangle as shown.

2. Cut out the triangle, cutting through both layers of the folded paper. This will create two congruent triangles.

3. Arrange the two triangles to form a parallelogram.

4. What is the height of the parallelogram? What is its base?

5. What is the area of the parallelogram?

6. How is the area of the triangle related to the area of the parallelogram?

Think and Discuss

7. Show how to write a formula for the area of a triangle with base b and height h.

8. Explain how to use your formula to find the area of this triangle.

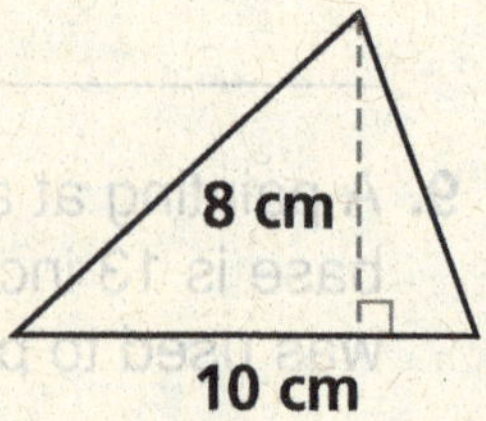

Name ______________________ Date __________ Class ________

LESSON **10B**

Review for Mastery

Area of Rectangles and Triangles

To find the area of a rectangle, multiply the length and the width.

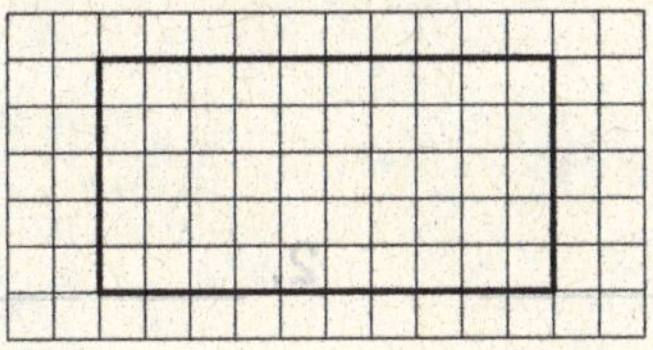

Find the area of the rectangle.

$A = lw$

$A = 5 \cdot 10 = 50$

So, the area of the rectangle is 50 square units.

Find the area of each rectangle.

1.

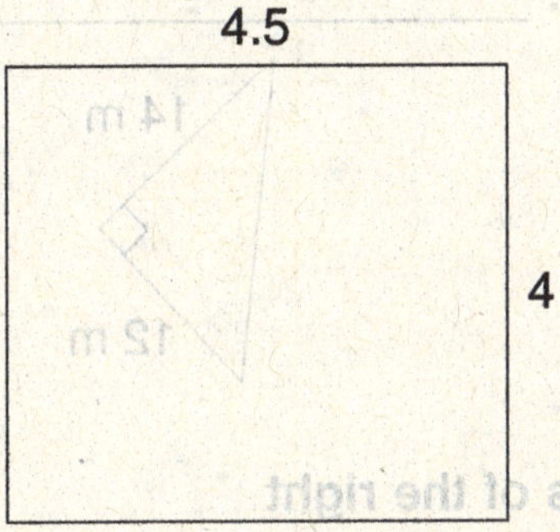

____

2.

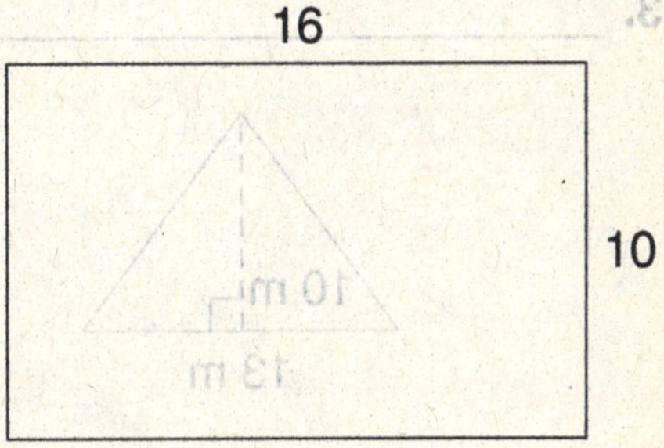

____

To find the area of a triangle, first turn your triangle into a rectangle.

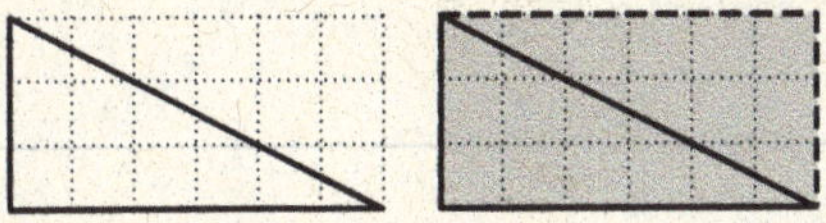

Next, find the area of the rectangle. 6 • 3 = 18

The triangle is half the area of the formed rectangle or $A = \frac{1}{2}bh$, so divide the product by 2.

$18 \div 2 = 9$ So, the area of the triangle is 9 square units.

Find the area of each triangle.

3.

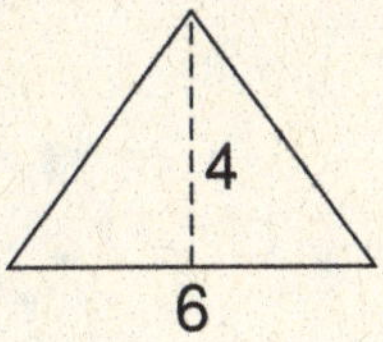

____

4.

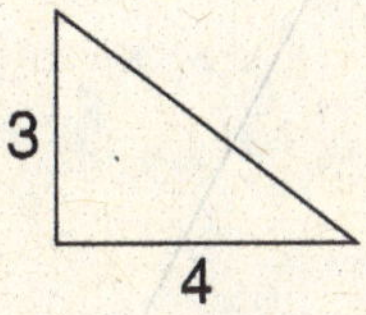

____

Name ______________________ Date __________ Class __________

LESSON **10B**

Homework and Practice

Area of Rectangles and Triangles

Find the area of each triangle.

1. ____________________

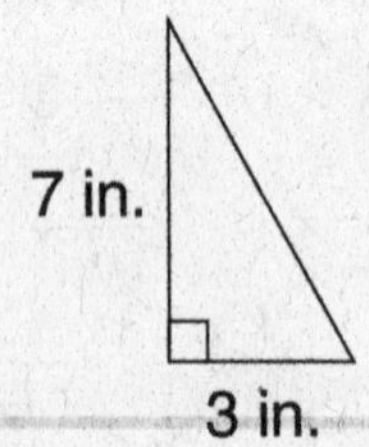

2. ____________________

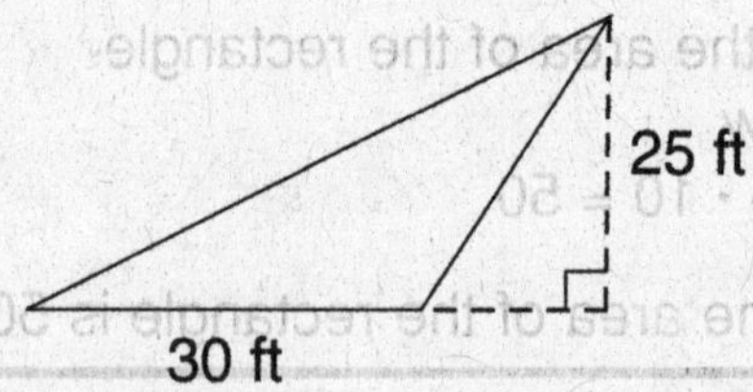

3. ____________________

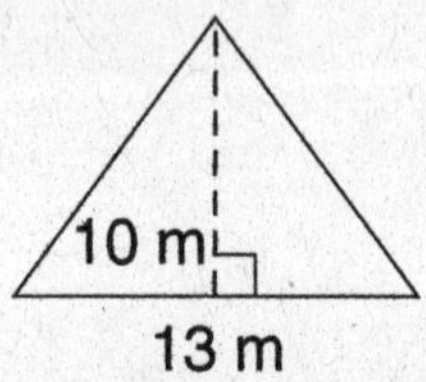

4. ____________________

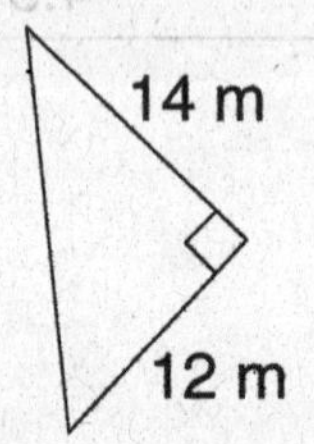

Use a ruler to measure the lengths of the legs of the right triangle. Then find its area. Select and use units of appropriate size and type.

5. ____________________

____

____

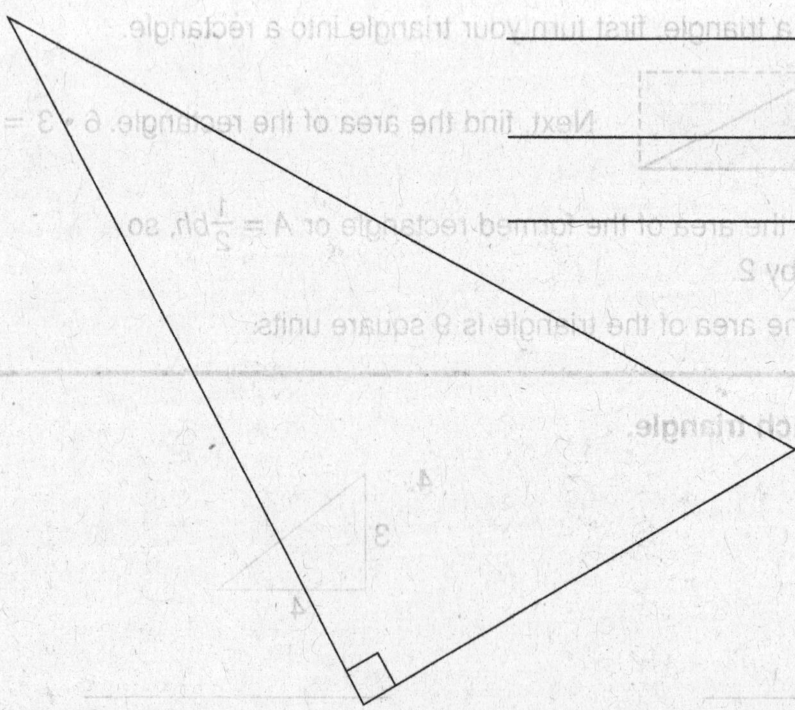

Name ________________ Date ________ Class ________

LESSON **10B**

Homework and Practice

Area of Rectangles and Triangles, continued

Find the area of each rectangle.

6. ____________________

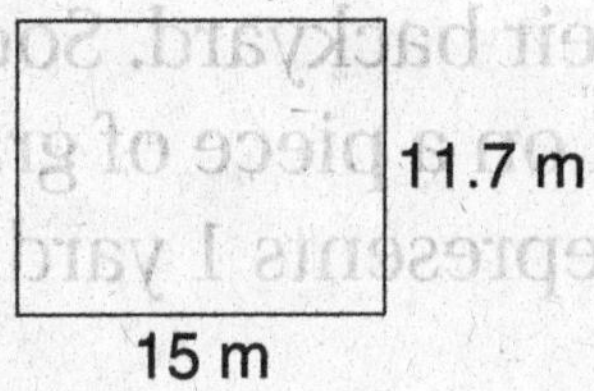

7. ____________________

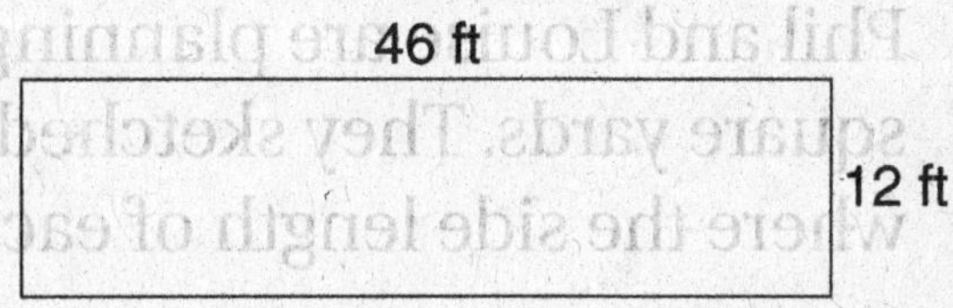

8. ____________________

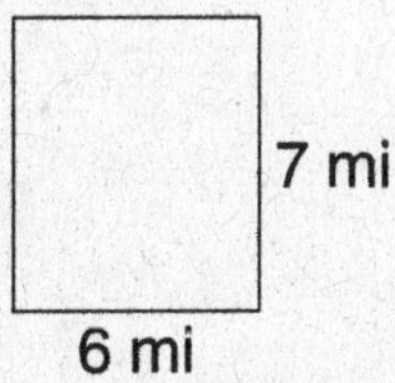

9. ____________________

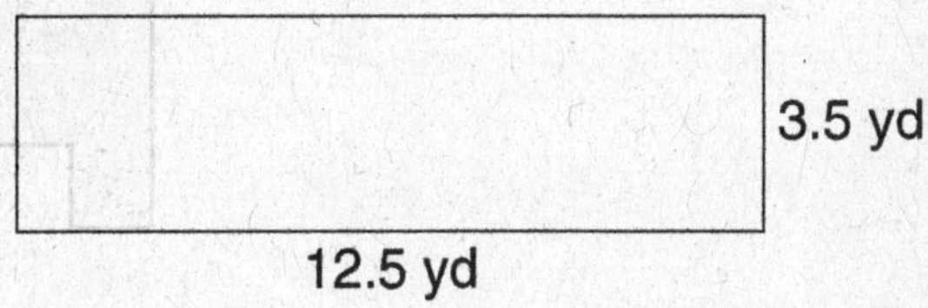

Use a ruler to measure the lengths of the sides of the rectangle. Then find its area. Select and use units of appropriate size and type.

10. ____________________

____

____

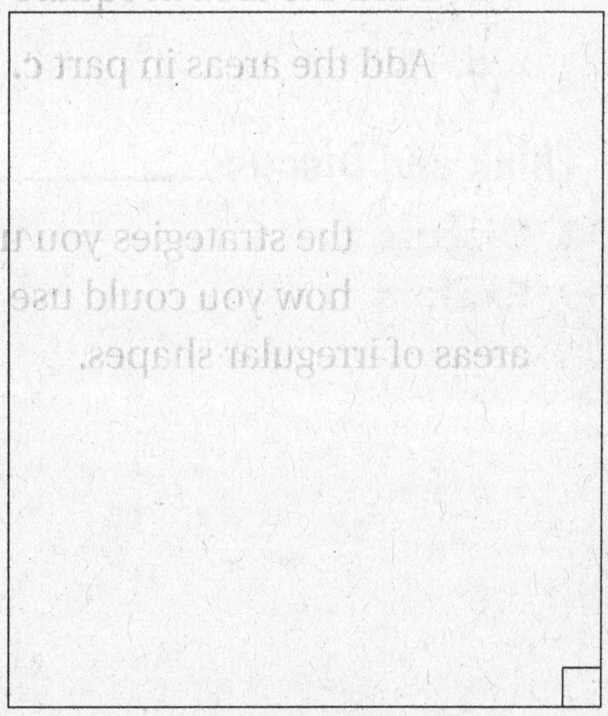

EXPLORATION

Program of Studies and Core Content
MA-6-M-S-MPA3,
MA-06-2.1.1

10C Area of Composite Figures

Phil and Louise are planning to sod their backyard. Sod is sold in square yards. They sketched their yard on a piece of graph paper, where the side length of each square represents 1 yard.

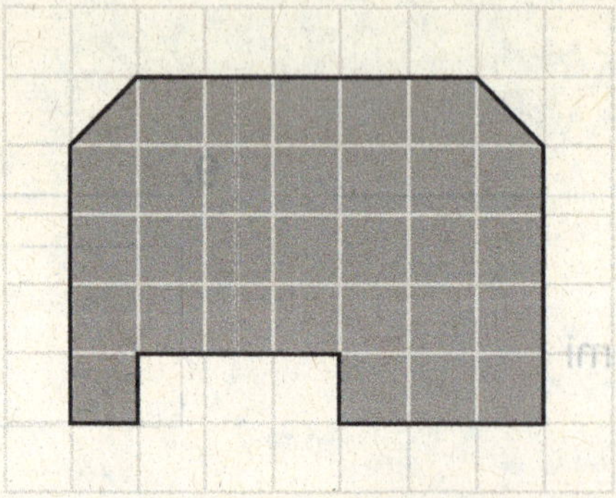

1. **a.** Divide the figure into several simpler figures.
   b. Name each figure from part **a.**
   c. Find the area in square yards of each figure.
   d. Add the areas in part **c.**

Think and Discuss

2. **Discuss** the strategies you used for finding area.
3. **Explain** how you could use squares to help you find the areas of irregular shapes.

Name ______________________ Date ____________ Class ____________

LESSON **10C**

Review for Mastery

Area of Composite Figures

Sometimes you can use area formulas you know to help you find the area of other figures.

To find the area of the figure below, first divide the figure into figures you know.

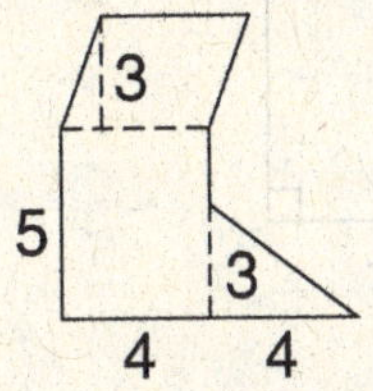

The figure is made up of a triangle, a parallelogram, and a rectangle.

Next, find the area of each figure.

Triangle	Parallelogram	Rectangle
$A = \frac{1}{2}bh$	$A = bh$	$A = \ell w$
$= \frac{1}{2}(3 \cdot 4)$	$= 3 \cdot 4$	$= 4 \cdot 5$
$= 6$	$= 12$	$= 20$

Then, find the sum of all of the areas.

$6 + 12 + 20 = 38$ The area of the figure is 38 square units.

Find the area of each figure.

1.

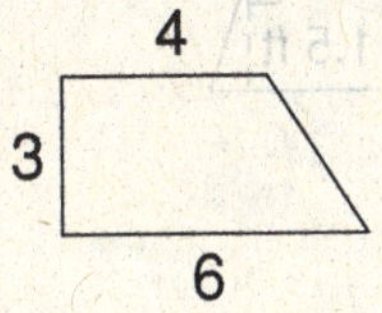

____

2.

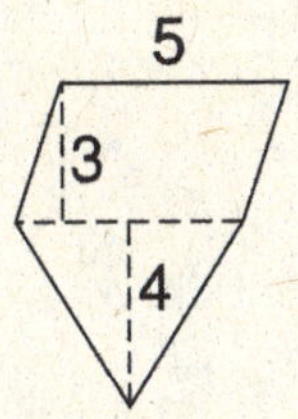

____

3.

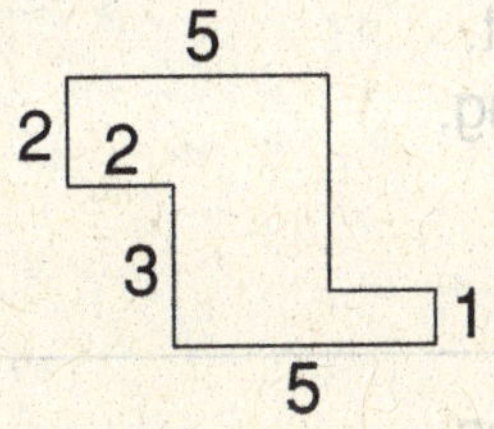

____

4.

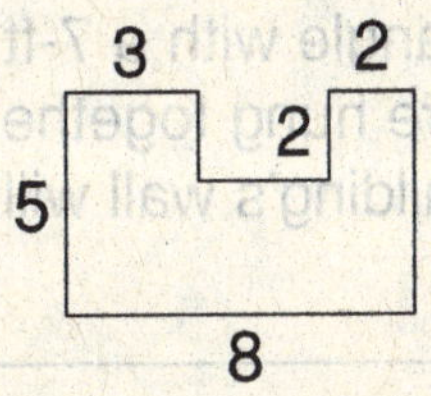

____

Name ______________________ Date ____________ Class ____________

LESSON 10C

Homework and Practice

Area of Composite Figures

Find the area of each polygon.

1.

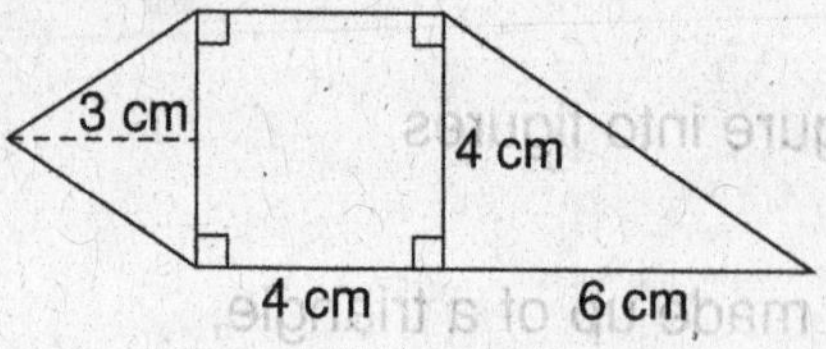

____

2.

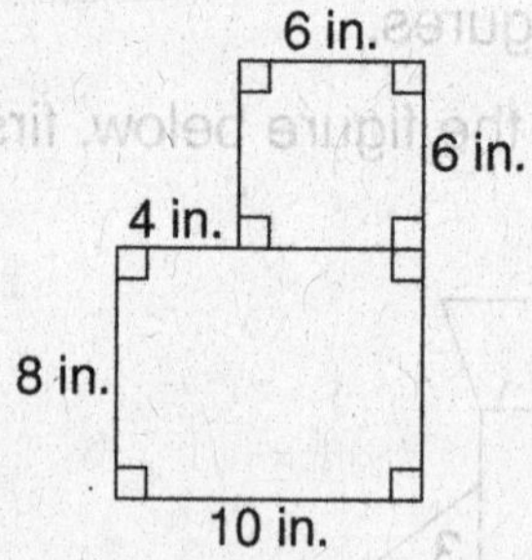

____

3.

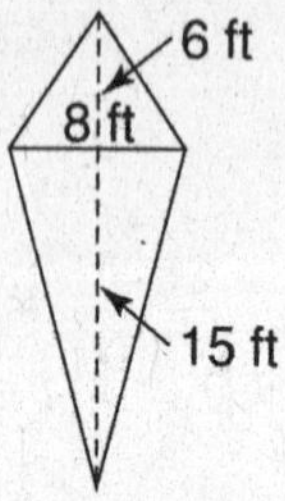

____

4.

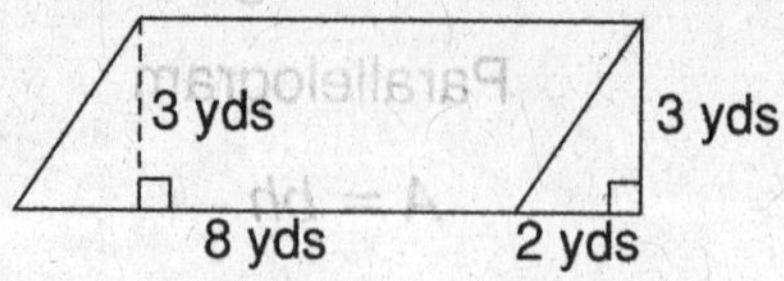

____

5.

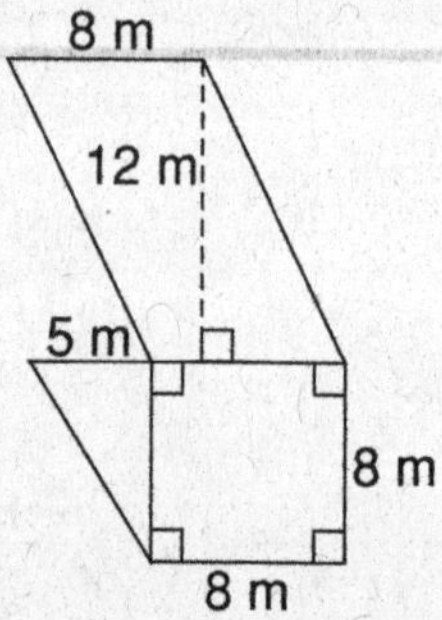

____

6.

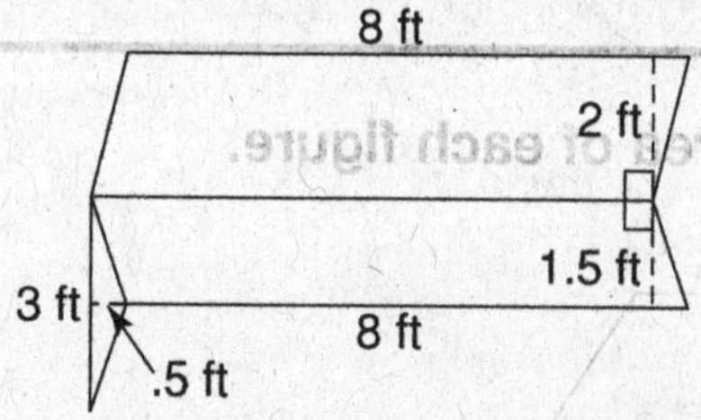

____

7. Three paintings are shaped like a 5-ft square, a 12 ft-by-8 ft rectangle, and a triangle with a 7-ft base and a height of 2 feet. If those paintings are hung together on the outside of a building, how much of the building's wall will they cover altogether?

____

8. Two diagonals divide a square carpet into 4 congruent triangles. The base of each triangle is 8 feet with a height of 4 feet. What is the area of the entire carpet?

____

EXPLORATION

Program of Studies and Core Content

MA-6-M-S-MPA1, MA-6-M-S-MPA2, MA-6-M-S-MPA6, MA-06-2.1.3

10D Comparing Perimeters and Areas

Suzanne is enlarging a color copy of a 3 in. by 5 in. photograph to 6 in. by 10 in. A model is shown below.

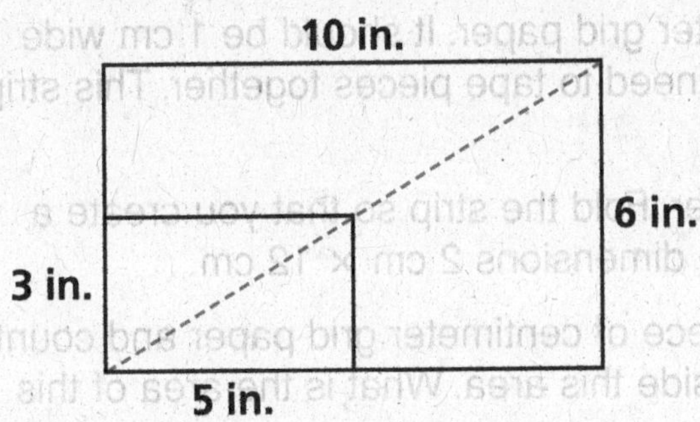

Find the perimeter of each color copy.

	Color Copy	Perimeter = 2 · Length + 2 · Width
1.	3 in. by 5 in.	
2.	6 in. by 10 in.	

3. How do the perimeters compare?

Find the area of each color copy.

	Color Copy	Area = Length · Width
4.	3 in. by 5 in.	
5.	6 in. by 10 in.	

6. How do the areas compare?

Think and Discuss

7. **Explain** how you compared the perimeters.

8. **Explain** how you compared the areas.

Name ______________________ Date __________ Class __________

LESSON **10D**

Hands-On Lab

Comparing Perimeters and Areas

You will use centimeter grid paper and scissors to determine and compare the perimeter and area of rectangles.

Activity 1

You are building a model of an enclosure for an animal with 28 cm meters of fencing. How large an enclosure can you build?

1. Cut out a strip of centimeter grid paper. It should be 1 cm wide and 28 cm long. You may need to tape pieces together. This strip represents the fence.

2. Tape the two ends together. Fold the strip so that you create a rectangular fence with the dimensions 2 cm × 12 cm.

3. Place this fence over a piece of centimeter grid paper and count the number of squares inside this area. What is the area of this enclosure?

____

4. Flatten and refold the strip to create four more fences:

6 cm × 8 cm 4 cm × 10 cm 5 cm × 9 cm 7 cm × 7 cm

Measure each area using grid paper. Enter the results in this table.

Dimensions of Enclosure	Area of Enclosure
2 cm × 12 cm	
6 cm × 8 cm	
4 cm × 10 cm	
5 cm × 9 cm	
7 cm × 7 cm	

Name ______________________________ Date ____________ Class ____________

LESSON 10D

Hands-On Lab

Comparing Perimeters and Areas, continued

Think and Discuss

1. What is the perimeter of each enclosure in the table?

____

2. Why does each enclosure have the same perimeter but a different area?

____

____

3. Enter the values from the table in the Activity into the table below, in order from smallest area to greatest area. What relationship do you see between the length and width of the enclosure and its area? Hint: Look at the difference between each enclosure's length and width, such as $12 - 2 = 10$.

Dimensions of Enclosure	Area of Enclosure

____

____

Try This

1. Cut out two strips of centimeter grid paper, one 44 cm long, and the other 40 cm long. Create two enclosures such that the enclosure you create with the 40-cm strip is larger than the enclosure you create with the 44-cm strip. What are each enclosure's dimensions?

____

____

____

Name ______________________ Date __________ Class __________

LESSON 10D

Hands-On Lab

Comparing Perimeters and Areas, continued

Activity 2

1. On centimeter grid paper, draw a rectangle that is 5 cm long and 6 cm wide. Label it *R*.
2. Calculate the area and perimeter of this rectangle. Include the units.

   Area of rectangle *R* = ____________

   Perimeter of rectangle *R* = ____________
3. Draw a rectangle that is 10 cm long and 12 cm wide. Label it *S*.

   Area of rectangle *S*: ____________

   What is the perimeter of rectangle *S*? ____________
4. Draw a rectangle that is 15 cm long and 18 cm wide. Label it *T*.

   Area of rectangle *T*: ____________

   Perimeter of rectangle *T*: ____________

Name ______________________ Date ____________ Class ____________

LESSON **10D**

Hands-On Lab

Comparing Perimeters and Areas, continued

Think and Discuss

1. Compare the areas and perimeters of rectangles *R* and *S*. What do you see?

____

____

2. Compare the areas and perimeters of rectangles *R* and *T*. What do you see?

____

____

3. Can a rectangle have the same value for its area and its perimeter? What is different about the two measurements even if the numerical value is the same? Explain.

____

____

Name ______________________ Date ____________ Class ____________

LESSON 10D

Review for Mastery

Comparing Perimeters and Areas

When the dimensions of a figure change, the perimeter and area of the figure change.

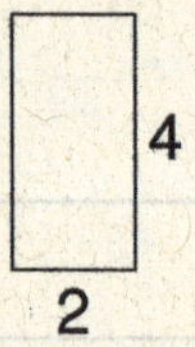

$P = 2\ell + 2w$ $\quad A = \ell w$

$= (2 \cdot 4) + (2 \cdot 2)$ $\quad = 4 \cdot 2$

$= 8 + 4$ $\quad = 8$ square units

$= 12$ units

The dimensions are $\ell = 4$ and $w = 2$.

If each dimension of the rectangle is divided by 2, the perimeter and area change.

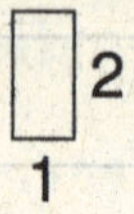

$P = 2\ell + 2w$ $\quad A = \ell w$

$= (2 \cdot 2) + (2 \cdot 1)$ $\quad = 2 \cdot 1$

$= 4 + 2$ $\quad = 2$ square units

$= 6$ units

The new dimensions are $\ell = 2$ and $w = 1$.

When the dimensions of a rectangle are divided by 2, the perimeter is divided by 2 and the area is divided by 4.

Write how the perimeter and area change when the dimensions change.

1.

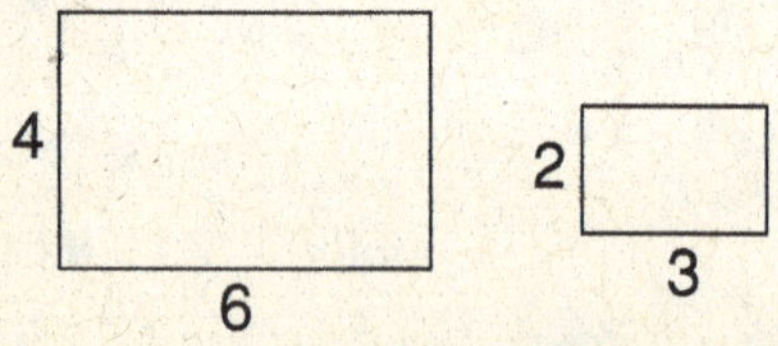

____

____

2.

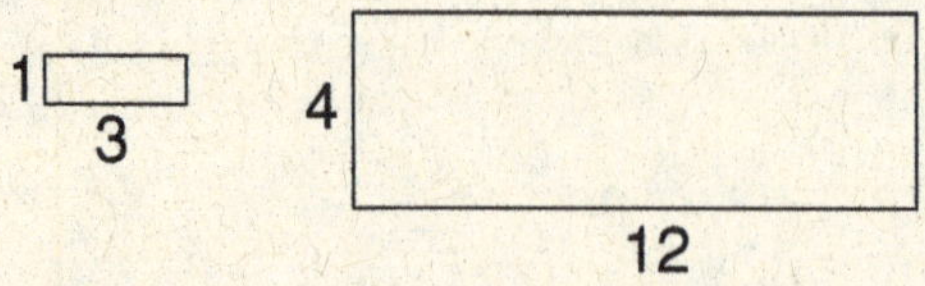

____

____

Name ______________________ Date __________ Class __________

LESSON **10D**

Homework and Practice

Comparing Perimeters and Areas

Find how the perimeter and the area of the figure change when its dimensions change.

1.

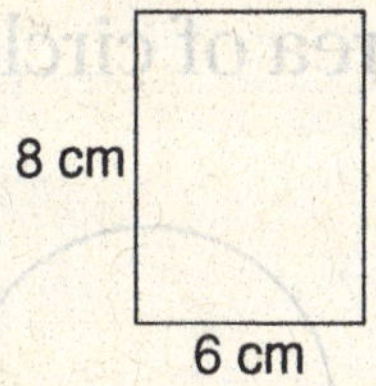

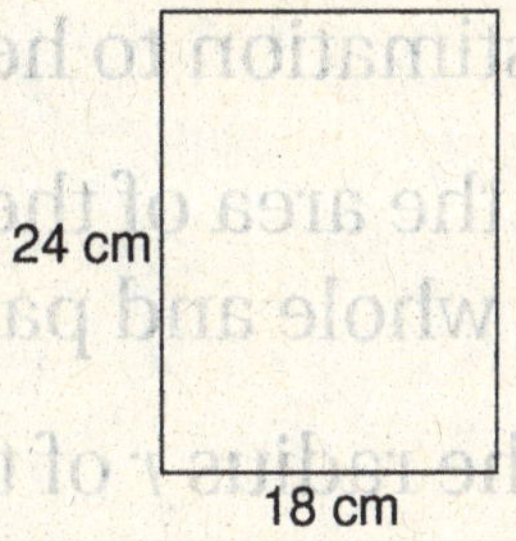

______________________ ______________________

____

____

2. Use a centimeter ruler to measure the triangle. Then draw another triangle with dimensions that are one-third as great as the given triangle. How do the perimeter and the area change when the dimensions change?

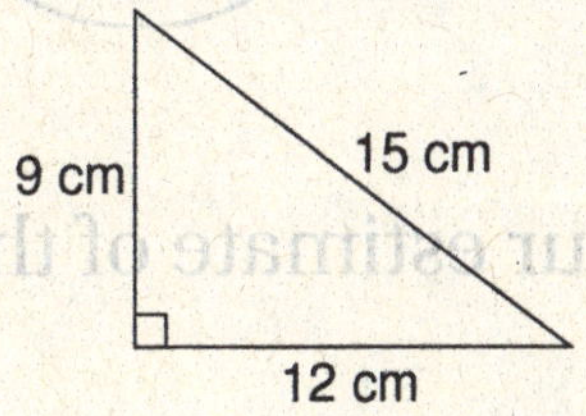

____

____

3. Bert wants to make a larger version of a drawing he saw in a magazine. The drawing fits into a square with each side measuring 7 cm. If Bert makes his copy twice the size of the square measured in the original drawing, how much space will it cover on the paper he uses?

____

4. How many centimeters of wood will Bert need to make a frame for his drawing from Exercise 3?

____

EXPLORATION

Program of Studies and Core Content
MA-6-AT-S-EI2

10E Area of Circles

You can use estimation to help you investigate the area of circles.

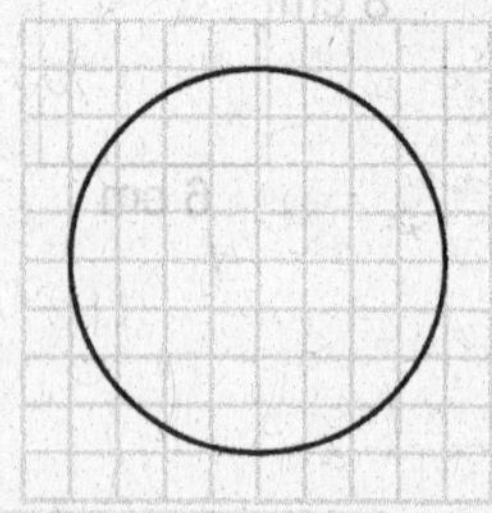

1. Estimate the area of the circle by counting whole and partial squares.
2. What is the radius r of the circle?
3. Calculate πr^2 using 3.14 for *pi.*
4. How does the value of πr^2 compare to your estimate of the circle's area?

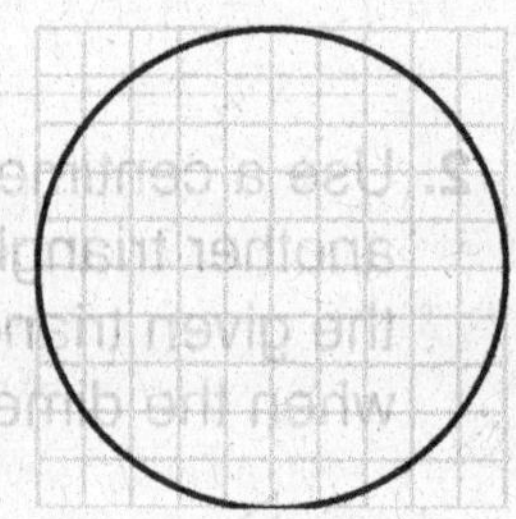

5. Repeat the process for this circle. First estimate the area by counting whole and partial squares.
6. What is the radius of the circle?
7. Calculate πr^2 using 3.14 for *pi.*
8. How does the value of πr^2 compare to your estimate of the circle's area?

Think and Discuss

9. **Describe** any shortcuts you found for counting the squares.
10. **Explain** how you can use what you discovered to write a formula for the area of a circle with radius r.

Name ______________________ Date __________ Class __________

LESSON **10E**

Review for Mastery

Area of Circles

To estimate the area A of a circle, use the formula $A = \pi r^2$ and 3 for π.

$A = \pi r^2$
$\approx 3 \cdot 2^2$
$\approx 3 \cdot 4$
≈ 12 square units.

The area is about 12 square units.

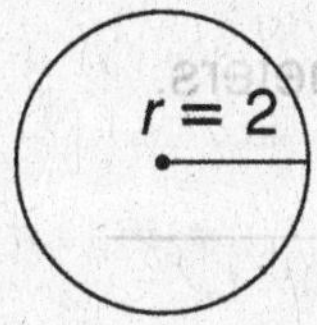

Estimate the area of each circle. Use 3 for π.

1.

____

2.

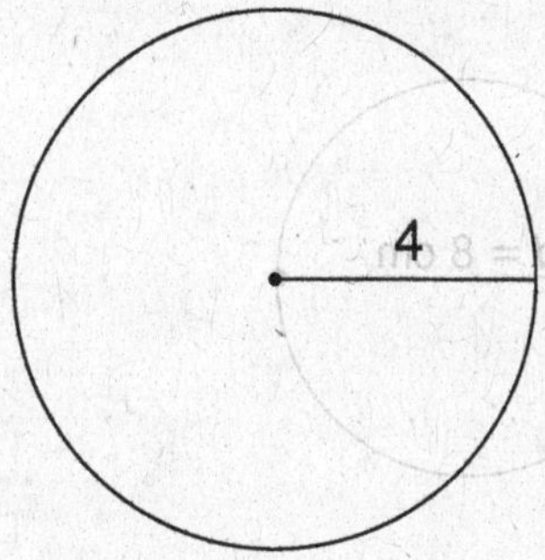

____

3.

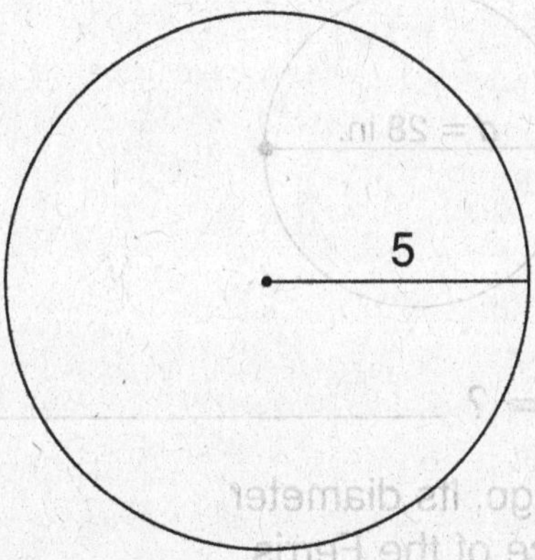

____

4.

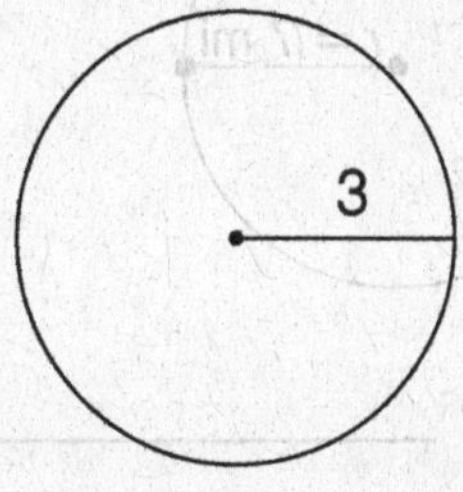

____

Name ______________________________ Date ____________ Class ____________

LESSON **10E**

Homework and Practice

Area of Circles

Use the circle to answer each question.

1. Name the circle. ____________________

2. Name two diameters.

____

3. Name four radii.

____

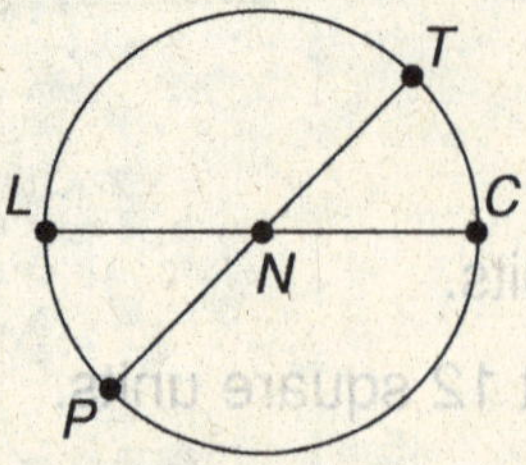

Find each missing value to the nearest hundredth. Use 3.14 for pi.

4.

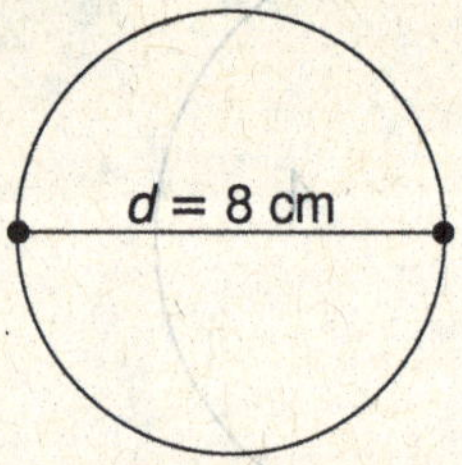

$C = ?$ ____________________

5.

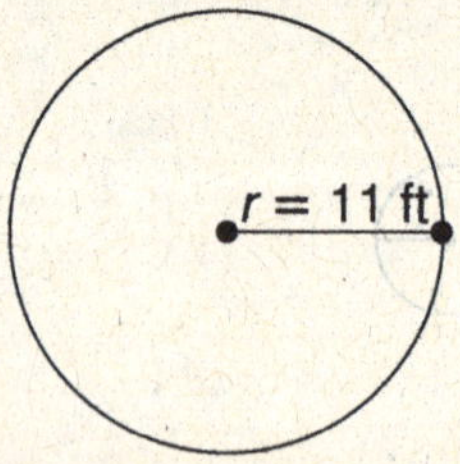

$C = ?$ ____________________

Find the area of each circle. Use $\frac{22}{7}$ for pi.

6.

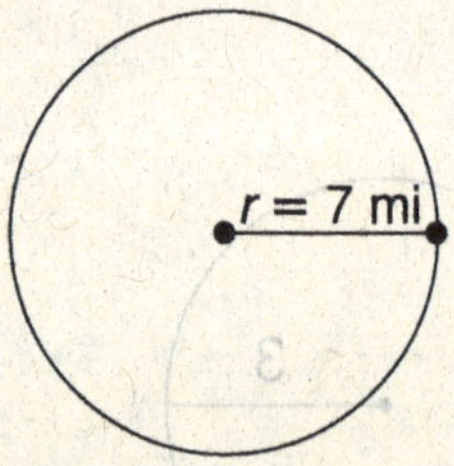

$A = ?$ ____________________

7.

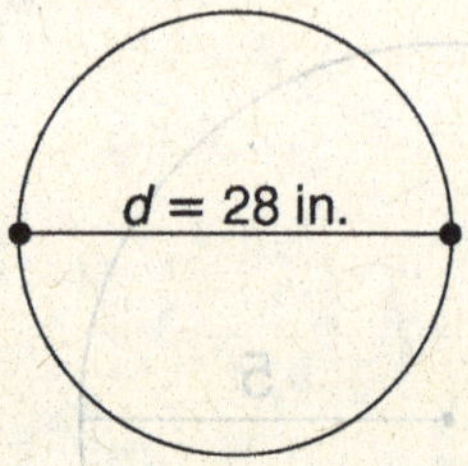

$A = ?$ ____________________

8. The first Ferris wheel was built in 1893 in Chicago. Its diameter was 250 feet. What was the actual circumference of the Ferris wheel? Use 3.14 for pi.

____

9. Stonehenge, a circle of large carved stones in England, was built more than 1,000 years ago. The circle of stones has a diameter of 108 feet. What is the area of Stonehenge in feet? Use 3.14 for pi.

____

EXPLORATION

Program of Studies and Core Content
MA-6-G-S-SR5,
MA-06-3.1.3

10F Three-Dimensional Figures

A cube is a solid figure with six faces, twelve edges, and eight vertices.

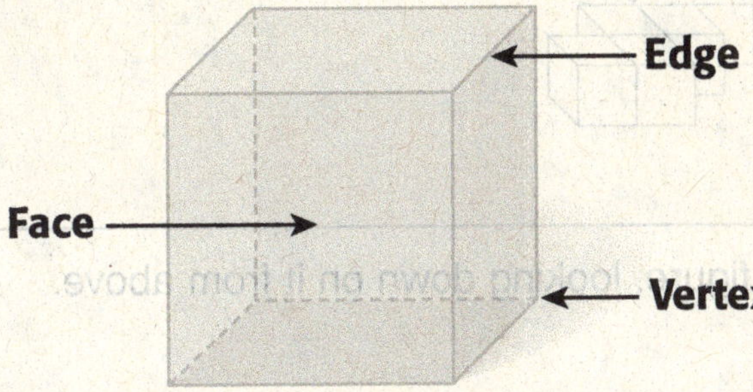

Determine how many faces, edges, and vertices each solid figure has.

		Faces	Edges	Vertices
1.				
2.				
3.				

Think and Discuss

4. Explain how many edges it takes to form a vertex.

5. Explain how many faces it takes to form an edge.

Name ______________________ Date __________ Class __________

LESSON 10F

Hands-On Lab

Drawing Three-Dimensional Figures

Drawing views of three-dimensional figures is easier if you can look at the actual models in three dimensions.

Activity

1. Use cubes to build the following figure.

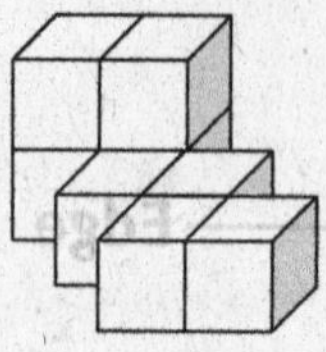

2. Draw the top view of the figure, looking down on it from above. Label your drawing "top."

3. Draw a front view of the figure looking at it from eye-level. Label your drawing "front."

4. Draw a side view of the figure, looking at it from your right. Label your drawing "side."

Name ______________________________ Date ___________ Class ___________

LESSON 10F

Hands-On Lab

Drawing Three-Dimensional Figures, continued

Think and Discuss

1. Is every cube seen in every view? Why or why not?

____

2. Build a figure using only the front and side views from the previous example. Will it necessarily match the original figure? Explain.

____

____

____

Try This

1. Use cubes to build the following figure. (Assume there is a cube in the back left corner.)

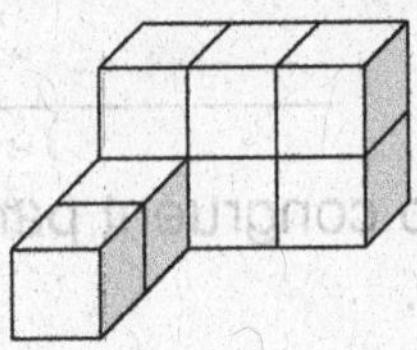

____

2. Draw the top view of the figure, looking down on it from above. Label your drawing "top."

3. Draw a front view of the figure looking at it from eye-level. Label your drawing "front."

4. Draw a side view of the figure, looking at it from your right. Label your drawing "side."

Name ______________________ Date __________ Class __________

LESSON **10F**

Review for Mastery

Three-Dimensional Figures

A polyhedron is a three-dimensional figure. A polyhedron has flat surfaces, called faces. Faces that share a side form an edge. The point at which three or more edges meet is called a vertex.

This polyhedron has 4 faces, 4 vertices, and 6 edges.

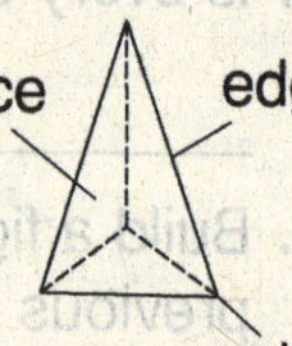

Identify the number of faces, edges, and vertices on each three-dimensional figure.

1.

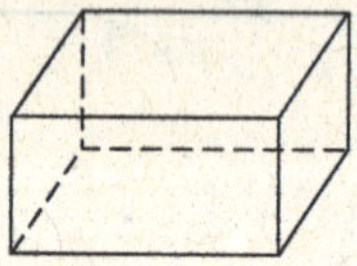

____

2.

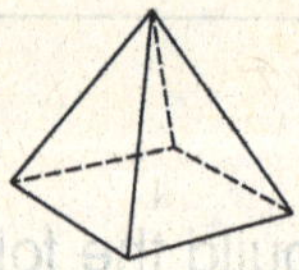

____

A prism is a three-dimensional figure with two congruent parallel bases. The other faces are all parallelograms.

A pyramid has one base. The other faces are triangles.

A cylinder has two circular parallel bases joined by a curved surface.

A cone has one circular base and a curved surface that comes to a point.

Pyramids and prisms are polyhedrons because each face is a polygon. Cylinders and cones are not polyhedrons because they have curved surfaces.

Tell whether each figure is a polyhedron and name each three-dimensional figure.

3.

____

4.

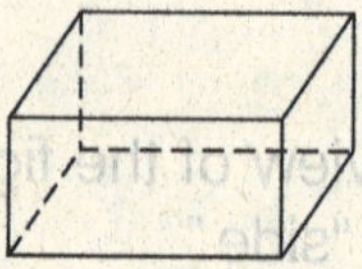

____

Name ______________________ Date __________ Class __________

LESSON 10F

Homework and Practice

Solid Figures

Identify the number of faces, edges, and vertices in each solid figure.

1.

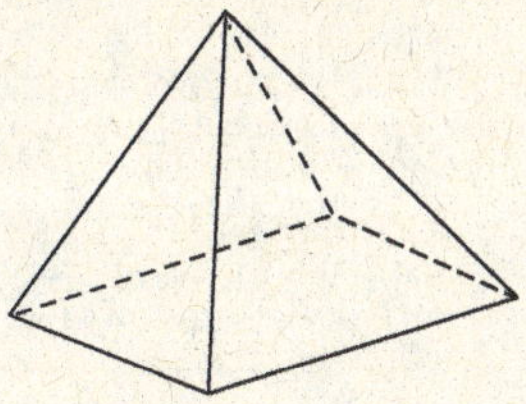

2.

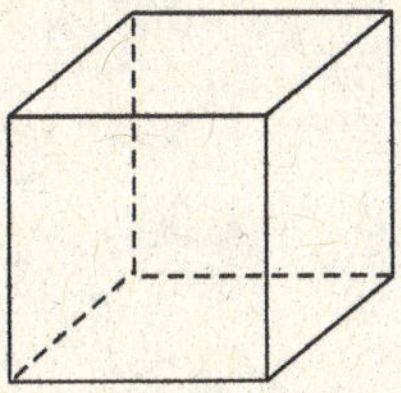

3.

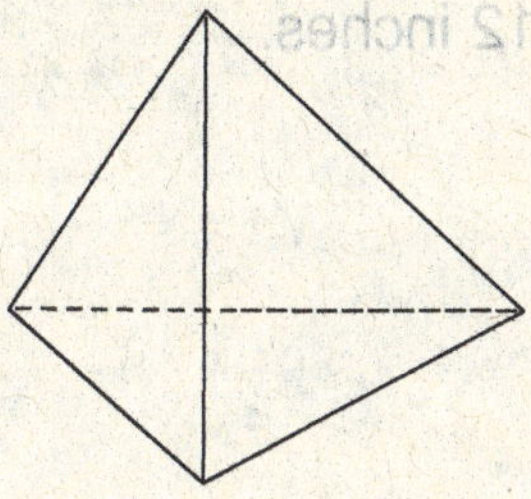

Tell whether each figure is a polyhedron and name the solid.

4.

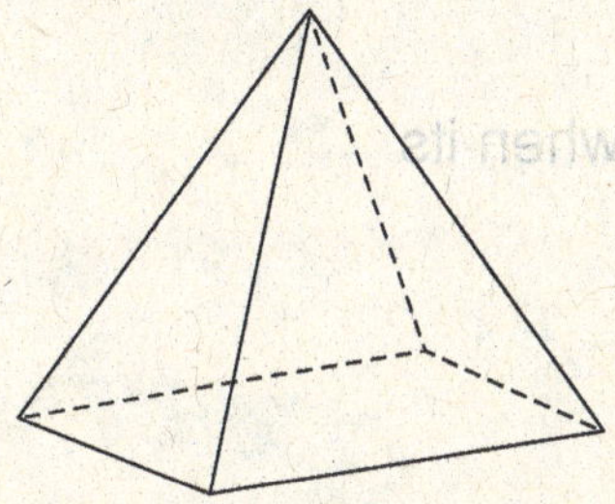

5.

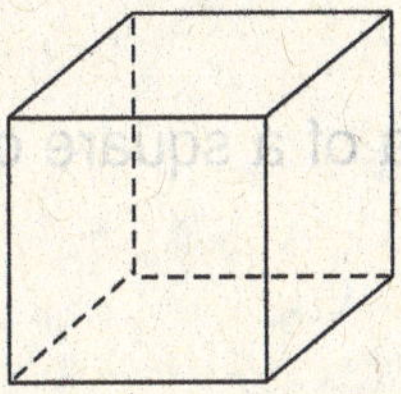

6.

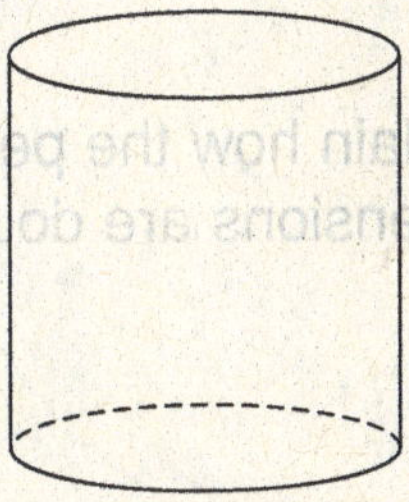

7. Jamie wants to make a box in the shape of a triangular pyramid. How many pieces of wood does she need? In what shape should she cut them? Explain.

8. A sphere is a three-dimensional figure shaped like a ball. Is a sphere a polyhedron? If so, how many faces does a sphere have? Describe a sphere in your own words.

CHAPTER 10

Chapter Summary

Answer these questions to summarize the important concepts from Chapter 10 in your own words.

1. Explain how to find the area of a triangle with base 13 inches and height 12 inches.

2. Explain how to find the area of a composite figure.

3. Explain how the perimeter and area of a square change when its dimensions are doubled.

4. Describe the difference between the side of a rectangle and a face of a rectangular prism.

5. Explain how to determine if a solid figure is a polyhedron.

For more review of Chapter 10:

- Complete the Chapter 10 Study Guide and Review on pages 590–592.
- Complete the Ready to Go On quizzes on pages 562 and 586.

Name ______________________ Date __________ Class __________

CHAPTER 10

KCCT Prep

Multiple Choice

1. A rectangle has a length of 3.2 in. and a perimeter of 10.4 in. What is the width of the rectangle?

Ⓐ 2 in. Ⓒ 4 in.
Ⓑ 3.2 in. Ⓓ 7.2 in.

2. The area of the figure shown is 60 cm^2. What is the value of x?

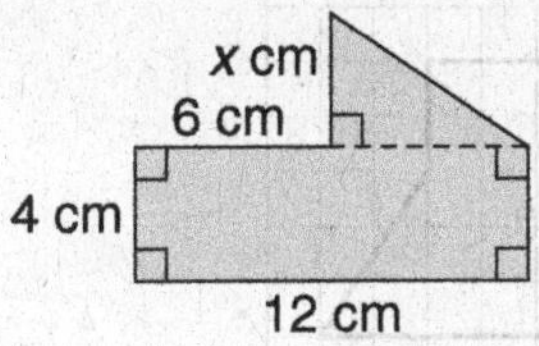

Ⓐ 2 Ⓒ 5
Ⓑ 4 Ⓓ 8

3. Which is the best way to classify an angle that is supplementary to $\angle ABD$?

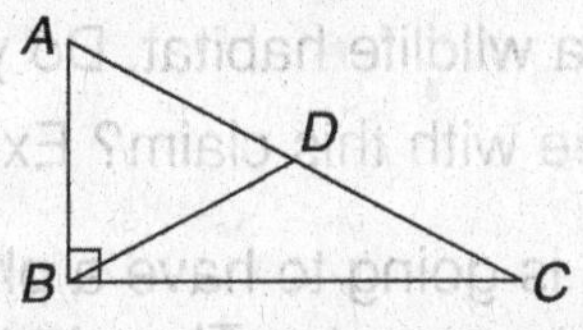

Ⓐ Acute Ⓒ Right
Ⓑ Obtuse Ⓓ Straight

4. The value of the expression ■ $\times 2 + 8 \div 2^2$ is 12. What is the missing number in the expression?

Ⓐ 2 Ⓒ 5
Ⓑ 4 Ⓓ 6

5. Which of these is closest to 0.7?

Ⓐ 0.08 Ⓒ $\frac{3}{4}$
Ⓑ $\frac{7}{100}$ Ⓓ 0.071

6. Which of these is the best estimate for the circumference of the circle?

Ⓐ 22 mm Ⓒ 88 mm
Ⓑ 44 mm Ⓓ 176 mm

7. A triangle has a base of 12 ft and an area of 48 ft^2. What is the height of the triangle?

Ⓐ 4 ft Ⓒ 8 ft
Ⓑ 6 ft Ⓓ 16 ft

8. There are 3,980 people in an arena. About 24% of the people took public transportation to the arena. Which is the best estimate of the number of people who took public transportation to the arena?

Ⓐ 250 Ⓒ 1,000
Ⓑ 500 Ⓓ 1,500

9. Which set of data is best displayed in a line graph?

Ⓐ The results of a survey in which 100 students were asked about their favorite type of music
Ⓑ The ages of people at an art gallery
Ⓒ The population of a town each year from 2000 to the present
Ⓓ The prices of 9 different brands of calculators

10. A map has a scale of 1 cm:12 km. The distance between two cities on the map is 3.85 cm. Which is the best estimate of the actual distance between the two cities?

Ⓐ 3 km Ⓒ 24 km
Ⓑ 4 km Ⓓ 48 km

11. Kyle takes a rectangular piece of wood and cuts a square from one of the corners, as shown. What is the area of the remaining piece of wood?

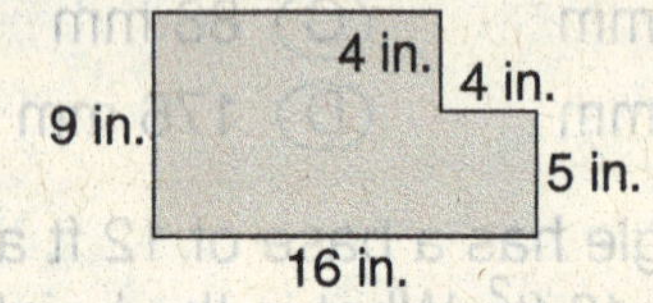

Ⓐ 128 in^2 Ⓒ 144 in^2
Ⓑ 132 in^2 Ⓓ 160 in^2

12. Latoya has 9 boxes of pencils. The solution of the equation $9b = 108$ tells how many pencils are in each box. How many pencils are there altogether in 4 boxes?

Ⓐ 24 Ⓒ 48
Ⓑ 36 Ⓓ 64

13. Alan has $5\frac{1}{3}$ pounds of tomatoes. It takes $1\frac{3}{4}$ pounds of the tomatoes to make a pot of sauce. How many full pots of sauce can he make from the tomatoes?

Ⓐ 2 Ⓒ 4
Ⓑ 3 Ⓓ 5

14. A square park has a perimeter of 181.2 meters. Which is the best estimate of the length of one side of the park?

Ⓐ 13 m Ⓒ 65 m
Ⓑ 45 m Ⓓ 90 m

Open Response

15. Diego plots the points (1, 3), (5, 3), (4, 5), and (2, 5). Then he connects the points to form a quadrilateral.

a. What type of quadrilateral does Diego form?

b. Diego reflects the quadrilateral across the line that passes through the points (1, 3) and (5, 3). Draw the result of the reflection.

16. The figure shows a map of a state park. Each square of the grid represents 1 mi^2.

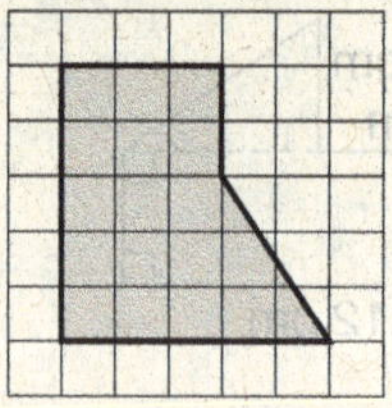

a. Find the area of the state park.

b. In the park, $\frac{9}{20}$ of the land is set aside as a wildlife habitat. A park ranger claims that nearly 12 mi^2 are a wildlife habitat. Do you agree with this claim? Explain.

17. Martina is going to have a photograph made into a poster. The photograph is 10 inches long and 8 inches wide.

a. The poster will have twice the length and twice the width of the photograph. How many times greater will the area of the poster be compared to the area of the photograph?

b. It costs \$0.04 per square inch to print the poster. How much will the poster cost?

c. Martina wants to order a frame for the poster. She guesses that the perimeter of the poster will be twice the perimeter of the photograph. Do you agree? Why or why not?

CHAPTER 11

Project

Graphs and Functions: Graphing the Earth

Career: Geographer

Geographers are interested in characteristics of our natural world, such as landforms, natural resources, and climate. Some geographers spend time in the field collecting information. Others can create maps, charts, and graphs. Geographers use integers to express information such as high and low temperatures and elevations above and below sea level.

Before you start the project, work through the Continental Ups and Downs worksheet on the next page.

In this chapter project . . .

You draw an elevation map of your town or county.

What to do:

- Learn about some of the highest and lowest points around the world.
- Research the highest and lowest elevations in your town or county. If available, find the elevations of several other nearby locations or landmarks.
- Make a scale drawing of your town or county and label the points of elevation. Label the highest and lowest points and add that information to your map.

What to present:

- An elevation map of your town or county.
- A table of locations or landmarks from your map, ranked by their altitude (from lowest to highest).

Where's the math?

- Use integers to represent altitudes above and below sea level.
- Use proportions to create a scale drawing.
- Use a coordinate grid to represent the positions of your locations and landmarks in your town or county.

go.hrw.com
Chapter Project Online
KEYWORD: MR7 Ch10

Name ______________________ Date __________ Class __________

CHAPTER 11

Project Recording Sheet

Graphs and Functions

Continental Ups and Downs

It has taken geographers many years to survey the Earth. Now in many places of each continent specific points have been marked so that other measurements can be made. The points, called benchmarks, often have a plaque listing the location, latitude, longitude and, in some cases, altitude above sea level.

Continent	**Highest Point** (m)	**Rank**	**Lowest Point** (m)	**Rank**
Africa	Mt. Kilimanjaro: 5,895		Lake Assal: −156	
Antarctica	Vinson Massif: 4,897		Bentley Subglacial Trench: −2,538	
Asia	Mt. Everest: 8,850		Dead Sea: −411	
Australia	Mt. Kosciusko: 2,228		Lake Eyre: −12	
Europe	Mt. Elbrus: 5,642		Caspian Sea: −28	
North America	Denali: 6,194		Death Valley: −86	
South America	Mt. Aconcagua: 6,960		Valdes Peninsula: −40	

1. Complete the chart by ranking the continents. First rank them by highest point, and then rank them by lowest point.
2. Create a scale drawing comparing the high and low points of each continent. Use sea level as the baseline on your drawing and show the high and low points above and below sea level.

Research: What are the High and Low points of each of the 50 states? Which state has the highest point? The lowest point?

What is the Bentley Subglacial Trench?

____

____

What is the deepest part of the ocean?

____

____

EXPLORATION

Program of Studies and Core Content
MA-6-G-S-CG2, MA-6-AT-S-PRF1, MA-6-AT-S-PRF2, MA-6-AT-S-PRF4, MA-06-4.1.2, MA-06-5.1.1, MA-06-5.1.2, MA-06-5.1.3, MA-06-5.1.4, MA-06-5.1.5

11A Tables and Functions

A school has scheduled a trip for 210 students to a theme park. The school can rent up to six buses for $200 each. Each bus seats a maximum of 60 students. Tickets to the theme park cost $25 per student.

1. Use the first example as a guide to complete the table.

Group Number	Number of Students	Cost of Bus Rental and Tickets	Total Cost
1	10	$200 + 25 \cdot 10 = 200 + 250$	$450.00
2	20		
3	30		
4	40		
5	50		
6	60		

2. What numbers in the table remain constant?

3. What numbers in the table vary?

4. Use the cost of the bus ($200), the cost of each ticket ($25), and the number of students (x) to write an equation for the total cost (c). (*Hint:* Look at the middle column of the table to write the equation.)

Think and Discuss

5. Explain what makes the total cost of a group vary.

6. Discuss possible ways of reducing the total cost of taking 210 students to the theme park.

Name ______________________________ Date __________ Class __________

LESSON 11A

Technology Lab 1

Tables and Functions

Use a spreadsheet to apply functions that you discover while examining data tables. A spreadsheet can interpret your equations and do the calculations with the data.

Activity

Write an equation for the function that is modeled by the data in the table below. Then, turn your equation into a spreadsheet-friendly formula and extend the data.

Input = ***i***	25	50	2	5
Output = ***o***	2	1	25	10

The rule for this data set's function is ________________

1. Put the Input label and data in the cells of Row 1.
2. Leave a row or two for a classmate to enter what they believe the function to be after seeing the output.
3. Put the Output label in cell A4 and the formula for the function in cell B4.

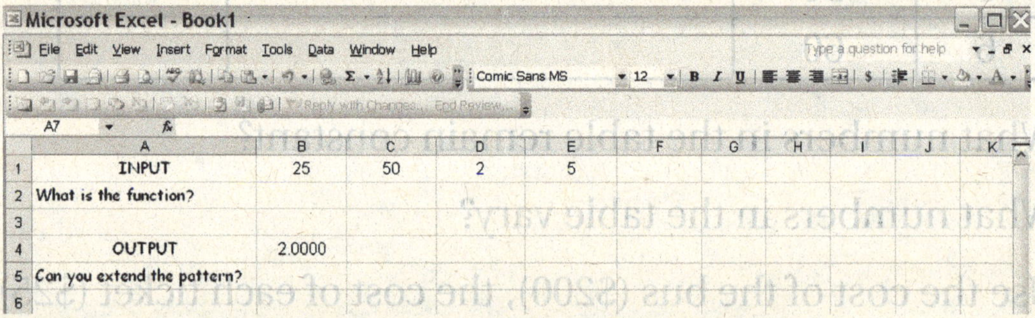

Remember, in a spreadsheet, math operations require special keystrokes. The equation for this data is **50 ÷ input = output**. Write the equation for the computer as = 50/B1. Enter the formula into cell B4.

4. Copy the formula all along row 4 so that every input will be computed into an output. Click on the cell with the formula and drag the small bold handle at the bottom of the cell along the row.

Name ______________________________ Date ____________ Class ____________

LESSON **11A**

Technology Lab 1

Tables and Functions, continued

Think and Discuss

1. Suppose the input in the first set of data increased by 5. How would that change the equation? How would the change affect the output?

____

____

____

2. Try this data in another section of the spreadsheet. Follow the same procedure.

Data = ***d***	4	6	9	12
Result = ***r***	32	34	37	40

The rule for this data set's function is ________________

Try This

For each set of data, do the following:

- **Determine the function rule for each table.**
- **Program the function into a spreadsheet.**
- **Check your rule by substituting the data given in the top row.**

1. Enter the data into the last two spaces.

Begin	8	15	17	20	
End	13	20	22		32

The rule for this data set is

____

2.

Data	5	20		40	4
Result	8		20	1	10

The rule for this data set is

____

Name ______________________________ Date ____________ Class ____________

LESSON **11A**

Technology Lab 2

Linear and Nonlinear Functions

You can get a general sense of whether a data set corresponds to a linear function by entering the data in an Excel worksheet and plotting the data as a chart.

Activity

The chart shows the open space observed by mapping from the air in Springvale County in five successive years.

Create an Excel chart that shows whether the data set in the following table has the form of a linear function.

	Open Space (1000s of acres)
2001	4.2
2002	3.75
2003	3.3
2004	2.85
2005	2.4

1. Enter the data in an Excel worksheet in the same form as above. Leave the top left cell (A1) blank as in the table.
2. Drag with the mouse from **A1** to **B6** to highlight the entire table.
3. Choose **Insert** from the main menu. From the Insert menu, choose **Chart.** The **Chart Wizard** dialog box appears.
4. From the **Chart Type** list at left, choose **Line**. From the **Chart Sub-Type** graphic thumbnails at right, choose **Line** (top left). Click the **Finish** button.

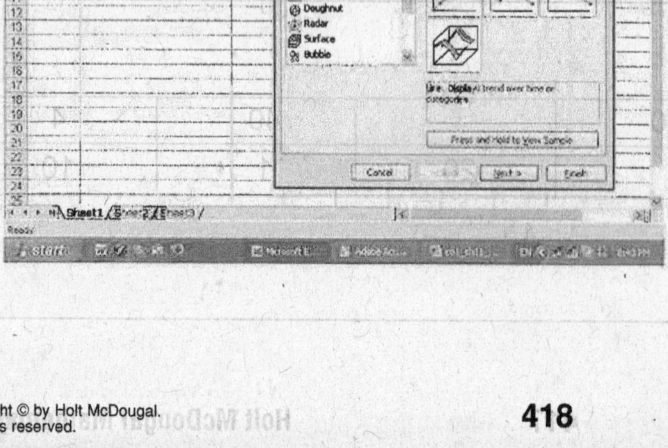

Name ____________________________ Date ____________ Class ____________

LESSON **11A**

Technology Lab 2

Linear and Nonlinear Functions, continued

5. A line graph appears displaying the data in the table. The data appear to form a straight line, indicating that open space in the county is decreasing at a steady rate from year to year.

Think and Discuss

1. If open space in the county were being lost at an accelerating rate, what might the graph look like? How would it look if the loss of open space was continuing but beginning to level off?

____

____

____

2. What would the graph look like if open space was declining at a steady rate, but more rapidly?

____

____

____

Try This

The chart above shows the amount of light measured at different distances (in feet) from a light source. Enter the above data in an Excel worksheet. Follow the same steps as before to create a chart from the data. Does the function appear to be linear?

____

Distance	Lumens
1	1000
2	250
3	111
4	62.5
5	40
6	27.8

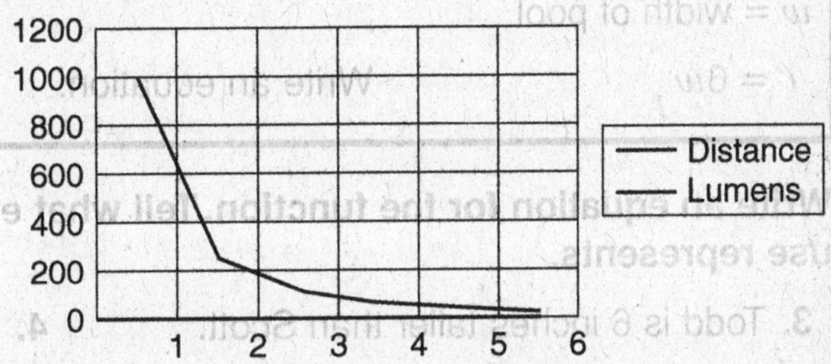

Name ________________________________ Date ___________ Class ___________

LESSON **11A**

Review for Mastery

Tables and Functions

A function is a rule that relates two quantities so that each input value corresponds to exactly one output value. In the table below, the *x*-values are the input and the *y*-values are the output.

x	0	1	2	3	4	5	6	7
y	4	5	6	7	8	9	10	?

To write an equation for a table of values, first compare the *x*- and *y*-values to find a pattern.

Each *y*-value is 4 more than its corresponding *x*-value.

Then use the pattern to write a rule for the table.

$y = x + 4$

You can use the rule to find a missing value in a table. To find the value of *y* in table above when $x = 7$, substitute 7 for *x* in the equation.

$y = x + 4$
$y = 7 + 4$
$y = 11$
So *y* is 11 when *x* is 7.

Write an equation for a function that gives the values in each table. Use the equation to find the value of *y* for the indicated value of *x*.

1.

x	1	2	3	4	5	6
y	3	6	9	12	15	?

____

2.

x	18	17	16	15	14	13
y	15	14	13	?	11	10

____

You can also write equations for functions that are described in words.

The length of the pool is 6 times the width of the pool.

ℓ = length of pool   Choose variables for the equation.
w = width of pool

$\ell = 6w$ Write an equation.

Write an equation for the function. Tell what each variable you use represents.

3. Todd is 6 inches taller than Scott.

____

____

____

4. Alana is 4 times as old as Tracey.

____

____

____

Name ____________________ Date __________ Class __________

LESSON 11A

Homework and Practice

Tables and Functions

Write an equation for a function that gives the values in each table. Use the equation to find the value of *y* for the indicated value of *x*.

1.

x	90	72	51	24	9
y	30	24	17	8	

2.

x	4	7	10	13	16
y	15	18	21	24	

3.

x	50	38	26	14	9
y	42	30	18	6	

4.

x	2	4	6	8	10
y	14	28	42	56	

Write an equation for the function. Tell what each variable you use represents.

5. Derek paid four times as much as Stephen for a hotel room.

6. A truck can travel 23 miles for each gallon of gas. Write an equation that shows how far the truck can go.

7. Write an equation that shows how many dollars are in a jar for the amount of nickels in the jar.

8. A doughnut shop is trying to find out how fast they sell a new type of doughnut. After an 8 hour day, they sold 24 boxes of the new doughnut, and after 3 hours the next day, they sold 9 boxes of the new doughnuts. Write an equation to show how many boxes of doughnuts are being sold.

EXPLORATION

Program of Studies and Core Content
MA-6-G-S-CG1, MA-6-AT-S-PRF3, MA-06-3.3.1

11B Ordered Pairs

The table shows the number of faces and vertices of the five regular polyhedrons.

Faces	4	8	20	6	12
Vertices	4	6	12	8	20

This data can be represented on a graph using ***ordered pairs.*** Each ordered pair is composed of the number of faces and the number of vertices of the polyhedron.

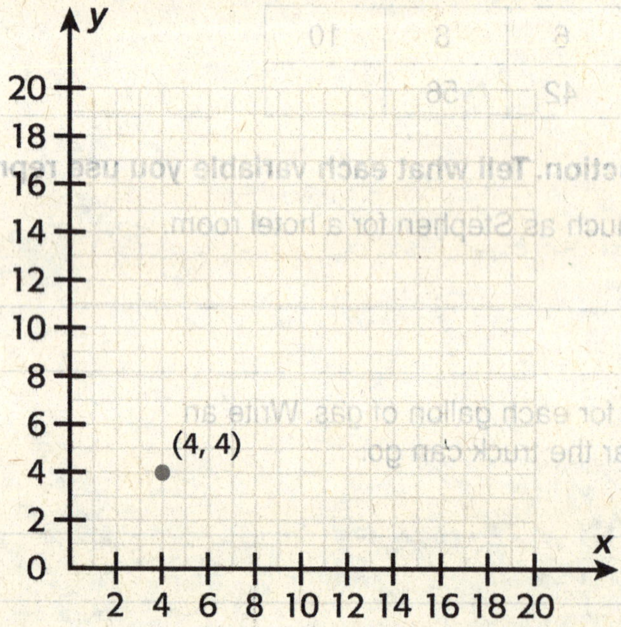

1. Plot and label the remaining ordered pairs. First find the number of faces on the horizontal line. From the number of faces, move up to find the number of vertices on the vertical line.

Think and Discuss

2. Explain how to plot ordered pairs.

3. Discuss what the point (6, 8) means.

Name ______________________ Date ____________ Class __________

LESSON **11B**

Review for Mastery

Ordered Pairs

A coordinate grid is formed by horizontal and vertical lines and is used to locate points.

An ordered pair names the location of a point by using two numbers.

The ordered pair (2, 5) gives the location of point *A* on the coordinate grid.

The first number, 2, tells the horizontal distance from the origin (0, 0).

The second number, 5, tells the vertical distance.

To find the ordered pair for point *B*, start at (0, 0). Then move 6 units right and 4 units up. The coordinates of point *B* are (6, 4).

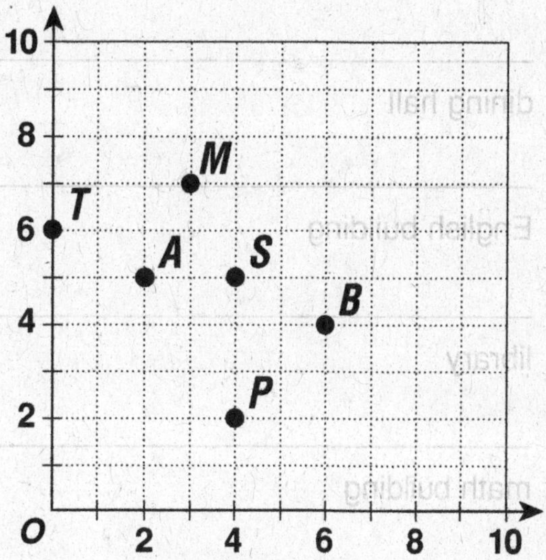

Give the ordered pair for each point shown on the coordinate grid above.

1. *P* ____________ **2.** *T* ____________ **3.** *M* ____________ **4.** *S* ____________

You can plot points in a coordinate grid.

To plot *F* (6, 4), start at (0, 0). Then move 6 units right and 4 units up.

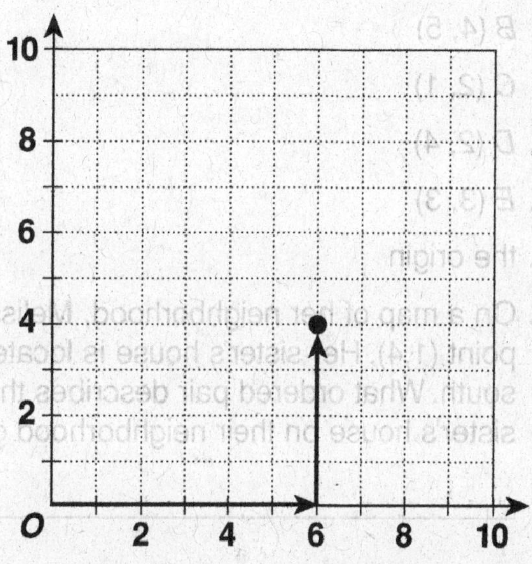

Plot each point in the coordinate grid above.

5. *V* (5, 6) **6.** *G* (3, 2) **7.** *K* (7, 0) **8.** *C* (1, 5)

Name ____________________ Date __________ Class __________

LESSON **11B**

Homework and Practice

Ordered Pairs

Name the ordered pair for each location on the grid.

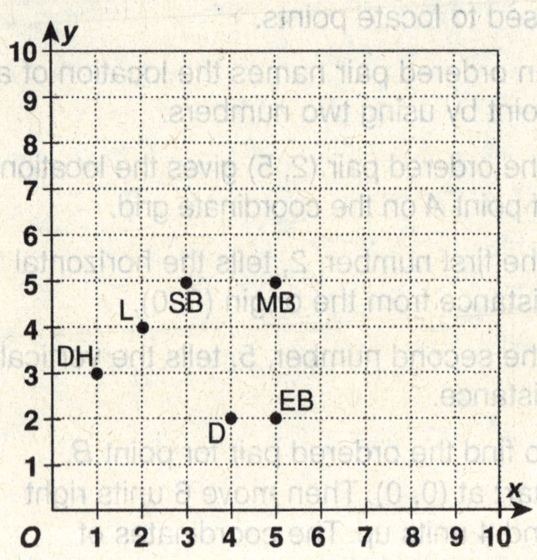

1. science building

____

2. dining hall

____

3. English building

____

4. library

____

5. math building

____

6. dormitories

____

Graph and label each point on the coordinate grid.

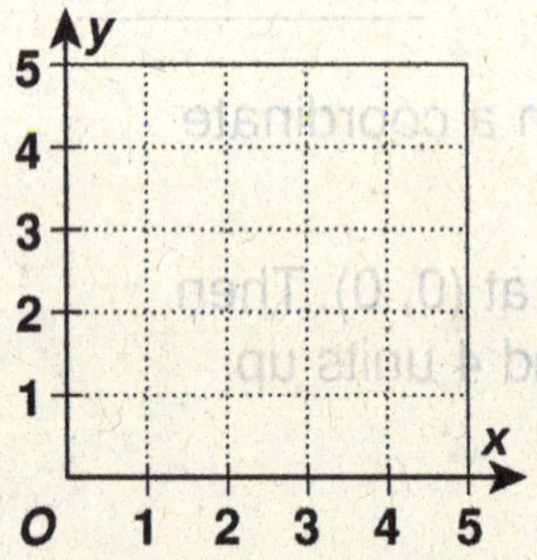

7. *A* (4, 2)
8. *B* (4, 5)
9. *C* (2, 1)
10. *D* (2, 4)
11. *E* (3, 3)
12. the origin
13. On a map of her neighborhood, Melissa's house is located at point (1,4). Her sister's house is located 5 units east and 3 units south. What ordered pair describes the location of Melissa's sister's house on their neighborhood grid map?

____

EXPLORATION

Program of Studies and Core Content
MA-6-G-S-CG2, MA-6-AT-S-PRF2, MA-6-AT-S-PRF3, MA-6-AT-S-PRF4, MA-06-4.1.2, MA-06-5.1.2, MA-06-5.1.3, MA-06-5.1.4, MA-06-5.1.5

11C Graphing Functions

To graph a ***linear equation,*** you can plot ordered pairs from a table of x- and y-values as points on a coordinate grid.

The table and graph at right model the function $y = x + 2$.

x	y
0	2
1	3
2	4
3	5

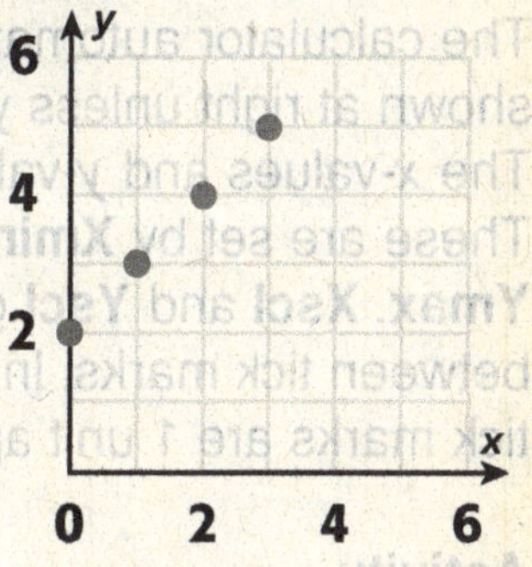

Graph each set of ordered pairs on the coordinate grid.

1. $y = x - 2$

x	y
2	0
3	1
4	2
5	3

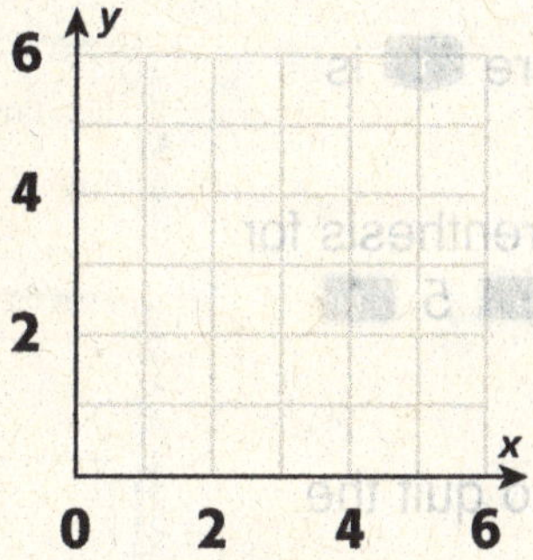

2. $y = x + 1$

x	y
1	2
2	3
3	4
4	5

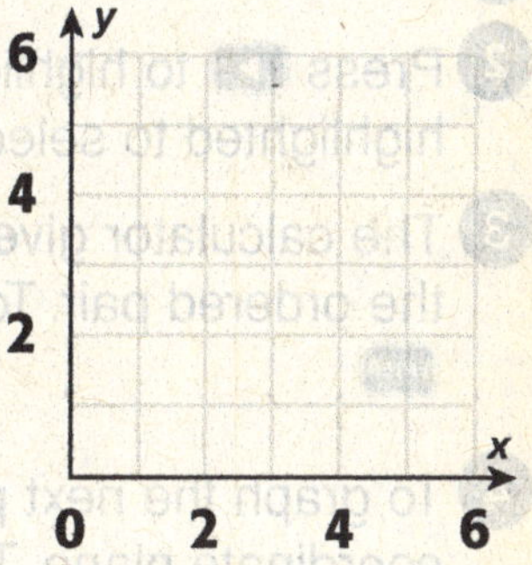

Use the points on each graph to complete each table.

3.

x	y

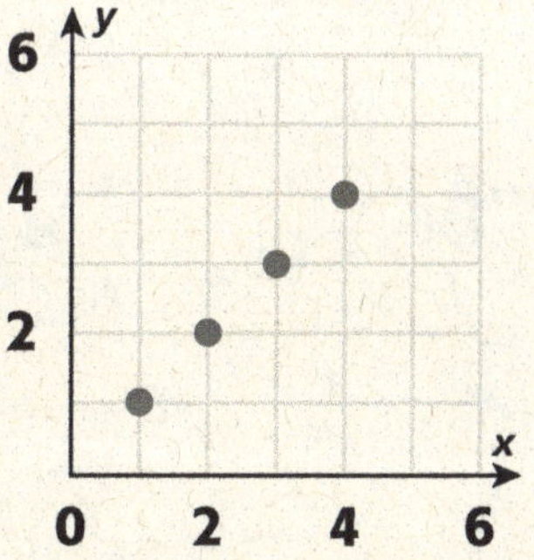

4.

x	y

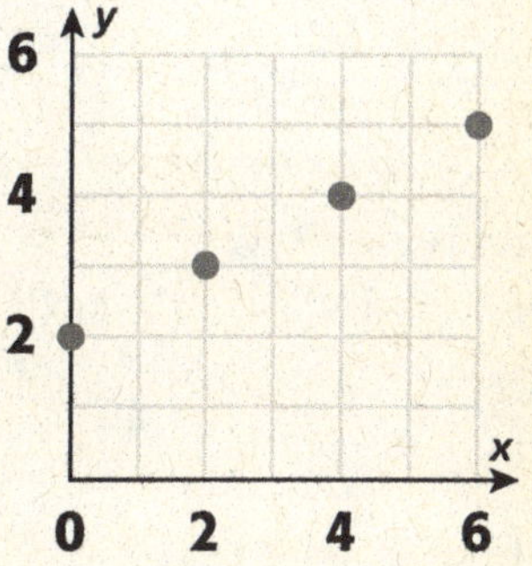

Think and Discuss

5. Explain how to use the points on the graph of a linear equation to write a table of ordered pairs.

Name ______________________________ Date __________ Class __________

LESSON 11C

Technology Lab 1

Graph Points

To graph on your calculator, you must first set the viewing window. To do this, press WINDOW.

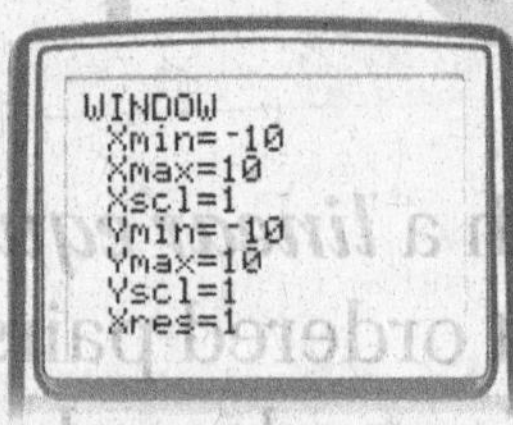

The calculator automatically uses the **standard window** shown at right unless you change the settings. The *x*-values and *y*-values go from −10 to 10. These are set by **Xmin, Xmax, Ymin,** and **Ymax**. **Xscl** and **Yscl** give the distance between tick marks. In the standard window, tick marks are 1 unit apart.

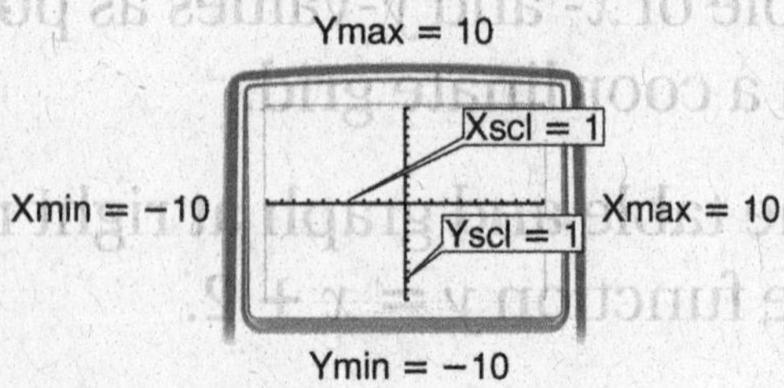

Activity

Graph the points (2, 5), (4, 3), (−4, 6) and (5, −7) on your calculator.

1. Access the **DRAW** menu by pressing 2nd PRGM (DRAW).

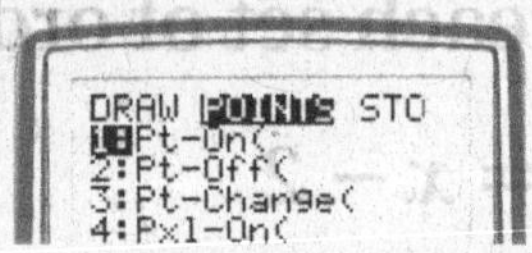

2. Press ▶ to highlight **POINTS**. Make sure 1 is highlighted to select **Pt-On**. Press ENTER

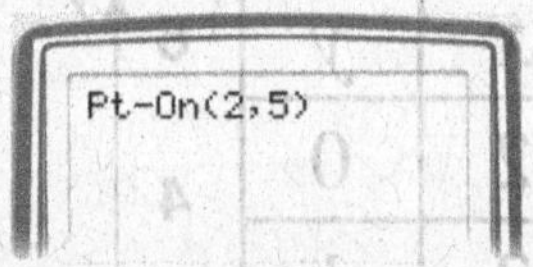

3. The calculator gives you the opening parenthesis for the ordered pair. To plot (2, 5), press 2 , 5) ENTER.

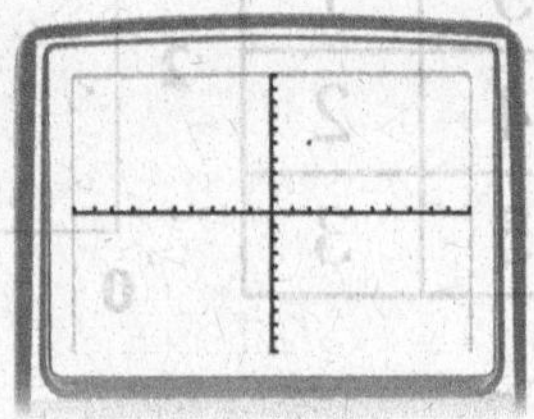

4. To graph the next point, press 2nd MODE (QUIT) to quit the coordinate plane. Then repeat steps 1–3.

Name ________________________________ Date ____________ Class ____________

LESSON 11C

Technology Lab 1

Graph Points, continued

Think and Discuss

1. Suppose you want to graph (9, 16) on a graphing calculator. Should you use the standard window? Why or why not?

____

____

Try This

1. Graph (1, 3), (5, 3), (−5, −2), (3, −6), and (−8, 1) on your calculator.

Name ______________________ Date __________ Class __________

LESSON 11C

Technology Lab 2

Graphing Functions

Japan is where high-speed railways were born. The Shinkansen (Bullet Train) travels at 220 km/h (about 130 mi/h).

Activity

Suppose a Shinkansen travels at 200 km/h. How far does the train travel in one hour? In two hours? In three hours? Make a table that represents the relationship between time and distance.

Time (h)	Distance (km)
1	200
2	400
3	600
4	800
5	1000

Express the relationship between time (t) and distance (d) with a function.

$d = 200t$

You can also make a graph to show this relationship. A graphing calculator or graphing software will draw this graph for you when you enter the function.

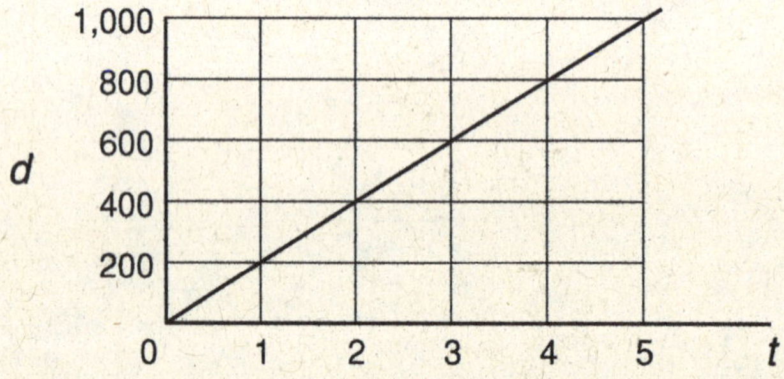

Use your graph to find the distance the train has traveled in one hour and 30 minutes.

The train has traveled 300 km: $d = 200 \cdot \frac{3}{2} = 300$ km

Name ________________________ Date ____________ Class ____________

LESSON **11C**

Technology Lab 2

Graphing Functions

Think and Discuss

1. What if the train leaves the station, and then stops on the tracks after only a few feet. One hour later, the train starts again and travels for 3 hours. How would the function change and how would that change affect the graph in the activity?

____

____

Try This

1. Use your graphing calculator or graphing software to draw the graph for the function: $y = \frac{2}{3}x$.

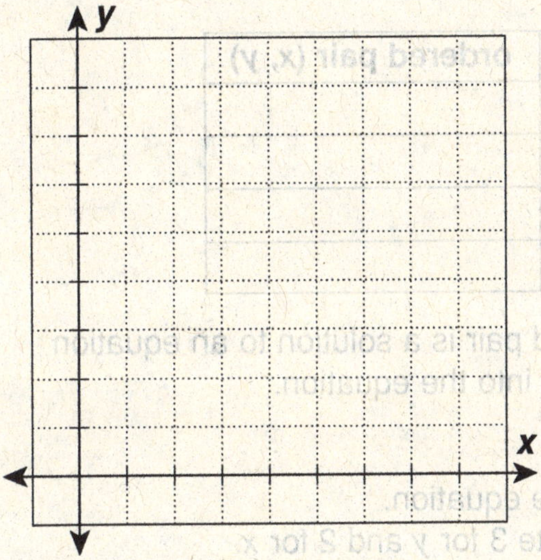

Name ______________________ Date ____________ Class ____________

LESSON 11C

Review for Mastery

Graphing Functions

You can express solutions to equations as ordered pairs.

To express solutions to $y = x - 3$ as ordered pairs, substitute the given values for x in the equation. Then write the solutions as ordered pairs.

x	$x - 3$	y	ordered pair (x, y)
9	9 − 3	6	(9, 6)
7	7 − 3	4	(7, 4)
5	5 − 3	2	(5, 2)
3	3 − 3	0	(3, 0)

Use the given x-values to write solutions to $y = 2x$ as ordered pairs.

1.

x	$2x$	y	ordered pair (x, y)
1			
2			
3			
4			

You can check whether an ordered pair is a solution to an equation by substituting the x- and y-values into the equation.

Is (2, 3) a solution to $y = x + 1$?

$y = x + 1$ — Write the equation.
$3 = 2 + 1$ — Substitute 3 for y and 2 for x.
$3 = 3$ ✓

So (2, 3) is a solution to $y = x + 1$.

Determine whether each ordered pair is a solution to the given equation.

2. (1, 4); $y = 4x$

____

3. (3, 1); $y = 2x - 3$

____

4. (2, 8); $y = 2x + 4$

____

Name ____________________ Date __________ Class __________

LESSON 11C

Homework and Practice

Graphing Functions

Use the given *x*-values to write solutions of each equation as ordered pairs.

1. $y = 9x + 1$ for $x = 0, 1, 2$

2. $y = 3x$ for $x = 2, 5, 8$

Determine whether each ordered pair is a solution of the given equation.

3. (7, 4); $y = x + 3$ ______

4. (2, 30); $y = 14x + 2$ ______

5. (3, 3); $y = 4x + 6$ ______

6. (2, 8); $y = 4x$ ______

Use the graph for exercises 7–11.

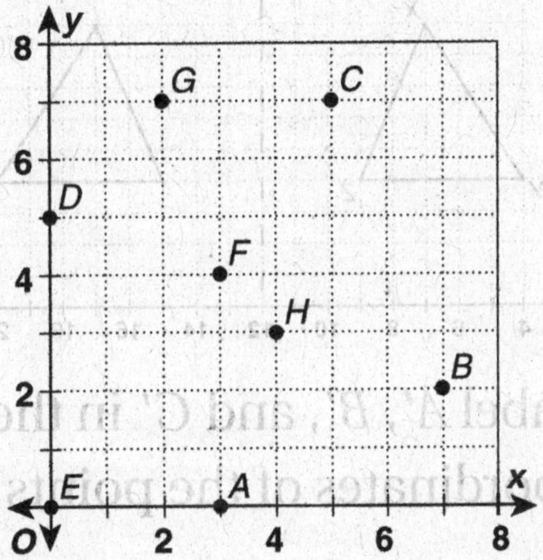

7. Point *D* lies on the ______-axis.

8. Point ____ has coordinates (4, 3).

9. Point *B* has coordinates ________.

10. Point ____ lies on the *x*-axis.

11. The origin is point ____.

Graph the function described by each equation.

12. $y = x + 3$

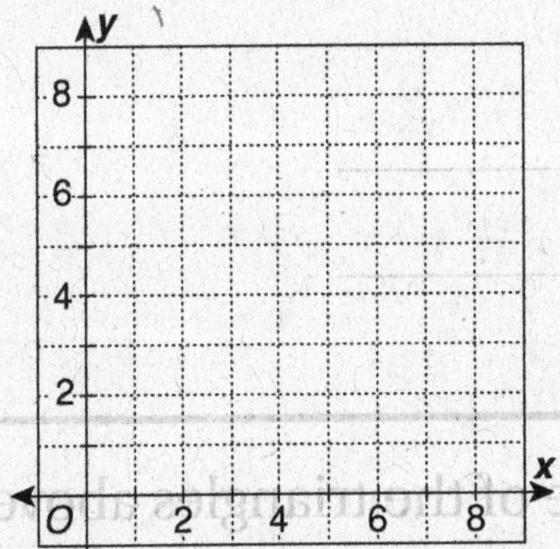

13. $y = 2x$

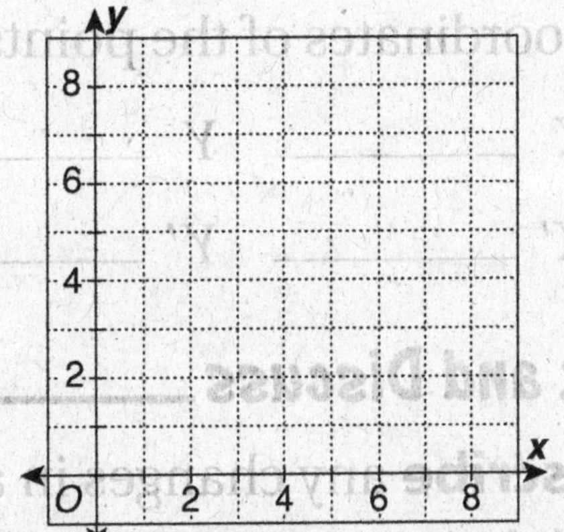

EXPLORATION

Program of Studies and Core Content
MA-6-G-S-TS2, MA-06-3.2.2

11D Reflections and Translations

You can transform figures in the first quadrant of the coordinate plane. The graph below shows a translation and a reflection.

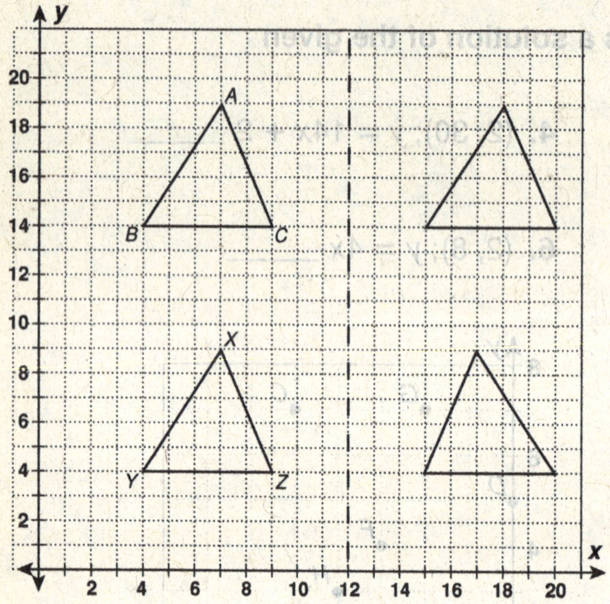

Triangle *ABC* was translated to the right 11 units. The new triangle is the image of triangle *ABC*. It has corresponding coordinates *A′*, *B′*, and *C′*.

Triangle *XYZ* was reflected across the vertical line. The new triangle is the image of triangle *XYZ*. It has corresponding coordinates *X′*, *Y′*, and *Z′*.

1. Label *A′*, *B′*, and *C′* in the graph above. Then write the coordinates of the points below.

A ________ *B* ________ *C* ________

A′ ________ *B′* ________ *C′* ________

2. Label *X′*, *Y′*, and *Z′* in the graph above. Then write the coordinates of the points below.

X ________ *Y* ________ *Z* ________

X′ ________ *Y′* ________ *Z′* ________

Think and Discuss

3. Describe any changes in appearance of the triangles above and their images. Which of the triangles are congruent?

Name ______________________ Date __________ Class __________

LESSON 11D

Review for Mastery

Reflections and Translations

When a figure is reflected across a horizontal line, the *x*-coordinates of the corresponding vertices of the image do not change.

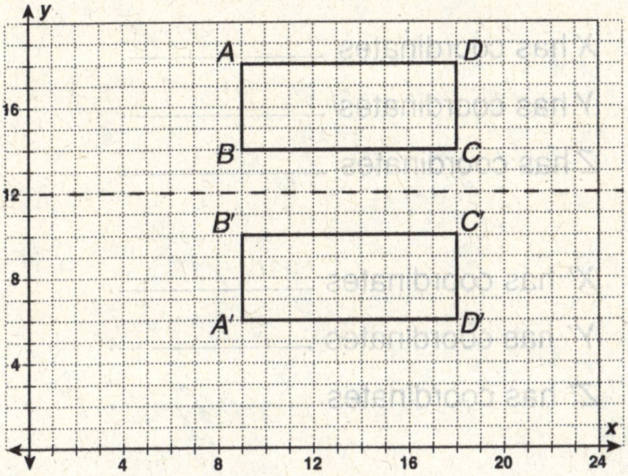

ABCD is reflected across the horizontal line shown.

A has coordinates (9, 18).
A′ has coordinates (9, 6).

B has coordinates (9, 14).
B′ has coordinates (9, 10).

C has coordinates (18, 14).
C′ has coordinates (18, 10).

D has coordinates (18, 18).
D′ has coordinates (18, 6).

When a figure is translated vertically, or slides up or down, the *x*-coordinates of the corresponding vertices of the image do not change.

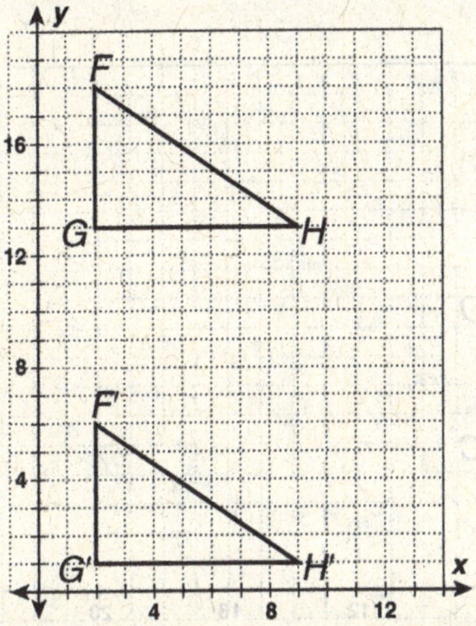

FGH is translated down 12 units.

F has coordinates (2, 18).
F′ has coordinates (2, 6).

G has coordinates (2, 13).
G′ has coordinates (2, 1).

H has coordinates (9, 13).
H′ has coordinates (9, 1).

For each transformation, describe what you can about the change in coordinates for the corresponding vertices of the image.

1. a translation 5 units left

____

2. a reflection over a vertical line

____

3. a translation 3 units right, and 2 units up

____

Name ______________________________ Date ______________ Class ______________

LESSON **11D**

Homework and Practice

Reflections and Translations

Sketch the reflection image of the figure over the line. Label the vertices of the reflection image, and indicate the coordinates of each vertex.

1.

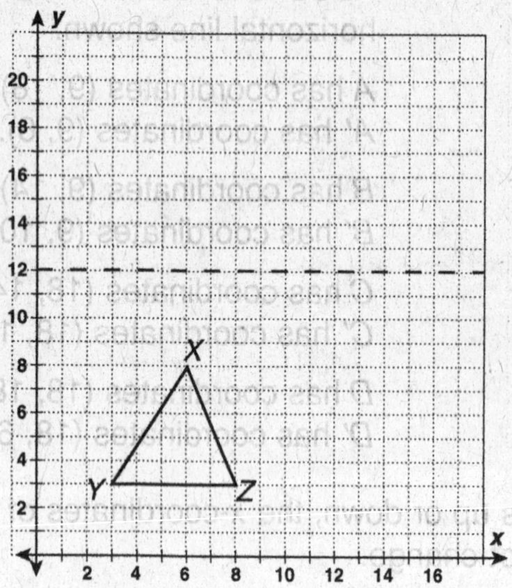

X has coordinates ______________.

Y has coordinates ______________.

Z has coordinates ______________.

X′ has coordinates ______________.

Y′ has coordinates ______________.

Z′ has coordinates ______________.

2. *A* has coordinates ______________.

B has coordinates ______________.

C has coordinates ______________.

D has coordinates ______________.

A′ has coordinates ______________.

B′ has coordinates ______________.

C′ has coordinates ______________.

D′ has coordinates ______________.

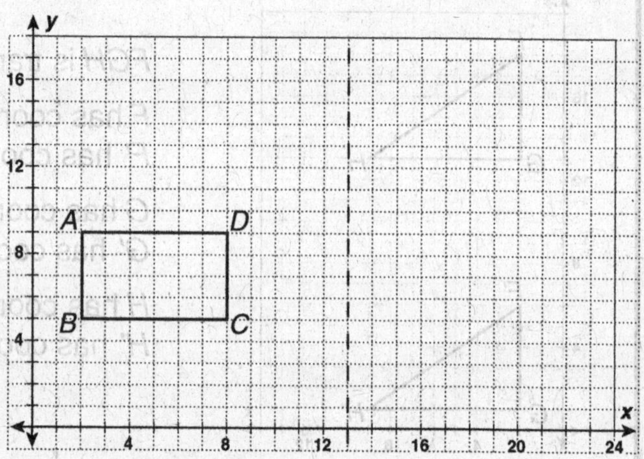

Rectangle *MNPQ* has coordinates *M*(5, 3), *N*(5, 5), *P*(7, 5) and *Q*(7, 3).

3. *MNPQ* is translated 2 units down and 3 units left. Find the coordinates of the vertices of the image *M′N′P′Q′*.

4. *MNPQ* is reflected over the horizontal line at $y = 6$. Find the coordinates of the vertices of the image *M′N′P′Q′*.

EXPLORATION

11E Rotations

Program of Studies and Core Content
MA-6-G-S-TS3, MA-06-3.2.3

You can explore rotations in the coordinate plane with the following activity.

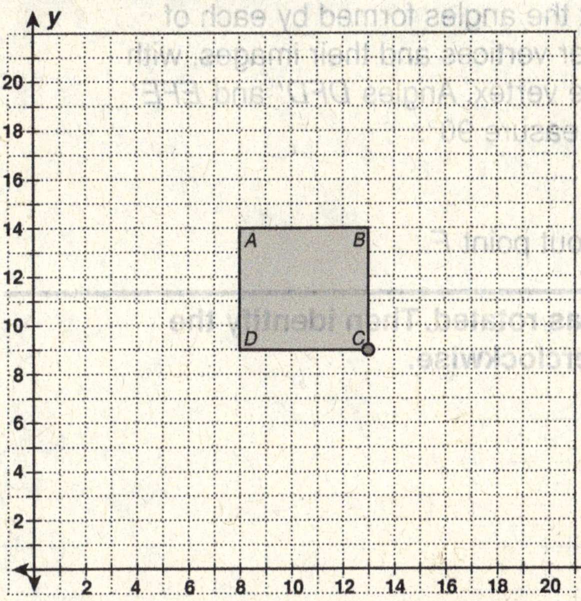

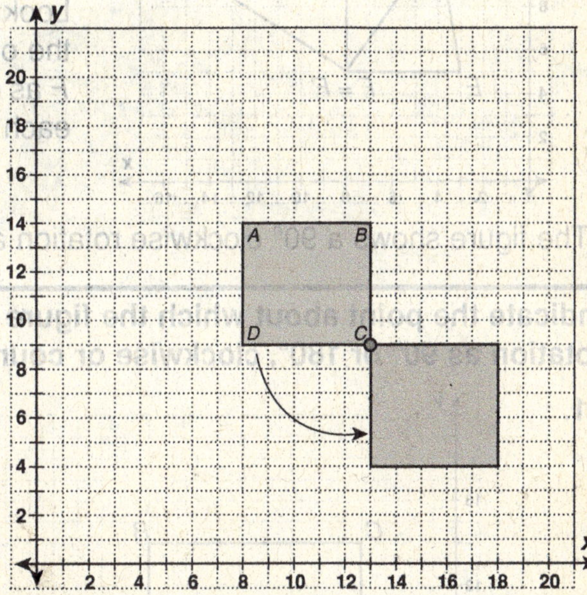

1. Use a pad of graph paper. Cut a piece of construction paper to cover the area shown above (left). Label the corners *A*, *B*, *C*, and *D* as shown.

2. Secure your figure with a tack at point *C*. Rotate the figure as shown above (right) 180° counterclockwise.

3. Rotate the figure an additional 90°. This represents a 270° counterclockwise rotation.

4. Suppose you had originally rotated the figure 90° clockwise. How would it compare to the other rotations?

Think and Discuss

5. **Discuss** the change of coordinates of the vertices after each rotation with classmates. Which point never moved?

Name ______________________________ Date ____________ Class ____________

LESSON **11E**

Review for Mastery

Rotations

You can identify a rotation by comparing the coordinates of the vertices of the original figure and its image.

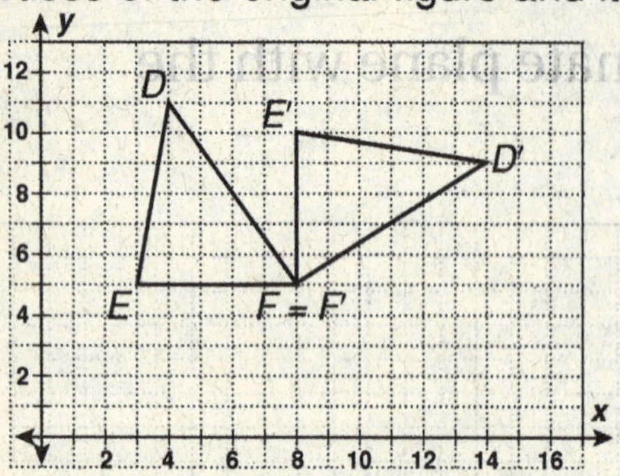

$F = F'$, so triangle DEF was rotated about point F.

Look at the angles formed by each of the other vertices and their images, with F as the vertex. Angles DFD' and EFE' each measure 90°.

The figure shows a 90º clockwise rotation about point *F*.

Indicate the point about which the figure was rotated. Then identify the rotation as 90° or 180°, clockwise or counterclockwise.

1.

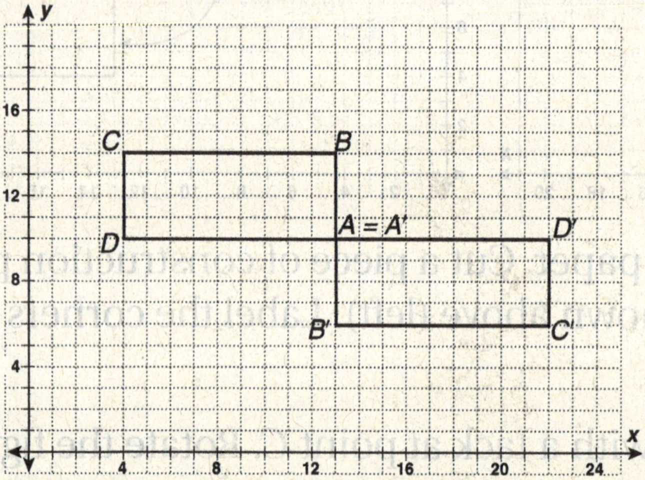

2.

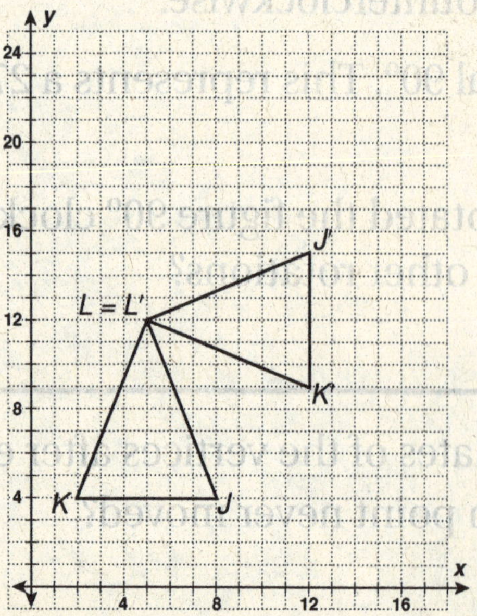

Name ______________________________ Date ____________ Class ____________

LESSON 11E

Homework and Practice

Rotations

Indicate the point about which the figure was rotated. Then identify the rotation as 90° or 180°, clockwise or counterclockwise.

1.

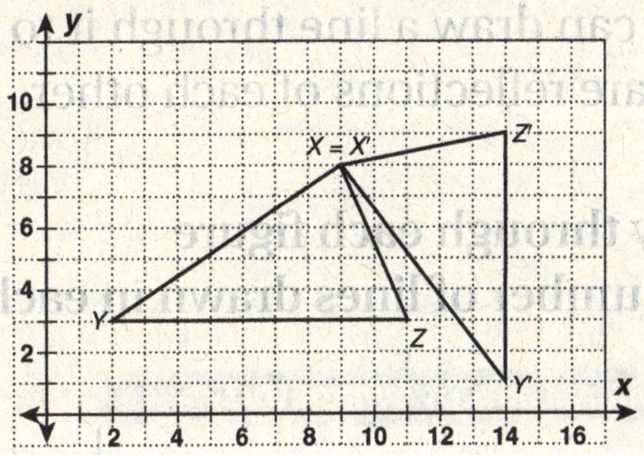

____

____

2.

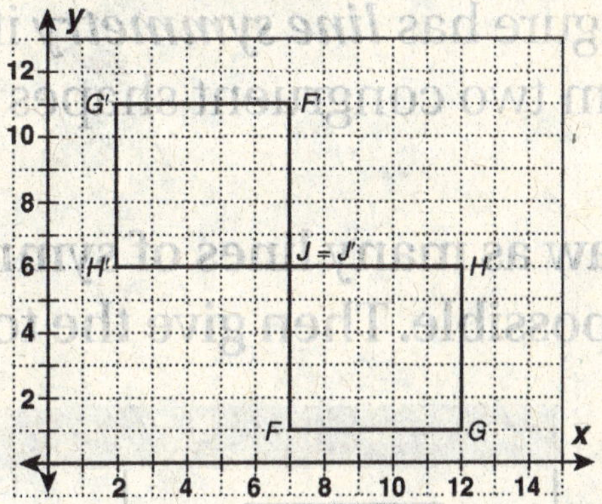

____

____

3.

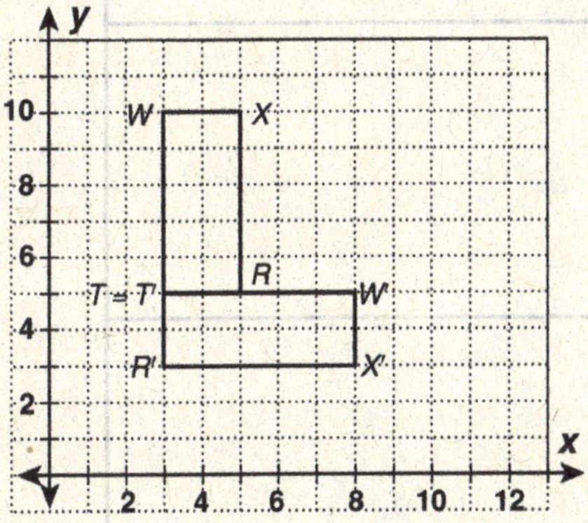

____

____

4.

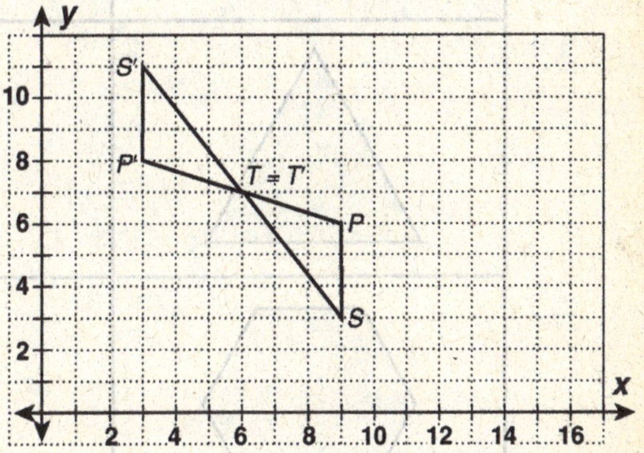

____

____

Triangle *MNP* has coordinates *M*(5, 6), *N*(5, 7), and *P*(7, 6). Use a graph to find the indicated coordinates.

5. *MNP* is rotated 180º clockwise about point *P*. Find the coordinates of the vertices of the image *M′N′P′*.

6. *MNP* is rotated 90° counterclockwise about point *M*. Find the coordinates of the vertices of the image *M′N′P′*.

7. *MNP* is rotated 180° counterclockwise about point *M*. Find the coordinates of the vertices of the image *M′N′P′*.

8. *MNP* is rotated 90° clockwise about point *N*. Find the coordinates of the vertices of the image *M′N′P′*.

EXPLORATION

Program of Studies and Core Content
MA-6-G-S-TS1,
MA-06-3.2.1

11F Line Symmetry

A figure has ***line symmetry*** if you can draw a line through it to form two congruent shapes that are reflections of each other.

Draw as many lines of symmetry through each figure as possible. Then give the total number of lines drawn in each.

		How Many Lines of Symmetry?
1.		
2.	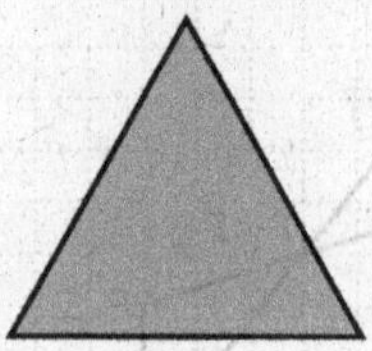	
3.	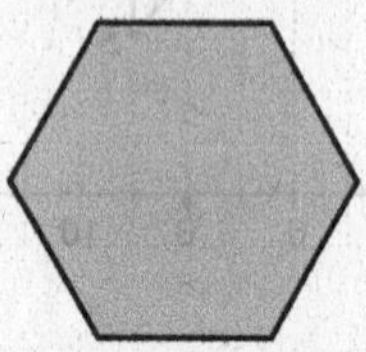	
4.	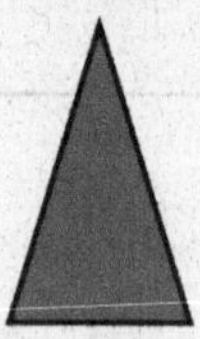	

Think and Discuss

5. Discuss the characteristics of the figures that have more than one line of symmetry.

6. Discuss the characteristics of the figures that have only one line of symmetry.

Name ______________________________ Date ____________ Class ____________

LESSON 11F

Review for Mastery

Line Symmetry

A figure has line symmetry if it can be folded or reflected so that the two overlapping parts of the figure are congruent. The line of reflection is called the line of symmetry.

To figure out if a dashed line is a line of symmetry, test to see if the two parts match exactly when folded or reflected across the line.

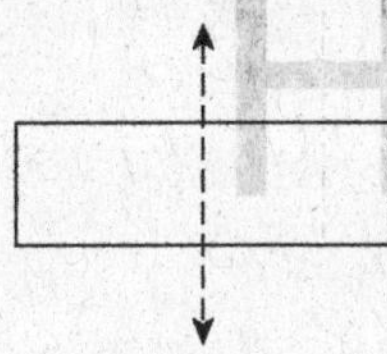

The two parts match, so the line is a line of symmetry.

Decide if each line is a line of symmetry.

1.

____

2.

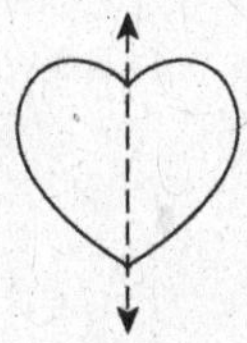

____

To find all of the lines of symmetry in the regular octagon, first trace the polygon and cut it out.

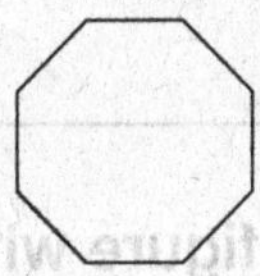

Then fold the polygon in half in different ways to find the lines of symmetry.

A regular octagon has 8 lines of symmetry.

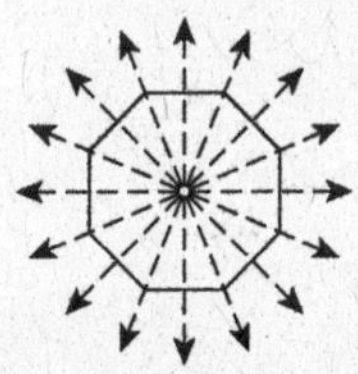

Find all of the lines of symmetry in each figure. Write how many lines of symmetry each figure has.

3.

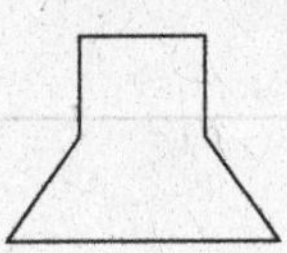

____

4.

____

Name ______________________ Date ____________ Class ____________

LESSON 11F

Homework and Practice

Line Symmetry

Determine whether each dashed line appears to be a line of symmetry.

1. 
____

2. 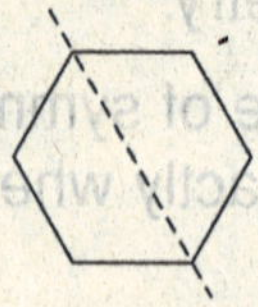
____

3. 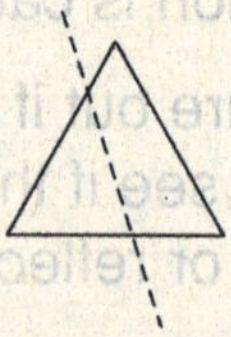
____

Find all the lines of symmetry in each regular polygon.

4. 
____

5. 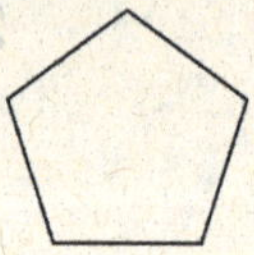
____

6. 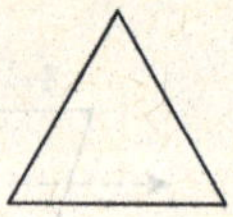
____

Sketch a plane figure with the indicated number of lines of symmetry.

7. two

8. one

9. zero

10. Which has more lines of symmetry, a regular pentagon or a square?

____

11. Of the letters A–L, which have lines of symmetry?

____

CHAPTER 11

Chapter Summary

Answer these questions to summarize the important concepts from Chapter 11 in your own words.

1. Explain why the coordinates of a point on the plane are called an ordered pair.

2. Explain how to graph the point (7, 9).

3. Explain how to graph $y = x - 7$.

4. How can you tell if a transformation is a reflection or a translation?

5. Explain how to rotate a figure 90° clockwise.

6. What is line symmetry?

For more review of Chapter 11:

- Complete the Chapter 11 Study Guide and Review on pages 656–658 of your textbook.
- Complete the Ready to Go On quizzes on pages 614, 632 and 650 of your textbook.

CHAPTER 11

KCCT Prep

Multiple Choice

1. Which point is located farthest from the *y*-axis?

Ⓐ (2, 7) Ⓒ (5, 0)
Ⓑ (3, 6) Ⓓ (6, 4)

2. Which table matches the function $y = 2x + 3$?

Ⓐ

x	1	3	5	7
y	5	7	9	11

Ⓑ

x	2	4	6	8
y	7	11	15	19

Ⓒ

x	1	2	3	4
y	4	5	6	7

Ⓓ

x	3	4	5	9
y	9	11	15	25

3. A rectangle has a perimeter of 18.6 m and a width of 4.1 m. What is the length of the rectangle?

Ⓐ 2.05 m Ⓒ 7.25 m
Ⓑ 5.2 m Ⓓ 14.5 m

4. Which fraction is closest to 0.6?

Ⓐ $\frac{3}{8}$ Ⓒ $\frac{3}{4}$
Ⓑ $\frac{9}{16}$ Ⓓ $\frac{15}{16}$

5. In the expression $(\square + 4) \times 3 + 1^3$, the missing number is a whole number. Which of these CANNOT be the value of the expression?

Ⓐ 13 Ⓒ 22
Ⓑ 19 Ⓓ 26

6. Which is the best estimate of $ab + 1$ when $a = 3.2$ and $b = 3.9$?

Ⓐ 10 Ⓒ 13
Ⓑ 12 Ⓓ 17

7. Which equation describes the function shown in the table?

x	1	2	3	5
y	4	7	10	16

Ⓐ $y = 4x$ Ⓒ $y = 3x$
Ⓑ $y = 4x - 1$ Ⓓ $3x + 1$

8. Given that $a > 0$ and $b > 0$, which point lies farthest from the *y*-axis?

Ⓐ $(a, 0)$ Ⓒ $(a + b, a)$
Ⓑ $(b, 0)$ Ⓓ $(0, a + b)$

9. Marianne earns money babysitting. When she babysits 3 hours, she earns \$21; for 5 hours, she earns \$35; and for 6 hours, she earns \$42. Suppose *h* is the number of hours Marianne babysits and *m* is the amount she earns. Which equation describes the function?

Ⓐ $m = 3h$ Ⓒ $m = 7h$
Ⓑ $m = h + 18$ Ⓓ $m = \frac{h}{7}$

10. Which is the best estimate of the area of the triangle?

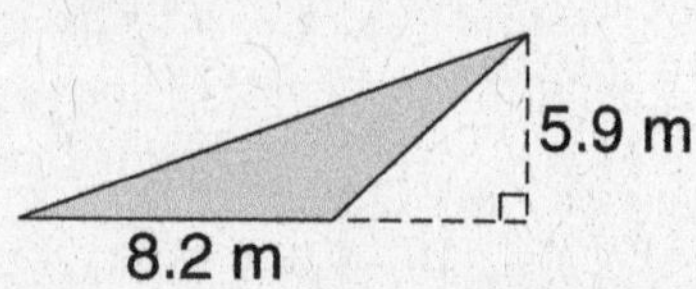

Ⓐ 7 m^2 Ⓒ 24 m^2
Ⓑ 14 m^2 Ⓓ 48 m^2

11. Carmen makes a table for the function $y = 2x + 3$. She uses whole numbers for the x-values. Which of the following does NOT appear as one of the y-values.

Ⓐ 9 Ⓒ 14
Ⓑ 11 Ⓓ 21

12. A vending machine contains 160 juice drinks. Of the 160 drinks, $\frac{2}{5}$ are orange flavored. How many drinks in the vending machine are NOT orange-flavored?

Ⓐ 32 Ⓒ 72
Ⓑ 64 Ⓓ 96

13. Which of these is closest to 9.6% of 51?

Ⓐ 0.5 Ⓒ 10
Ⓑ 5 Ⓓ 50

14. Each square of the grid represents 1 cm^2. What is the area of the shaded figure?

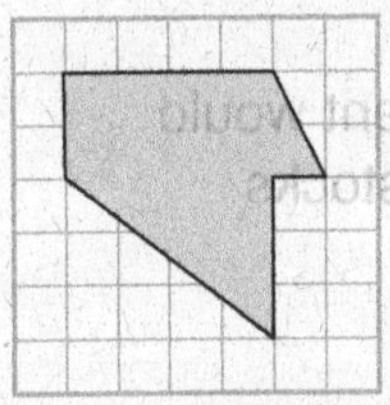

Ⓐ 14 cm^2 Ⓒ 21 cm^2
Ⓑ 15 cm^2 Ⓓ 22 cm^2

15. What is the missing value in the table?

x	1	3	4	5
y	6	■	21	26

Ⓐ 8 Ⓒ 15
Ⓑ 11 Ⓓ 16

16. How many lines of symmetry does a square have?

Ⓐ 1 Ⓒ 4
Ⓑ 2 Ⓓ 8

Open Response

17. Denzel is paid a weekly salary, plus an extra amount for any overtime hours he works. The table shows the total amount he earns for different numbers of overtime hours.

Overtime Hours	0	1	2	3
Total Earnings	\$500	\$520	■	\$560

a. What is the missing value in the table?

b. Suppose Denzel works h hours of overtime. Write an equation that gives his total earnings t.

c. What is the total amount Denzel earns if he works 5.6 hours of overtime? Explain.

18. Jenna plots the following points on a coordinate plane: (2, 1), (6, 1), (6, 3), (2, 3).

a. Jenna connects the four points. What type of quadrilateral does she form?

b. The figure is translated 2 units right and 1 unit up. What are the coordinates of the translated image?

c. The figure is rotated 180° clockwise about the point (6, 3). What are the coordinates of the rotated image?

CHAPTER 12

Project

Probability: Financial Planning

Career: Financial Advisor

We all must decide how much money to spend and how much to invest and save for the future. Financial advisors help people make these decisions. Financial advisors must understand the relationship between risk and earnings. An investment with a high probability of returning a profit is less risky than an investment with a lower probability of returning a profit. However, riskier investments may return larger profits.

Before you start the project, work through the Financial Planning worksheet on the next page.

In this chapter project . . .

You make a "virtual" investment and track your losses and earnings over time.

What to do:

- Learn about interest and how it is calculated.
- Start with an investment of $1,000 and choose stocks or investment funds for your virtual investment. Research various stocks and funds and decide how risky you would like your investment to be.
- After several months, calculate how much money your investment would be worth based on the real-world performance of the funds or stocks you chose.

What to present:

- A data display showing how your investments performed over several months.
- The earnings from your stocks or investment funds. Compare your earnings with your classmates'.

Where's the math?

- Calculate the value of your investment, and the interest earned.
- Discuss how probability relates to investment opportunities.

go.hrw.com
Chapter Project Online
KEYWORD: MR7 Ch12

Name ______________________ Date __________ Class __________

CHAPTER 12

Project Recording Sheet

Probability

Financial Planning

When you save money in a bank, you are actually lending your money to the bank. When you finance a purchase with a loan or a credit card, you are borrowing money. In either case, whenever a loan is made, the borrower has to pay a price, called interest, to the lender, for the opportunity to borrow the money.

Saving or borrowing money involves:

- Principal: The amount of money being borrowed or lent
- Interest: The price paid by the borrower to the lender

Interest Earned
Interest can be compounded at various times. The chart on p. 664 of your textbook shows money compounded annually.

Interest Paid
A credit card company charges interest compounded daily.

You purchase a new stereo for $200. Your finance charge is 19.98% per year. The minimum payment required is $5.00.

Use the following conditions to complete the chart to see how much you actually pay for a $200 stereo.
1. You pay only the minimum payment required.
2. You pay $25 a month.

Monthly Statement

Balance	Finance Charge	Minimum Payment	Your Payment	New Balance
$200.00				

Tip: Use your calculator. Change the percentage to a decimal. To find the monthly finance charge, multiply the balance × 0.1998 ÷12.

What did you learn about interest rates and minimum payments?

____

____

Extension: Research some "safe" investments and some "risky" investments. Watch the value of these investments for several weeks. Which type of investment do you think is better to make? Why?

EXPLORATION

Program of Studies and Core Content
MA-6-DAP-S-P5

12A Introduction to Probability

Probability describes how likely it is that an event will occur. For example, it is likely that a dog will eat a treat, and it is unlikely that a human being will live 200 years.

Tell whether each event is likely or unlikely to happen.

	Event	Probability
1.	Having a blackout during a thunderstorm	
2.	Losing your money in an old vending machine	
3.	Winning the lottery	
4.	Finding a $100 bill on the street	
5.	Passing a test for which you studied very hard	
6.	Being in a traffic jam at 5:00 P.M. on Monday	
7.	Hearing your favorite song when you first turn on the radio	

Think and Discuss

8. Describe an event that is impossible.

9. Describe an event that is certain to happen.

Name ______________________________ Date __________ Class __________

LESSON **12A**

Review for Mastery

Introduction to Probability

You can use what you know about fractions, decimals, and percents to compare probabilities.

When you roll two standard number cubes, the chance that you roll a sum of 2 is $\frac{1}{36}$. The chance that you roll a sum of 7 is $\frac{1}{6}$.

To figure out which event is more likely to occur, compare the probabilities of the events.

First make sure the fractions have the same denominators.

$\frac{1}{36} = \frac{1}{36}$

$\frac{1}{6} = \frac{6}{36}$ Find an equivalent fraction.

$\frac{6}{36} > \frac{1}{36}$

So rolling a sum of 7 is more likely to occur than rolling a sum of 2.

Compare the probabilities to decide which event is more likely to occur.

1. The probability that a month of the year ends with R is $\frac{1}{3}$. The probability that a month of the year ends with E is $\frac{1}{12}$. Is it more likely that a month of the year will end with R or E?

____

2. When you spin a spinner, there is a 20% chance of it landing on 1, a 30% chance of it landing on 2, and a 50% chance of it landing on 3. Is it more likely that the spinner will land on 1 or 2?

____

3. When you pick a marble from a bag, the chance that the marble is blue is 0.3, the chance that is it purple is 0.4, and the chance that it is red is 0.3. Is it more likely that the marble you pick will be red or blue?

____

Name ______________________________ Date ____________ Class ____________

LESSON **12A**

Homework and Practice

Introduction to Probability

Write impossible, unlikely, as likely as not, likely, or certain to describe each event

1. landing on blue

____

2. landing on orange

____

3. landing on red

____

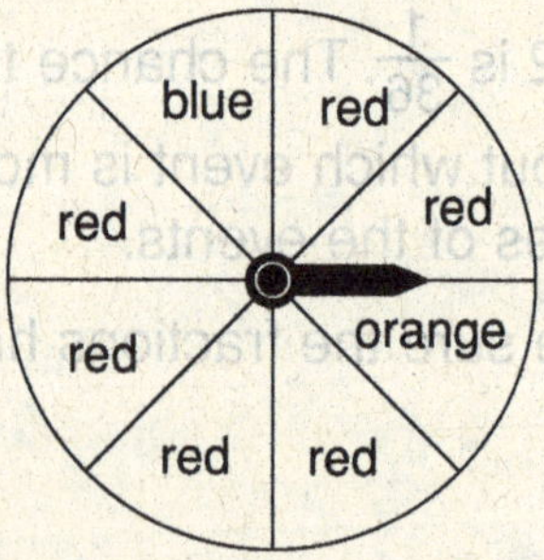

4. landing on blue, red, or orange

____

5. You will spin the spinner clockwise.

____

Write each probability as a decimal and as a fraction.

6. There is a 65% chance of rain tomorrow. _______________________

7. There is a 5% chance of snow tomorrow. _______________________

8. There is a 50% chance of hail tomorrow. _______________________

Compare probabilities.

9. Are you more likely to win a DVD player or a stereo?

____

10. Are you more likely to win a diamond ring, a watch, or a stereo?

____

11. Which item are you most likely to receive? Which are you least likely to receive?

____

Prize Winning Probabilities	
Color TV	28%
DVD player	15%
Watch	35%
Stereo	16%
Diamond ring	6%

12. A spinner has 8 equal sections marked with the numbers 1 through 8. Are you more likely to land on an odd or even number?

____

EXPLORATION

Program of Studies and Core Content

MA-6-DAP-S-P4, MA-06-4.4.2

12B Experimental Probability

You can find the ***experimental probability*** of an event by dividing the number of times an event occurs by the total number of times the experiment is performed.

$$\text{probability} = \frac{\text{number of times an event occurs}}{\text{total number of trials}}$$

1. The data in the table show the number of free throws five players made in a season. Find the $\frac{\text{made}}{\text{attempts}}$ ratio for each player.

	Bo	Jack	Ali	Kim	José
Free Throws Made	30	32	15	36	24
Attempts	48	64	25	48	49
$\frac{\text{Made}}{\text{Attempts}}$					

2. Which player has the best chance of making a free throw?

3. Which player has the worst chance of making a free throw?

Think and Discuss

4. Discuss how you determined the answers for numbers 2 and 3.

5. Explain how to write each $\frac{\text{made}}{\text{attempts}}$ ratio as a percent.

Name ______________________ Date __________ Class __________

LESSON 12B

Technology Lab

Experimental Probability

Probability is the likelihood that an event will occur.

Activity

There are three boxes containing colored tiles. One box contains two green tiles; one box contains two red tiles; the third box contains one red tile and one green tile. The boxes are shuffled, and a student picks one tile out of one box; the tile is green. How often do you expect a second chosen tile also to be green? Use the table below to record experimental data after repeating the experiment 50 times.

Total # of Experiments	# of Times the First Tile is Green	# of Times Both Tiles are Green	Experimental Probability of the First Tile Being Green	Experimental Probability of Both Tiles Being Green	Experimental Probability of Both Tiles Being Green, Given That the First Tile is Green
50	23	16	$\frac{23}{50} = 0.46$	$\frac{16}{50} = 0.32$	$\frac{16}{23} \approx 0.7$

Formula to Calculate Experimental Probability:

Experimental Probability (*EP*) = (Number of cases when the event occurs)/(Total number of cases)

The experiment is repeated 50 times; the first tile was green 23 times. Using your calculator, perform the following function:

23 [÷] 50 [ENTER] 0.46

The number of times the first 2 tiles were green was 16 out of the 50 trials. Using your calculator, perform the following function:

16 [÷] 50 [ENTER] 0.32

The number of times both tiles were green after the first one was green was 16. Using your calculator, perform the following function:

16 [÷] 23 [ENTER] 0.6956521739

Name ______________________________ Date ____________ Class ____________

LESSON 12B

Technology Lab

Experimental Probability, continued

Think and Discuss

1. What would you expect the probability of drawing one green tile to be? Why?

____

____

____

____

Try This

1. Roll a six-sided die twice. Mark down your results in a chart. Repeat this same experiment 50 times. Make a chart of the outcomes like the one in the activity above. Mark down how many times you get an even number on one roll. Mark how many times you get an even number on the second roll. Figure out the experimental probability using your calculator.

Total # of Experiments	# of Times the First Roll is Even	# of Times Both Rolls are Even	Experimental Probability of the First Roll Being Even	Experimental Probability of Both Rolls Being Even	Experimental Probability of Both Rolls Being Even, Given That the First Roll is Even

Name ____________________ Date __________ Class __________

LESSON 12B

Review for Mastery

Experimental Probability

An experiment is an activity involving chance that can have different results. Rolling a standard number cube is an experiment.

The different results that are possible are called the outcomes of an experiment. If you are rolling a standard number cube, 5 is one possible outcome.

Identify the outcome shown.

1. spinning a spinner

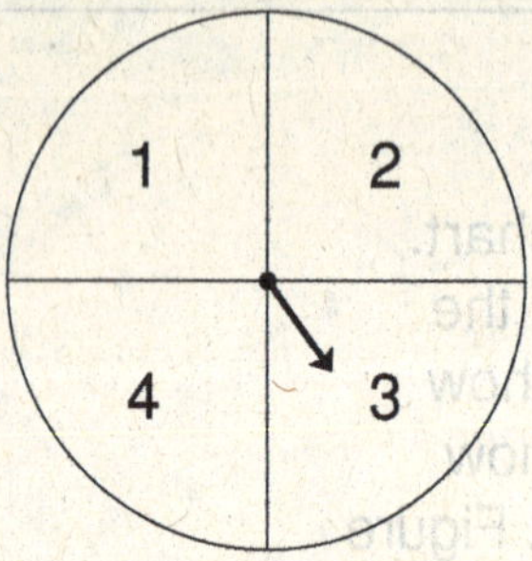

____

2. tossing a coin

____

If you perform an experiment many times, you can estimate the probability of an event.

To find the experimental probability of an event, use a formula.

$$\text{probability} \approx \frac{\text{number of times the event occurs}}{\text{total number of trials}}$$

Suppose you rolled a standard number cube 20 times and it landed on 5 six times. The probability of rolling a 5 is

$$P(\text{rolling } 5) \approx \frac{6}{20} = \frac{3}{10}$$

Find the experimental probability of each event.

Pulling Marbles Out of a Bag

Red	Blue	Green
𝍸 IIII	𝍸 I	𝍸

3. The experimental probability of pulling out a red marble. __________

4. The experimental probability of pulling out a green marble. __________

Name ______________________ Date __________ Class __________

LESSON **12B**

Homework and Practice

Experimental Probability

For each experiment, identify the outcome shown and the sample space.

1.

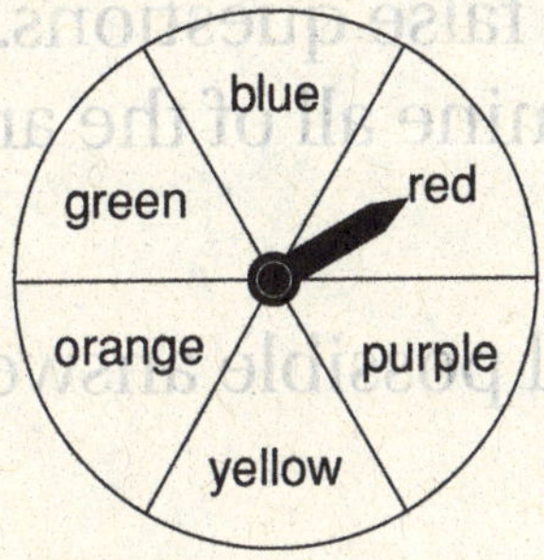

outcome: ______________________

sample space: ______________________

____

2.

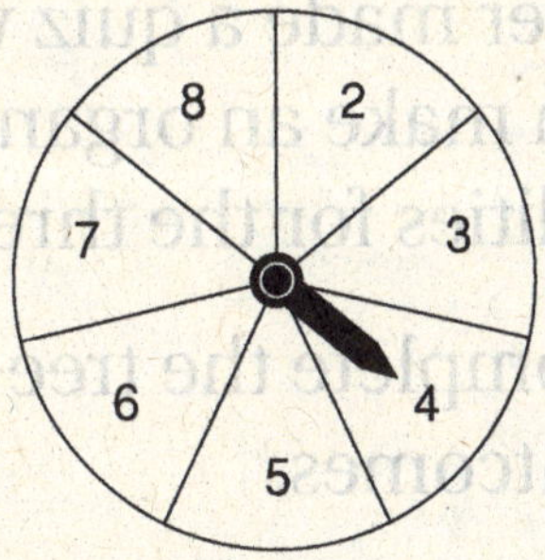

outcome: ______________________

sample space: ______________________

____

Adam has a standard deck of playing cards. He picked one card, recorded the suit, and placed it back in the deck. He repeated this process several times and recorded his results in the table.

Heart	IIII
Diamond	卌 II
Spade	卌 I
Club	III

3. Find the experimental probability that a card selected from the deck will be a club.

____

4. Find the experimental probability that a card selected from the deck will be spade.

____

5. Based on Adam's experiment, which card suit is he most likely to select from the deck?

____

6. Based on Adam's experiment, which card suit is he least likely to select from the deck?

____

7. In 100 coin tosses, John got tails up 50 times. What is the experimental probability that John will get tails up on his next toss?

____

EXPLORATION

Program of Studies and Core Content

MA-6-DAP-S-P1, MA-6-DAP-S-P2, MA-6-DAP-S-P5, MA-06-4.4.1

12C Counting Methods and Sample Spaces

A teacher made a quiz with three true or false questions. You can make an organized list to determine all of the answer possibilities for the three questions.

1. Complete the tree diagram to list all possible answer outcomes.

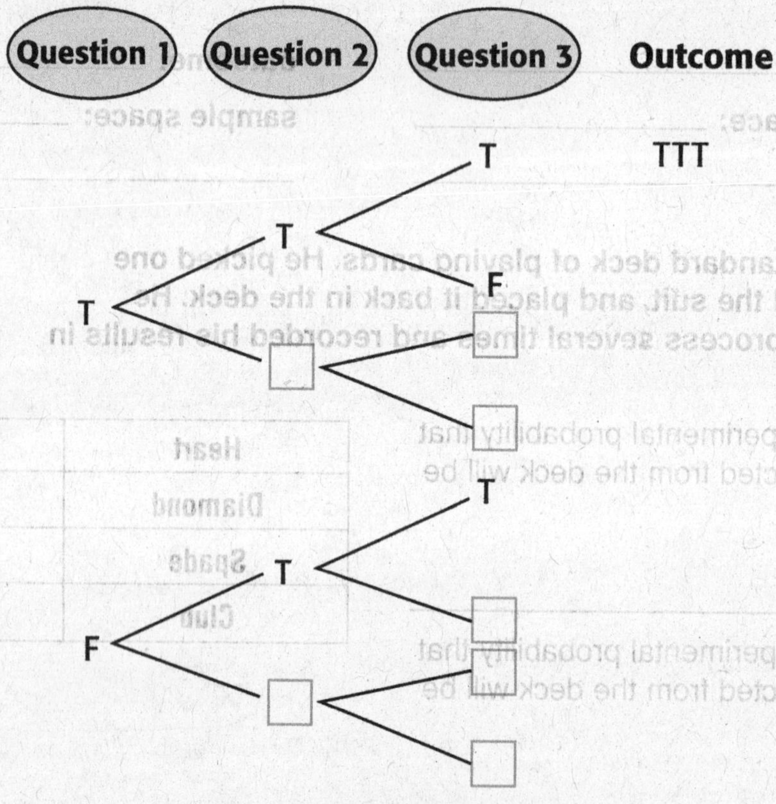

2. How many outcomes are possible?

Think and Discuss

3. Discuss your method for organizing the list of outcomes.

4. Explain how you could determine the number of outcomes possible with four questions.

Name ______________________ Date ____________ Class ____________

LESSON 12C

Hands-On Lab

Market Research

Before a company begins to develop a new product, the company will usually conduct research to determine whether or not the product is likely to be successful. In this lab, you will collect information from your classmates to determine if a new ready-made lunch pack will be a hit.

The **population** you are interested in is all students in your school. However, you probably will not be able to ask everybody in your school about the new ready-made lunch pack. So you must choose a sample to ask. A **sample** is a group of individuals selected from a population. (Students you survey are a sample of all students in your school.) You will also need to be sure that the students you do ask represent the opinions of all the students in your school. So you will need to use a random sample. **Random samples** are samples in which everyone in the population (students at your school) has an equal opportunity of being chosen.

Activity

1. Plan how you will choose your random sample. Write out your plan in the space below.

____

____

____

____

2. Design your new ready-made lunch pack. Try to stay close to some existing product, but with a new twist.

____

____

____

____

____

____

Name ______________________ Date __________ Class __________

LESSON 12C

Hands-On Lab

Market Research, continued

3. Design your survey and collect your information. This must be a question that you can ask which students can answer quickly and easily. The answer must be numerical. In the space below, design your survey.

____

____

____

____

____

____

____

Think and Discuss

1. Examine your data. Can you determine if students would buy the ready-made lunch pack?

____

____

____

2. Construct a bar graph of the responses. Construct a frequency table of the responses. Which data display helps you to best determine if students would buy your ready-made lunch pack? Explain.

____

____

Show your work.

Name ______________________ Date __________ Class __________

LESSON **12C**

Hands-On Lab

Market Research, continued

3. Find the mean of the data. Does the mean help you to determine if students would buy your ready-made lunch pack? Explain.

____

____

____

4. If your sample space was not a random selection of the population, how might this affect your survey results? Provide an example.

____

____

____

Try This

1. A class surveyed their classmates to determine whether or not they liked deep-dish pizza. This was the question that was asked.

Do you enjoy eating deep-dish pizza? Select a number from 1 to 5 where 1 means that you do not enjoy eating deep-dish pizza and 5 means you enjoy eating deep-dish pizza very much.

Do not enjoy		Somewhat enjoy		Enjoy very much
1	2	3	4	5

The table shows the results of the survey

a. Find the mean. __________

b. What does the mean tell you?

____

____

Student Number	Response
1	3
2	5
3	5
4	2
5	3
6	2
7	5
8	4
9	2
10	4
11	4
12	5
13	1
14	3
15	1
Mean	

Name ______________________________ Date ____________ Class ____________

LESSON 12C

Hands-On Lab

Market Research, continued

2. Write a new question that could be asked in a survey regarding deep-dish pizza. What random sample population would you use to conduct your survey?

____

____

____

____

____

____

____

3. Conduct a survey and make a table of your results. Find the mean and write one sentence explaining the results of your survey.

____

Student Number	Result	Student Number	Result

Mean ____________ Results ______________________

Name ______________________ Date __________ Class __________

LESSON **12C**

Review for Mastery

Counting Methods and Sample Spaces

You can make an organized list to help you solve problems.

Look at the sandwich choices posted at the deli.

Bread	Meat
Wheat	Ham
Rye	Turkey
White	

To find the possible sandwiches that can be made with one type of bread and one type of meat, use a tree diagram to make an organized list of the sandwiches.

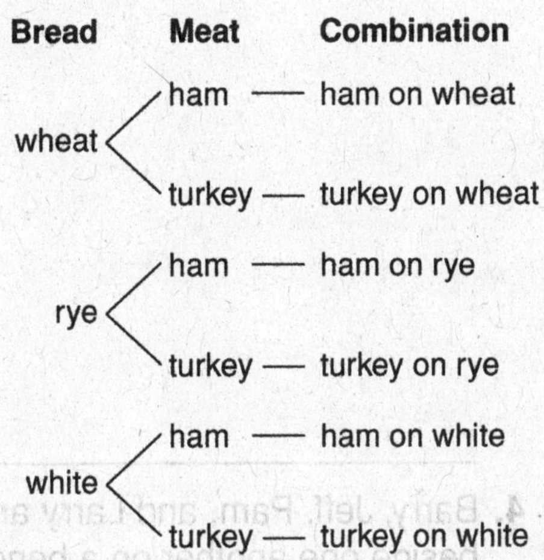

Then follow the branches to find the total number of outcomes.

There are 6 possible sandwiches.

Solve.

1. Erin can choose a waffle cone, sugar cone, or chocolate-dipped cone at the carnival. Her choices for ice cream are chocolate, vanilla, strawberry, or cookies and cream. How many different ice cream cones combinations are available to her?

____

2. Mrs. Baylus is a parent involved in a neighborhood car pool. She can drive on Monday, Wednesday, or Friday, before school or after school. How many different times can Mrs. Baylus drive for the car pool?

____

Name ______________________ Date ____________ Class ____________

LESSON **12C**

Homework and Practice

Counting Methods and Sample Spaces

Make an organized list to answer each question.

1. Jason needs to pick out a shirt for his graduate dinner. He can choose form two styles: long sleeves and short sleeves. In each of the shirts, he has the various colors blue, black, beige, and green. How many different shirts can Jason actually choose?

____

2. Mr. Johnson will attend a conference on one of four days. he can take the bus, drive his own car, or take a taxi cab. how many different ways on four days can he get to the conference?

____

3. Bruce stops at the store and wants to buy a snack. He is undecided as to whether he wants something sweet, something sour, or something chewy. There are two types of sweet candies: strawberry and cherry. There are three types of sour candies: grape, orange, and raspberry. There are two types of chewy candies: lemon, blueberry. How many different choices does Bruce have to choose from?

____

4. Barry, Jeff, Pam, and Larry are sitting beside one another on a bench. In how many different orders could they possibly be sitting from left to right, with Larry always sitting on the far left?

____

EXPLORATION

Program of Studies and Core Content
MA-6-DAP-S-P5,
MA-06-4.4.3

12D Theoretical Probability

When you flip a fair coin, the ***theoretical probability*** of getting tails is 50% and the theoretical probability of getting heads is 50%. These two outcomes are equally likely to occur.

Determine whether the outcomes in each experiment are all equally likely to occur.

	Equally Likely	Not Equally Likely
1. 		
2. 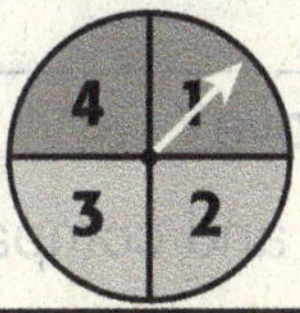		
3. 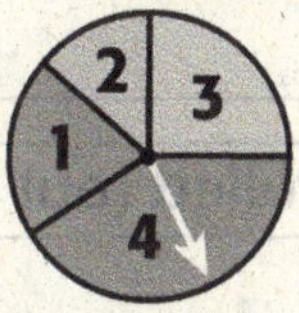		

Think and Discuss

4. Explain how you determined whether the outcomes in numbers 1–3 were equally likely or not.

Name ______________________________ Date __________ Class __________

LESSON **12D**

Hands-On Lab

Testing Theoretical Probability

You can test the relation between experimental and theoretical probability. Experimental probability is the chance that an event will occur, based on past experience.

Materials: number cubes, spinners, coins

Activity 1

Do this activity in groups of four.

1. First, as a group, calculate the theoretical probability of rolling a single number cube and rolling a 3. How many possible outcomes are there? __________

P(rolling a 3) (expressed as a fraction) = ____________

Express this result as a percent: ____________

2. As individuals, do an experiment to test this theoretical probability. Each student rolls a number cube 6 times and records the number of times a 3 is faceup. Record your experiment results.

	Number of times I rolled a 3	Total number of rolls
Me		

How close were you to the theoretical probability you calculated above? __________

Now enter the results of all the members of your group in this table and compare the results. How close were these results to the theoretical probability?

____

	Number of times a 3 was rolled (out of 6)
Me	
Student 2	
Student 3	
Student 4	
TOTAL	

3. Add up the results for all the students in your group. You will have a result out of 24 rolls. Express this result as a percent. ________________

How close is it to the theoretical probability? ______________

Name ______________________________ Date __________ Class __________

LESSON 12D

Hands-On Lab

Testing Theoretical Probability, continued

4. Each student now rolls the number cube 12 times. Record the number of times each student rolls a 3 in this table. Add up the results. Then express the final ratio as a percent.

	Number of times a 3 was rolled (out of 12)
Me	
Student 2	
Student 3	
Student 4	
TOTAL	
PERCENT	

Were the results of this larger experiment closer to the theoretical probability you calculated above?

____

Think and Discuss

Compare the results from the first individual experiments (6 rolls) to the final group experiment (48 rolls). Why does experimental probability differ from theoretical probability in small experiments?

____

Why is experimental probability close to or the same as theoretical probability in large experiments?

____

Try This

The theoretical probability of a student being a boy or a girl is 1/2 (or 50%). Test this theoretical probability by recording the number of girls and boys in your group, your class, and your school.

	Number of Girls	Number of Boys	Total Number	Express as a Probability Ratio	Express as a Percent
Group					
Class					
School					

Which result is closest to 50%? _______________________________

Name ____________________ Date __________ Class __________

LESSON **12D**

Hands-On Lab

Testing Theoretical Probability, continued

Activity 2

1. In groups of four, each student flips a coin 10 times and counts the number of times the result is *heads*. Record your results. Express the result as a probability ratio.

	Heads	Total Flips	Ratio
Me			
Student 2			
Student 3			
Student 4			
TOTAL			

2. Calculate the *theoretical probability* of flipping *heads*.

P (heads) = (as a fraction) __________ = (as a percent) __________

How close are your results to the theoretical probability? ____________________

Is the total for your group closer to the theoretical probability? ____________________

3. Using the blackboard, list the totals for all the groups in the class. Record them in this table. Add up the results to make a grand total.

	Heads	Total Flips	Ratio
Our Group			
Group 2			
Group 3			
Group 4			
Group 5			
Group 6			
Group 7			
Group 8			
TOTAL			

How many flips are included in the class total? ________ How close are the results to the theoretical probability? ________

Name ______________________________ Date ____________ Class ____________

LESSON **12D**

Hands-On Lab

Testing Theoretical Probability, continued

Think and Discuss

If you could double the number of students in your class and repeat this experiment, do you think the results would be closer or farther from the theoretical probability? Explain.

____

____

Try This

Create a circular spinner for a turtle race with five turtles (Blazer, Whitie, Secret, Spuds, and Tame). Based on the results of the past week's races, you know that Blazer is the fastest turtle (having won twice as many previous races as Whitie and Secret). Spuds and Tame are the slowest turtles (winning only half as often as Whitie and Secret). Create *unequal* sectors in this circular spinner to represent the different track records of each turtle. (Hint: Divide the circle into 10 equal sectors. Give Spuds and Tame one sector each. Give Whitie and Secret double sectors. Give Blazer the last four sectors.)

Given the way you have drawn the spinner, what is the theoretical probability for each turtle to win?

	Number of Sectors	Theoretical Probability of Winning	Experiments
Blazer	4		
Whitie	2		
Secret	2		
Spuds	1		
Tame	1		

Spin the spinner 20 times. See if your predictions match the experimental results.

Name ______________________________ Date ___________ Class ___________

LESSON 12D

Review for Mastery

Theoretical Probability

You can use theoretical probability to estimate the probability of an event.

To find the theoretical probability of an event, first find the number of ways the event can occur. Then divide that number by the total number of possible outcomes.

Think about a standard number cube. To find the theoretical probability of rolling a number greater than 2, find the number of possible outcomes that are greater than 2.

3, 4, 5, and 6 are greater than 2. So there are 4 outcomes that are greater than 2.

There are 6 possible outcomes, so divide 4 by 6.

$$\text{Probability} = \frac{\text{number of ways event can occur}}{\text{total number of possible outcomes}}$$

$$P(\text{rolling a number greater than 2}) = \frac{4}{6} = \frac{2}{3}$$

Find the theoretical probability of each event.

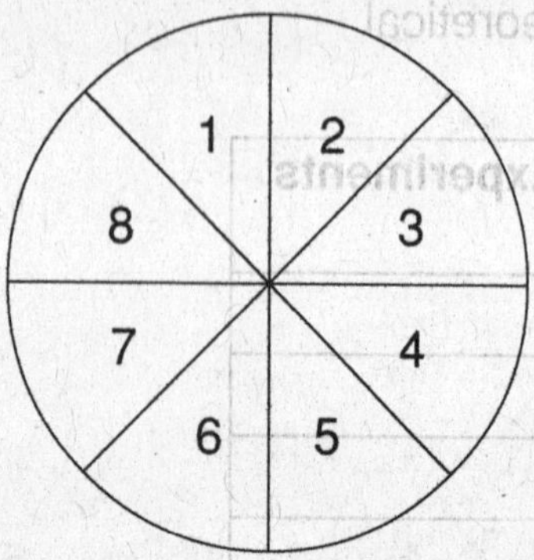

1. *P*(landing on an even number) ___________

2. *P*(landing on a prime number) ___________

3. *P*(landing on a number divisible by 3) ___________

4. *P*(landing on a number with 3 factors) ___________

5. *P*(landing on a number greater than 8) ___________

6. *P*(landing on a number less than 9) ___________

Name ______________________ Date __________ Class __________

LESSON **12D**

Homework and Practice

Theoretical Probability

Find the probability of each event using the spinner.

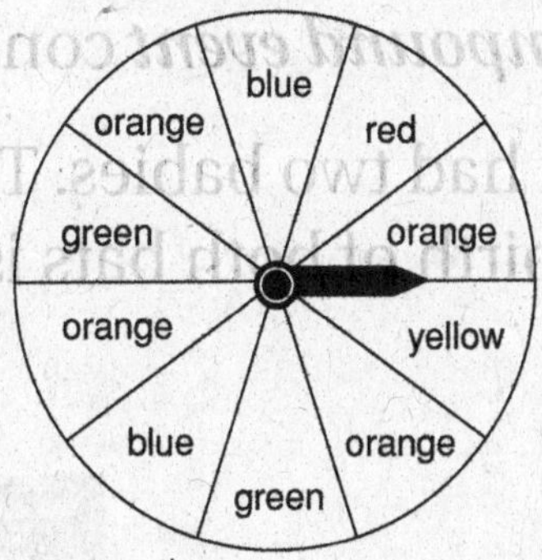

1. landing on blue ______________

2. landing on red, yellow, or orange

____

3. landing on red ______________

4. not landing on yellow ______________

Find the probability of each event using the bag of marbles.

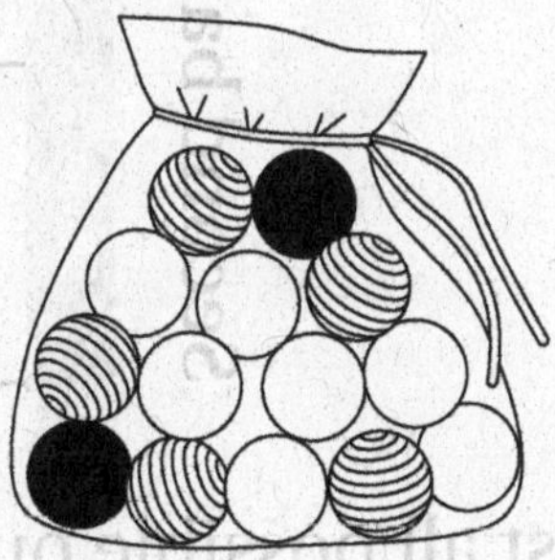

5. picking a black marble ______________

6. picking a striped marble ______________

7. picking a grey marble ______________

8. not picking a white marble ______________

A standard number cube is rolled. Find each probability.

9. *P*(1 or 2 or 3) ______________ **10.** *P*(less than 3) ______________

11. *P*(6) ______________ **12.** *P*(greater than 2) ______________

13. Out of 10 fair coin tosses, a coin landed tails up 9 times. How does this experimental probability of a fair coin landing tails up compare to the theoretical probability of the same event?

____

____

14. The probability of a spinner landing on blue is $\frac{2}{5}$. What is the probability of it not landing on blue, written as a percent?

____

EXPLORATION

Program of Studies and Core Content

MA-6-DAP-S-P1, MA-6-DAP-S-P2, MA-6-DAP-S-P5, MA-06-4.4.1

12E Compound Events

A ***compound event*** consists of two or more single events.

A bat had two babies. The birth of each baby bat is a single event. The birth of both bats is a compound event.

Second bat born \ First bat born	Female	Male
Female	FF	MF
Male	FM	MM

1. List all possible outcomes for two offspring.

2. Divide 1 (FF) by the total number of outcomes to find the probability that both bats are female.

3. Divide 1 (MM) by the total number of outcomes to find the probability that both bats are male.

4. Divide 2 (MF and FM) by the total number of outcomes to find the probability that one bat is female and the other is male.

Think and Discuss

5. Explain how to find the probability that three males would be born among four births.

Name ______________________________ Date ____________ Class ____________

LESSON **12E**

Technology Lab

Compound Events

> You will need access to a computer with Excel software.

Activity

A compound event involves the use of two or more items such as:

- two cards being drawn
- three coins being tossed
- two dice being rolled
- four people being chosen

Since there are six faces on one die, one of which is a 3, there is a 1 in 6 chance of getting a 3 with one roll of one die.

What is the probability of throwing two 3's with two dice? You can calculate the probability on an Excel spreadsheet.

Step 1 Open a new spreadsheet.

Step 2 In cell A1, type the heading "Calculation"; in cell B1, type the heading "Die One"; in cell C1, type "Math"; in cell D1, type "Die Two"; and in cell E1, type the heading "Probability".

Step 3 In cell A2, type "Fractional"; in cell A3, type "Decimal".

Step 4 Go to cell B2, click on "Format" on the toolbar; select "Cells"; select "Fraction", then select "Up to 1 digit", and type "1/6" in the cell; repeat for cell D2, and type "1/6" in the cell; repeat for cell E2, except select "Up to 2 digits" (do not type anything in the cell. This result will be calculated).

Step 5 Go to cells C2 and C3 and type "Multiply" in each cell.

Step 6 Go to cell B3, click on "Format" on the toolbar: select "Cells", then select "Percentage" and select "1 decimal place" and type "16.7%" in the cell; repeat for cell E3, but select "2 decimal places" (do not type anything in the cell. This result will be calculated).

Step 7 In cell E2, select the "=" on the toolbar and type the formula "=(B2*D2); in cell E3, select the "=" on the toolbar and type the formula "=(B3*D3)".

Name ______________________________ Date ____________ Class ____________

LESSON 12E

Technology Lab

Compound Events, continued

The spreadsheet will look like this:

Calculation	*Die One*	*Math*	*Die Two*	*Probability*
Fractional	*1/6*	*Multiply*	*1/6*	*1/36*
Decimal	*16.7%*	*Multiply*	*16.7%*	*2.78%*

The probability of rolling two 3's with two dice is 1 chance in 36, or about 2.78%. You could use the same format for other probabilities with dice.

Think and Discuss

1. A die is rolled. What is the probability that the number rolled is greater than 2 and even? What are the steps in arriving at the answer?

____

____

____

____

____

Try This

1. A student has 25 T-shirts, 13 pairs of jeans and 4 pairs of tennis shoes. How many different outfits can be made of these items, each consisting of 1 T-shirt, 1 pair of jeans and one pair of tennis shoes? Use the spreadsheet to calculate your answer.

____

____

Name ______________________________ Date ___________ Class ___________

LESSON **12E**

Review for Mastery

Compound Events

A compound event consists of two or more single events.

To find the probability of landing on a prime number and black, first make a list of all the possibilities.

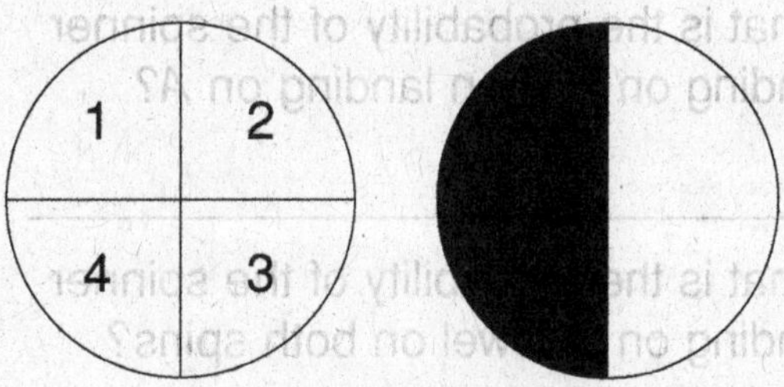

There are 8 possible outcomes.

Number Spinner (rows: **Color Spinner**)

	1	**2**	**3**	**4**
Black	1, B	2, B	3, B	4, B
White	1, W	2, W	3, W	4, W

Next, find the number of outcomes that have a prime number and black.

There are two outcomes that have a prime number and black.

Number Spinner (rows: **Color Spinner**)

	1	**2**	**3**	**4**
Black	1, B	(2, B)	(3, B)	4, B
White	1, W	2, W	3, W	4, W

Then find the probability.

$$P(\text{prime number, black}) = \frac{\text{2 ways the event can occur}}{\text{8 possible outcomes}}$$

$$= \frac{2}{8} = \frac{1}{4}$$

Use the spinners to find the probability of each compound event.

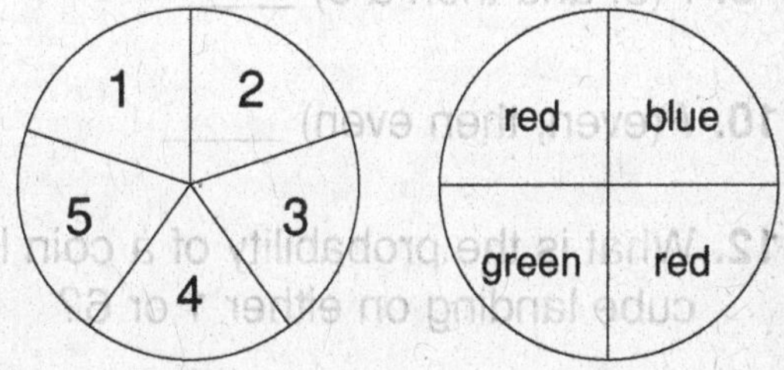

1. *P*(2, red) __________

2. *P*(odd number, blue) __________

3. *P*(even number, red) __________

4. *P*(number greater than 4, green) __________

5. *P*(number less than 4, red) __________

Name ________________________ Date ___________ Class ___________

LESSON **12E**

Homework and Practice

Compound Events

The spinner is spun twice.

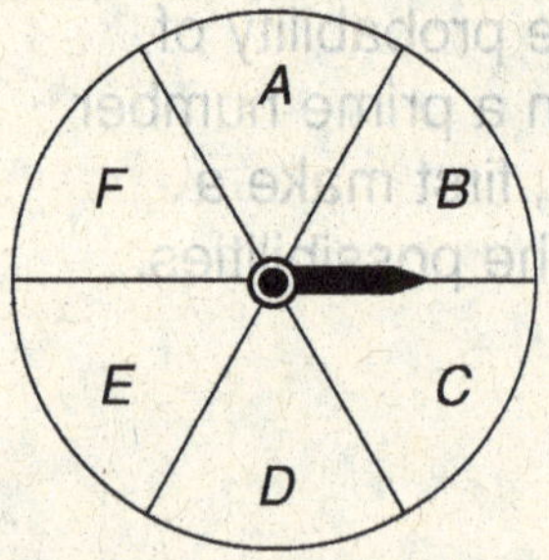

1. What is the probability of the spinner landing on *B*, then landing on *A*?

   ____

2. What is the probability of the spinner landing on a vowel on both spins?

   ____

3. What is the Probability of the spinner not landing on *E* the first times and not landing on *A* the second time?

   ____

4. What is the probability of spinning a *D* on the second spin?

   ____

A coin is tossed three times.

5. What is the probability of the coin landing tails up three times? _____

6. What is the probability of the first toss landing heads up? _____

7. What is the probability of the coin landing tails up twice and heads up once? _____

There are five cards, numbered 1, 2, 3, 4, and 5, in a bag. Each time a card is drawn, it is replaced. Find the probability of each event.

8. *P*(3, and then a 5) _____

9. *P*(4 or a 5, then 1) _____

10. *P*(even, then even) _____

11. *P*(greater than 2, then less than 3) _____

12. What is the probability of a coin landing on heads and a number cube landing on either 1 or 6?

    ____

13. If a couple has three children, what is the probability of two children being girls and one a boy?

    ____

EXPLORATION

Program of Studies and Core Content

MA-6-DAP-S-CD1, MA-6-DAP-S-P3, MA-6-DAP-S-P5, MA-06-4.4.2

12F Making Predictions

You can use probabilities to make ***predictions***. For example, the probability of rolling a 1 on a number cube is $\frac{1}{6}$. If you roll the cube 12 times, how many times do you predict you will roll a 1?

Each time you roll the cube, the probability of rolling a 1 is $\frac{1}{6}$.

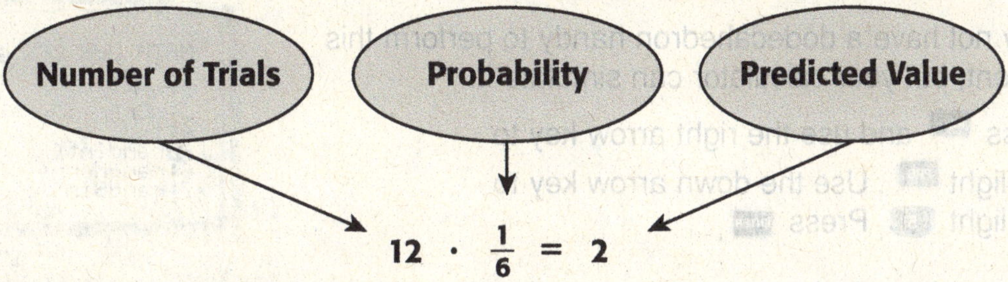

So you expect to roll a 1 twice when you roll a cube 12 times.

Use the model as a guide to make each prediction.

	Event	Probability	Number of Trials	Predicted Value
1.	Getting heads when flipping a coin	$\frac{1}{2}$	Flip the coin 20 times.	
2.	Spinning a 2 on a spinner divided into 4 equal sections	$\frac{1}{4}$	Spin the spinner 24 times.	

Think and Discuss

3. **Explain** how to find a predicted value.
4. **Discuss** whether it is certain that you will roll a 1 twice when you roll a number cube 12 times.

Name ______________________ Date __________ Class __________

LESSON **12F**

Technology Lab

Random Numbers

Your calculator can randomly generate numbers using the **randInt** function. This is helpful when you cannot actually perform an experiment. Instead, you can use random numbers to simulate it.

Activity

A dodecahedron is a 12-sided solid figure in which all faces are congruent. Imagine a dodecahedron whose faces are numbered from 1 to 12. Roll this dodecahedron 25 times and give your experimental probability of rolling a 6.

You may not have a dodecahedron handy to perform this experiment, but your calculator can simulate it.

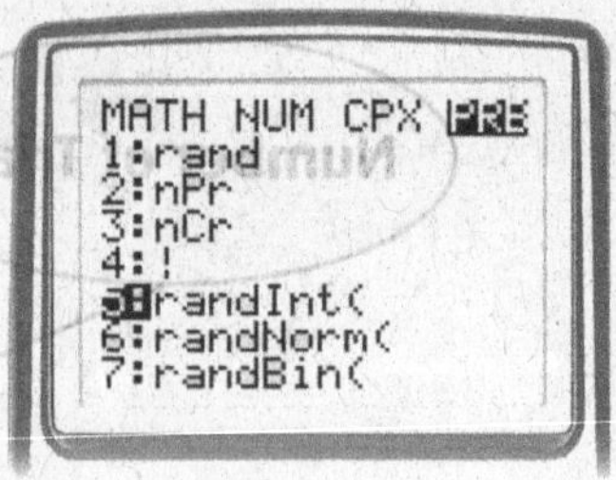

1. Press [MATH] and use the right arrow key to highlight [PRB]. Use the down arrow key to highlight [5]. Press [ENTER].

2. To have the calculator give you a number from 1 to 12, press 1 [,] 12 [)]. [ENTER]. The number that appears represents the number you have "rolled" on the dodecahedron.

3. To generate another random number, press [ENTER]. Continue to press [ENTER] as many times as you wish to represent rolling the dodecahedron.

Think and Discuss

1. How could you use your calculator to simulate rolling a standard number cube? flipping a coin?

____

____

2. What kinds of experiments do you think random numbers would be most useful to simulate?

____

Name ______________________________ Date ____________ Class ____________

LESSON **12F**

Technology Lab

Random Numbers, continued

Try This

1. Simulate rolling a dodecahedron 50 times, and give the experimental probability of rolling a 9.

____

2. Design your own experiments, and describe how you would use your calculator to simulate it. Then perform the simulation several times, and give your experimental probabilities of each outcome.

____

Name ______________________ Date __________ Class __________

LESSON 12F

Review for Mastery

Making Predictions

A prediction is a guess about something in the future. You can use theoretical probability to make predictions.

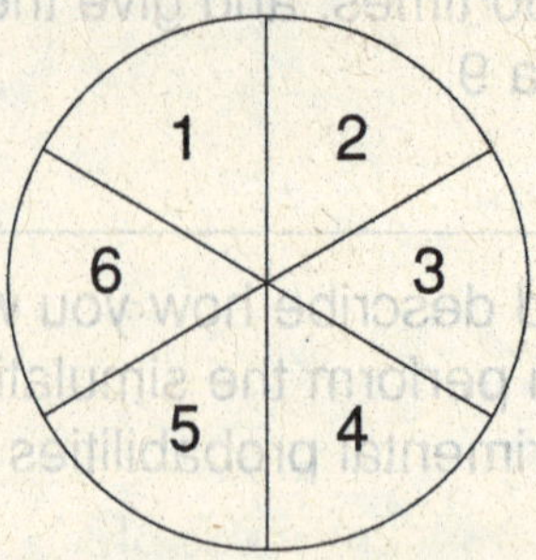

Look at the spinner above. To find the number of times you expect to land on a composite number in 27 spins, first find the theoretical probability of the event.

$P(\text{composite number}) = \frac{2}{6} = \frac{1}{3}$

This means that the spinner will land on a composite number about $\frac{1}{3}$ of the time.

Then multiply the theoretical probability by the number of times that you carry out the experiment.

Find $\frac{1}{3}$ of 27 times.

$\frac{1}{3} \cdot 27$

$= \frac{27}{3}$

$= 9$

You can predict that the spinner will land on a composite number about 9 times in 27 spins.

Look at the spinner above. Use theoretical probability to make a prediction.

1. Landing on an odd number in 42 spins.

____

2. Landing on a number that is neither prime nor composite in 30 spins.

____

3. Landing on a number that is less than 5 in 18 spins.

____

Name ______________________ Date __________ Class __________

LESSON **12F**

Homework and Practice

Making Predictions

Use the sample survey to answer exercises 1–5.

Favorite Foods	
Foods	**Number of Students**
Chicken Fingers	18
Hamburger	26
Hot Dog	12
Pizza	32
Salad	10
Other	7

1. If you randomly selected a person, what is the probability that his or her favorite food is pizza?

   ____

2. In a group of 210 people, how many do you predict would choose hamburgers as their favorite food?

   ____

3. In a class of 35 students, how many students do you predict would choose hot dogs as their favorite food?

   ____

4. In a group of 21 people, how many do you predict would choose salad as their favorite food?

   ____

5. Based on the sample survey, what do you think is the favorite food for most students at the school? Must this be the favorite food? Explain.

   ____

   ____

6. If you toss a fair coin 90 times, how many times would you expect a heads to be face up on the toss?

   ____

7. If you toss a fair die 188 times, how many times do you predict it will land on an odd number?

   ____

EXPLORATION

Program of Studies and Core Content
MA-6-DAP-S-ES2, MA-06-2.1.1

12G Sample Size and Reliability

The size of a sample you use to predict probabilities affects the reliability of your prediction. Similarly, the expected outcome of an experiment is affected by the number of trials.

Number of Flips	Heads	Tails	*P*(heads)
4			
10			
20			
40			
80			

1. Flip a coin to fill out the "heads" and "tails" columns in the table. You may want to keep track of your flips on a separate piece of paper before filling out the table.
2. Calculate the experimental probability of flipping heads for each number of flips. Write the results in the table to the nearest hundredth.
3. How many heads do you expect to get for each number of flips? What do expect the probability to be for each number of flips?
4. Describe any patterns that you see in the table. Compare your table with those of your classmates.

Think and Discuss

5. **Discuss** what you think would happen to the probability in the last column if you performed more and more flips.

Name ______________________ Date ____________ Class ____________

LESSON 12G

Review for Mastery

Sample Size and Reliability

A sample is often chosen from a population to make predictions about the population. To ensure your prediction is reliable, there are several factors to consider. The most fundamental of these are:

- The entire sample must be members of the population.
- The sample should be chosen at random to eliminate bias.
- The sample size must be large enough.

Generally, the larger the sample size, the more confident you can be about statements you make about the population.

Tom and Shandi each want to determine the probability that a randomly selected 11th-grade student in their high school owns a car. There are 107 students in 11th grade. Identify any poor sampling methods in each of their samples.

Tom is in 11th grade. So, he asks all 14 students in his algebra class whether or not they own a car.	Shandi takes a survey in the library. She asks 40 students whether or not they own a car.

Tom's sample has only 11th-grade students. However, a sample of 14 students is too small for the population he is studying. Also, limiting his poll to one algebra class is not very random.

Shandi's sample is large, since 40 students represent nearly half of the 11th grade. Unfortunately, the library likely has students from all grades at the high school. So, her sample may have members that are not in the population being studied.

For each of the following, write a sampling method that would result in a good sample. Explain why you think the sample size is adequate.

1. A restaurant owner wants to know how many customers order dessert after dinner in a week. She gets about 100 customers for dinner per week.

____

____

2. The town of Wilton has 38,512 residents. The town manager wants to know the percentage of residents that have internet access.

____

____

Name ______________________________ Date ____________ Class ____________

LESSON **12G**

Homework and Practice

Sample Size and Reliability

Read the sampling methods for each situation. Then decide if the sample represents the population, is random, and is large enough.

1. Claire and Justin both work at a movie theater. The corporate headquarters want to know the average age of customers that spend money on snacks.

Claire has the ticket booth person ask every other customer their age. She does this for all day and evening shows, weekdays and weekends, for one month.	Justin has the usher checking tickets for screen 3 to ask customers carrying snacks their age. The usher does this on Friday and Saturday evenings for two weeks.

____

____

____

2. Mr. Wilkes wants to start a Thursday evening kickboxing class at his local gym. The class will take the place of one of the aerobics classes. He wants to know if most of the 572 members are in favor of this.

Mr. Wilkes hands out surveys to those who come to his Tuesday evening kickboxing class, over a 4-week period.	Mr. Wilkes surveys all members that come to the gym on Friday. He does this over a two-month period.

____

____

____

3. A light bulb manufacturer claims that 91% of their bulbs last for at least 12 hours. Fatima bought a 4-pack of these light bulbs, but only one lasted more than 12 hours. Can she conclude the manufacturer's claim was false? Explain.

____

____

CHAPTER 12

Chapter Summary

Answer these questions to summarize the important concepts from Chapter 12 in your own words.

1. Explain how to find the experimental probability of an event.

2. Explain how to use the Fundamental Counting Principle to find the total number of possible outcomes for an event.

3. Explain how to find the theoretical probability of rolling a number greater than 2 on a fair number cube.

4. Explain how to find the complement of an event.

5. Explain how to predict the number of times you can expect to roll a 3 when rolling a number cube 210 times.

For more review of Chapter 12:

- Complete the Chapter 12 Study Guide and Review on pages 704–706 of your textbook.
- Complete the Ready to Go On quizzes on pages 686 and 698 of your textbook.

Name ______________________ Date __________ Class __________

CHAPTER 12

KCCT Prep

Multiple Choice

1. Vickie surveys a sample of customers at a deli. The survey shows that there is a 45% probability that a customer lives in the neighborhood. Out of 500 customers, how many would you predict live in the neighborhood?

Ⓐ 111 Ⓒ 275
Ⓑ 225 Ⓓ 450

2. Molly spins a spinner a total of 20 times. She finds that the experimental probability of landing on a red section is 0.6. How many times did the spinner land on a red section?

Ⓐ 6 Ⓒ 12
Ⓑ 8 Ⓓ 15

3. Which of these is closest to $\frac{1}{6}$ of 55?

Ⓐ 8 Ⓒ 10
Ⓑ 9 Ⓓ 11

4. Lori spins this spinner and another one at the same time. There are 20 outcomes in the sample space. How many outcomes does the other spinner have?

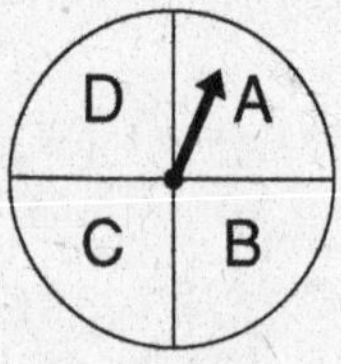

Ⓐ 4 Ⓒ 16
Ⓑ 5 Ⓓ 80

5. Dewayne has 7 notebooks. Each notebook weighs $\frac{5}{8}$ lb. Which is the best estimate of the total weight of the notebooks?

Ⓐ 3 lb Ⓒ 5 lb
Ⓑ 4 lb Ⓓ 6 lb

6. Which of these is NOT a term in the following pattern? 4, 8, 12, 16, 20, ...

Ⓐ 500 Ⓒ 512
Ⓑ 506 Ⓓ 528

7. Which equation describes the function in the table?

x	1	2	3	4
y	4	9	14	19

Ⓐ $y = x + 3$ Ⓒ $y = 3x + 1$
Ⓑ $y = 4x$ Ⓓ $y = 5x - 1$

8. Juan flips a coin and rolls a standard number cube. How many outcomes are in the sample space?

Ⓐ 2 Ⓒ 8
Ⓑ 7 Ⓓ 12

9. In Elmwood, it rains 48 days out of 160 days. What is the experimental probability that will rain on any given day in Elmwood?

Ⓐ 3% Ⓒ 33%
Ⓑ 30% Ⓓ 48%

10. The solution of $16 = \frac{y}{2}$ gives the number of granola bars in a box. How many granola bars are in 6 boxes?

Ⓐ 32 Ⓒ 128
Ⓑ 48 Ⓓ 192

11. Which point lies closest to the x-axis?

Ⓐ (7, 2) Ⓒ (0, 5)
Ⓑ (2, 3) Ⓓ (6, 6)

12. The area of the figure is 36 m^2. What is the value of x?

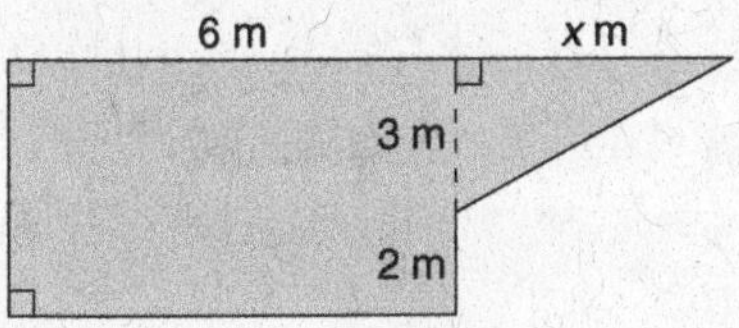

Ⓐ 2 Ⓒ 4
Ⓑ 3 Ⓓ 8

13. Kai rolled a number cube and recorded his results in the table. What is the experimental probability of rolling a 1?

Outcome	Frequency
1	12
2	10
3	8
4	16
5	12
6	10

Ⓐ $\frac{1}{12}$ Ⓒ $\frac{3}{17}$
Ⓑ $\frac{1}{6}$ Ⓓ $\frac{12}{17}$

14. The mean score on Roberto's three math quizzes is 8. Two of his scores are 6 and 8. What is the third score?

Ⓐ 6 Ⓒ 9
Ⓑ 8 Ⓓ 10

15. Which of these is closest to 55%?

Ⓐ $\frac{2}{5}$ Ⓒ 0.6
Ⓑ $\frac{3}{50}$ Ⓓ 0.06

Open Response

16. Cal has 40 songs on his MP3 player, as shown in the table.

Songs on Cal's MP3 Player	
Type of Song	**Number of Songs**
Rock	8
Hip-hop	20
R&B	12

a. What fraction of the songs are rock songs? hip-hop? R&B?

b. Write the fractions as a percents.

c. Cal chooses a song at random from his MP3 player. What is the probability that it will be a rock song?

d. Cal is going to download 40 new songs. When he is done, he wants the probability of hearing a rock song to be the greatest, and he wants there to be an equal chance of hearing a hip-hop song or an R&B song. Show how Cal can choose the 40 new songs.

17. A copy machine makes 4 copies in 12 seconds and 7 copies in 21 seconds.

a. Predict how long it takes the copy machine to make 9 copies. Explain your prediction.

b. Make a table showing the number of copies and the time it takes to make the copies. Include at least 4 pairs of values.

c. Write an equation that describes the function in the table. Tell what the variables stand for.

d. It takes Veronica 84 seconds to make some copies using this machine. Write and solve an equation to find out how many copies Veronica makes.